THE SCIENCE OF ABOLITION

The Science of Abolition

How Slaveholders Became
the Enemies of Progress

ERIC HERSCHTHAL

Yale
UNIVERSITY PRESS

NEW HAVEN AND LONDON

Published with assistance from the Annie Burr Lewis Fund.
Published with assistance from the Mary Cady Tew Memorial Fund.

Yale University Press books may be purchased in quantity for educational, business,
or promotional use. For information, please e-mail sales.press@yale.edu (U.S. office)
or sales@yaleup.co.uk (U.K. office).

Set in Electra type by IDS Infotech Ltd., Chandigarh, India.
Printed in the United States of America.

Library of Congress Control Number: 2020945668
ISBN 978-0-300-23680-4 (hardcover : alk. paper)

A catalogue record for this book is available from the British Library.

This paper meets the requirements of ANSI/NISO Z39.48-1992
(Permanence of Paper).

10 9 8 7 6 5 4 3 2 1

For Ilana
&
Mom, Dad, Seth, and Jordie

CONTENTS

CONTENTS

Should the Acknowledgments Be Longer or Short? Should There Be A Limit Before The Section Is Out-of-Control?   _Fri 6/18/21_

ACKNOWLEDGMENTS

Do Authors Give Free Books To Those Acknowledged In Their Books? Do Authors Expect Those Acknowledged To Buy Their Books? Is The Acknowlegment Section of A Book Supposed to Represent A Word-of-Mouth Endorsement? Is Name-Dropping A Form of Advertising?

This book would not have been possible without the support of remarkable teachers, and one in particular deserves special mention, Christopher L. Brown. He shepherded this project from its inception and has also been, since day one of graduate school, something more: a role model, intellectually and otherwise. So, first and last, thank you, Chris. A mighty thank-you to my dissertation committee as well—Eric Foner, David Waldstreicher, James Delbourgo, and Matthew Jones—who, through their close readings of each chapter, shaped this book into what it is today. Each one of them, along with Chris, also provided me with essential support at critical moments in my nascent academic career, and without them I'm not sure I would have made it through the long, tortuous slog of peer review, revising, and general job market misery. Technically, we're colleagues now. But actually, they're still my heroes.

Research grants—and the scholars, librarians, and staff who run them—have also made this book possible, and for them I am eternally grateful. A special thank-you to Daniel Richter for a priceless yearlong fellowship at the McNeil Center for Early American Studies at the University of Pennsylvania, where a fair amount of this book was written, and to my fellowship colleagues and the center's incomparable staff leaders, Amy L. Baxter-Bellamy and Barbara Natello, who make it all work. Another special thank-you to Brent Hayes Edwards and Sylviane Diouf for a truly magical year at the Schomburg Center for Research in Black Culture–New York Public Library, on a Lapidus Center for the Historical Analysis of Transatlantic Slavery Postdoctoral Fellowship. The collegial, rigorous—and fun!—culture Brent created in the Scholars' Center; the wisdom of Sister Aisha al-Adawiya; the support of the librarians; the advice of my fiercely intelligent cohort, Yuko Miki, Brian Jones, Imani Owens, Ansley Erickson, Anthony Bayani Rodriguez, Ayesha Hardison, Tyesha Maddox, and Hisham Aidi; and the help of the center's assistants-celebrities Margaret Odette and Naomi Lorrain—all shaped this book and my understanding of its place within African American scholarship. So again, thank you.

Revisions to this book were made possible by a two-year postdoctoral fellowship at the Ohio State University, and without the support of Simone Drake, chair of the African American and African Studies Department, this undertaking would have been difficult beyond words. Simone made sure my time was protected, treated me like a genuine colleague, and encouraged me to think like an interdisciplinary Black Studies scholar; this book— and the next—will forever reflect that influence. Other OSU colleagues and mentors who deserve heartfelt thanks include Quinn Capers IV, for selecting me for the fellowship and for being such an inspiring leader in the world of academic medicine; John Brooke, for mentoring me despite my being, let's face it, a random postdoc from another department; Margaret Ellen Newell and Joan Cashin, for similar reasons; and Valerie Lee and Kelly Jo Fulkerson-Dikuua for selecting me for the fellowship. In (or affiliated with) my own AAAS Department, a special thank-you for supporting me and this project: Lupenga Mphande, Judson Jeffries, Tiyi Morris, Ryan Skinner, Sarah Van Beurden, Linda Myers, Kwaki Korang, Franco Barchiesi, Molly Reinhoudt, Scopas Poggo, Kenneth Goings, Hasan Kwame Jeffries, Stephanie Shaw, Leta Hendricks, Ioanna Kipourou, Jerrell Beckham, and Candace Gaiters. The final touches on this book were completed at my new academic home—the University of Utah—so a very hearty thank-you to all my new colleagues in the History Department, for believing in this project, and for welcoming me to the U.

Other institutions were vital to the research and writing of this book. For research grants, thank you to the American Philosophical Society, the Huntington Library, the Omohundro Institute of Early American History and Culture, the University of Miami (Florida) Library, Columbia University's

History Department, and the History of Science Society / National Science Foundation. For archival material, thank you to the librarians at the Historical Society of Philadelphia, the Maryland Historical Society, Columbia University Library, Amherst University Library, the British Library, the Library of the Society of Friends in London, the GlaxoSmithKline archives, the Wedgwood Museum Library, the Royal Society of London Library, Emory University Library, the Uppsala University Library, the New York Public Library, the New-York Historical Society, the Connecticut Historical Museum, Yale University Library, and the Library of Congress.

Several parts of this book were presented at seminars and conferences, and the critiques I received at each of them greatly improved this project. A special thank-you for invitations to present research from this book at Johns Hopkins University's History of Science Colloquium, the Charles Warren Center for Studies in American History at Harvard University, the McNeil Center at the University of Pennsylvania, and Princeton University's Law, Difference, and Healthcare Workshop. And thank you to the conference organizers of the Omohundro Institute, the Society for Historians of the Early American Republic, the History of Science Society, the Association for the Study of African American Life and History, the Southern Historical Association, the American Historical Association, the Organization of American Historians, Emory University's Critical Junctures conference, Yale University's Critical Histories and Activist Futures conference, and the International Congress for the History of Science and Technology.

Anonymous peer-reviewers, journal editors, and numerable scholars who've been kind enough to comment on my ideas also indelibly shaped this project. My debts to Catherine E. Kelly, Joshua Piker, Zara Anishanslin, Sari Altschuler, Christopher Bilodeau, David I. Spanagel, Deirdre Coleman, Katy Lasdow, Evan Haefeli, Herb Sloan, David Greenberg, Kate Masur, Leslie Harris, Kirsten Fischer, George Aumoithe, Chris Blakely, Hugh Cagle, Craig Steven Wilder, Joris Mercelis, María M. Portuondo, James Green, Jessica Linker, Hilary Hallett, Peter Shulman, Paul Polgar, Nic Wood, Jane Landers, Beverly Tomek, Joyce Chaplin, Deirdre Cooper Owens, Edward Baptist, Sharla Fett, Walter Johnson, Andrew Lipman, Nathan Perl-Rosenthal, Ashli White, Patrick Spero, Caitlin Fitz, Hannah Farber, Craig

Hollander, Christopher Willoughby, Urmi Engineer Willoughby, Cameron Strang, Rana A. Hogarth, François Furstenberg, S. Max Edelson, Whitney Stewart, Benjamin Wright, Britt Rusert, Manisha Sinha, Sasha Turner, Kathryn Olivarius, Laura Rosanne Adderley, Nicholas Guyatt, and the one and only Stephen Koeth (rhymes with faith!). And a special mention to the historians who nurtured me well before the PhD, namely, Carol Berkin, Kathleen Brown, Anthony Grafton, and Olga Litvak. Finally, my enormous gratitude to all the people at Yale University Press: Adina Popescu Berk, my remarkable editor, who saw something worthwhile in this project early on; the Yale editorial staff, especially Eva Skewes, Ann Twombly, and Margaret Otzel, for their expertise with production and marketing; my anonymous peer-reviewers, whose critiques made the book's arguments much stronger; and Gerry Krieg, my cartographer, for his perfectly realized maps. Suffice it to say that any mistakes in this book are entirely my own.

My family also deserves special thanks; directly and indirectly, they have made this book possible. To Seth, Jordana, Poppy, and Nathan, for being lifelong champions of my intellectual endeavors, including the dumber ones. Ditto to my in-laws, Lisa, Marshall, Emily, and Andrew (and Lisa and Marshall, for a few crutch edits); I love you all. Mom, Dad: in many ways, this book is yours. Your love, support—and the premium you've always placed on education—form the deepest well from which this book springs. You've always supported my writing habit, and without that, not a word in these pages would exist. Thank you, and I love you. And finally, Ilana. You've suffered through too many years of hearing about this book, so for more than one reason, you should be happy it's done. Your advice and edits—and what you've taught me about real-world activism, social justice, and racial health disparities—have also profoundly shaped this project. But the most important way you've influenced this book is this: you've made my life better. I love you, and thank you for all the love you are forever giving back.

THE SCIENCE OF ABOLITION

Introduction

Throughout the summer of 1851, thousands of reporters, politicians, and ordinary people waited for hours to enter London's Crystal Palace. Visitors withstood sweltering heat to glimpse the marvelous new inventions on display, which organizers believed would inspire hope for humanity's future, showcasing how science and technology would transform the world into a more humane and civilized place. The event came at an auspicious time. Less than fifteen years earlier, the British had abolished slavery throughout its colonies and had taken the lead promoting emancipation worldwide. Britain portrayed its technologically advanced economy as the next phase in mankind's progress. Yet the dark reality was that Britain's economy still depended on slave labor. By the late 1850s, 77 percent of the raw cotton Britain imported to feed its steam-powered factories was being produced by nearly four million enslaved Black Americans.[1]

One of the most talked-about inventions at the Great Exhibition was a chemical processing machine for flax. Its inventors, a Brazilian chemist and a professor of chemistry at London's Royal Polytechnic Institute, wrote that their machine would lead to the "extinction of human slavery, with all its cruel and debasing horrors." If farmers in the American Midwest and in Ireland grew flax, the thinking went, their machine would turn free-grown flax into a cheaper alternative to slave-grown cotton, putting planters out of business. News of the flax machine hit America like an earthquake. Opponents of slavery, Black and white, touted the invention as a death knell for slavery. The editors of *Scientific American,* a popular scientific magazine based in New York, inspected the machine and gave it their full endorsement. Henry Bibb, a Black abolitionist and fugitive enslaved man living in Canada, wrote that "fugitive slaves should be standing in the front ranks of this experiment." Frederick Douglass declared that the invention would lead to a "considerable diminution in the wealth and power of the Southern States." Even in early 1861, when a civil war seemed inevitable, some white Northerners still held out hope for a scientific solution to slavery. "SCIENCE is king—not *cotton,*" wrote the *New England Farmer* in January 1861, upon hearing news of a new flax invention.[2]

Since the birth of the organized antislavery movement at the outbreak of the American Revolution, and through British emancipation and the American Civil War, antislavery advocates who were both men of science and not offered countless scientific arguments against slavery. Chemists contended that new fertilization techniques would make plantation soil more fertile, reducing the need for slave labor. Geologists, botanists, and explorers claimed that the soil and climate of new settlements in the American west, West Africa, and Southeast Asia were more fertile and conducive to free labor than those of the Caribbean and southern slave plantations. Engineers and inventors argued that steam engines and animal-powered machinery would make enslaved labor unnecessary. Antislavery editors and radical abolitionists, whether they were men of science or not, embraced and amplified these views: "In no one instance had slavery and the use of mechanical inventions existed together," wrote the *Anti-Slavery Monthly Reporter,* a leading mouthpiece for Britain's abolitionist movement, in 1825.[3]

Though the proslavery racial science of this period is well known, few scholars have explored the vast array of scientific knowledge *beyond* racial science that also bore on the question of slavery—from ideas rooted in chemistry and geology to those based on medicine, demography, and engineering. More to the point, few have realized that antislavery advocates, as much as their proslavery adversaries, relied on scientific discourse to defend their views. Taken together, this antislavery scientific discourse amounts to what I call the science of abolition—a wide range of scientific arguments that helped legitimate the antislavery movement and that ultimately cast slaveholders as unscientific and premodern: the enemies of progress.

Elite men of science who opposed slavery—chemists and geologists, explorers and professors of medicine—offered critiques of slavery in scientific journals, university lecture halls, natural history books, and the corridors of Congress and Parliament. Free and enslaved Black people, though largely excluded from elite scientific institutions, found ways to shape this antislavery scientific narrative, too. Before the 1830s, some served as unacknowledged scientific partners, finding botanical evidence or conducting agricultural experiments for elite naturalists, botanists, and explorers. By the antebellum period, free Black abolitionists, increasingly writing in their own newspapers, popularized the idea that new scientific inventions could destroy slave labor, and they encouraged Black people throughout the Atlantic world to engage in scientific work. Acquiring scientific expertise would prove Black people's value to white society and facilitate integration, many argued, while others contended that it would help sustain autonomous Black communities. Black and white antislavery advocates often wed their scientific arguments to different antislavery agendas: whites generally favored a more moderate agenda, and Blacks a more radical one. Nonetheless, by the 1850s, antislavery advocates of all persuasions had begun to solidify a public image of slaveholders as the enemies of science.[4]

Scholars have recently shown this image to be a myth, demonstrating that slaveholders throughout the Atlantic world in fact were deeply invested in new technologies, were engaged in scientific observation and experimentation, and generated new forms of medical knowledge. In other words, they are showing that, in reality if not in perception, slavery was modern. But what is remarkable is the degree to which antislavery men of science and abolitionists obscured this truth. By generating a wide-ranging antislavery scientific discourse—a science of abolition—they created an image of slaveholders not as modern, but as just the opposite: the enemies of progress.

This book began with a simple question: What role did men of science and scientific ideas play in the antislavery movement? Answering that question proved difficult. One of the first challenges was coming to a historically accurate understanding of what it meant to do science, and be a scientific practitioner, in the late eighteenth and early nineteenth centuries, the period in which this book is set. The precise meaning of *science*, and who could be a man of science, differed considerably from today's understanding. For one thing, the term *scientist* did not exist until 1834, and most of the elite scientific practitioners in this book would have been called *men of science*, the term I use throughout these pages. Becoming a man of science required little formal training and a fair amount of networking and self-fashioning: gaining entry to the right social networks; adopting the proper tone of deference and gentlemanly decorum; publishing in respectable journals; finding wealthy patrons. In addition, before the 1830s, being a man of science was rarely a full-time job and more a learned hobby. The most prominent men of science in this period, several of whom appear in this book, supported themselves with other activities: Benjamin Franklin with his printing business, Joseph Priestley as a tutor and clergyman, Erasmus Darwin as a physician.[5]

Yet this was also a period in which science was becoming recognizably modern. Universities and governments increasingly saw the value in scientific expertise and began funding scientific research. The British government paid for scientific expeditions into the African interior; federal and state legislatures in the United States commissioned chemists, geologists, and engineers to build roads, survey new territories, and investigate the nation's agricultural industries. Universities also began to create professorships in modern scientific disciplines: in 1802 Benjamin Silliman, an important figure in this study, became Yale's first professor of chemistry. In addition, new disciplines such as chemistry, geology, zoology, biology, and engineering began to emerge in this period, and a few figures in these pages made important discoveries in some of these fields. Priestley discovered the element we now call oxygen. Franklin discovered the nature of electrical currents. Darwin devised new theories of human evolution that would influence his grandson Charles. The emergence of recognizable disciplines, the increasing institutionalization of science, and the sheer number of scientific discoveries still with us—all explain why scholars can plausibly call the science in this era "modern," and even why it deserves its own moniker: the "second scientific revolution."[6]

Though modern disciplines began to emerge in this period, the boundaries between them remained porous. The absence of rigid disciplinary borders meant that men of science rarely specialized in any single field and often dabbled in many. This was particularly common in the eighteenth century, when men like Franklin wrote on everything from electricity to political arithmetic (an antecedent of demography), and men like Darwin authored texts on zoology, botany, and medicine. The porousness of these disciplines, and the fact that men of science often engaged in multiple disciplines at once, explains why this book takes a broad definition of science—one that includes chemistry, botany, natural history, and geology, but also medicine, engineering, mechanical inventions, and political arithmetic. What unifies the men of science in this book is that nearly all of them were members of scientific societies or wrote for scientific and medical journals. And on an epistemological level, all of them were engaging in a similar set of practices: they were observing, experimenting, categorizing, quantifying, or theorizing about the natural world—and that meant they were doing science.[7]

Despite their elite status, their authority was hardly uncontested. These men were living in a period not only in which the precise meaning of science was in flux, but also when the authority of elites—be it in government, religion, or science—was being challenged by newly empowered citizens and subjects. Therefore, we should not assume that their scientific arguments against slavery were unanimously embraced, either by their fellow men of science or by the broader, less genteel public.[8] But nor should we dismiss their contributions as irrelevant or without consequence. As this book demonstrates, many antislavery men of science made their arguments directly to political figures. Carl Wadström, an antislavery naturalist who explored West Africa in the 1780s, testified before Parliament about the feasibility of establishing a slave-free colony in Sierra Leone. Benjamin Rush, a professor of medicine at the University of Pennsylvania, made medical arguments against slavery in private letters to Thomas Jefferson. In 1831 Silliman published an official report commissioned by Congress that subtly challenged the need for slave labor in light of advancements in chemistry. It may be impossible to know whether these efforts changed politicians' minds. But at the very least we can say that their scientific standing gave them access to like-minded people in power, and political figures listened to what these men said.

In any event, elite men of science were not the only ones to use scientific arguments to attack slavery. Nonscientific antislavery editors and activists often published the antislavery arguments of men of science in their own newspapers and pamphlets, or they relied on scientific evidence to fashion their own antislavery arguments. Frederick Douglass routinely ran scientific articles against slavery in the *North Star* and his eponymous paper. Mainstream publications like *Harper's Magazine*, the *New York Tribune*, the *London Chronicle*, and *Scientific American* did much the same. Moreover, elite men of science were not the only ones to create antislavery scientific knowledge. This book also draws attention to the many scientific practitioners who, on account of their race or gender, were barred from formal scientific institutions, but who nonetheless helped produce or circulate antislavery scientific ideas. Throughout the 1790s, Benjamin Banneker, a free Black astronomer, published dozens of almanacs—popular pamphlets that predicted the sunrise and sunset for each day, among other useful astronomical information—that were also covered with antislavery missives. Frances Wright, a women's rights advocate and wealthy white abolitionist—who established a lyceum, or public lecture hall, for scientific lectures in 1829 in New York City—relied on scientific educational and managerial theories to promote her gradual emancipation program. Free Black women in the antebellum North, such as Maria W. Stewart, encouraged Black Americans to study scientific subjects in order to prove Black people's value to the nation. In Sierra Leone, free Black settlers conducted scientific expeditions for the colony's white abolitionist patrons, though their contributions were rarely acknowledged.

Indeed, in this book we see Black figures routinely engaging in scientific practice and discourse and using it for their own ends. This depiction fits uneasily with the dominant paradigms we have of Black people's relationship to Western science, particularly in the era of slavery. Traditionally, scholars have focused either on Black people's exclusion from scientific institutions, or on the ways in which scientific racism and new technologies prolonged slavery and sanctioned anti-Black violence. More recently, historians have highlighted the indigenous scientific cultures of enslaved people, from African-derived healing traditions to rice irrigation techniques transferred from West Africa to the New World. Some scholars even contend that enslaved and free Black people, in different times and places, developed alternative epistemological traditions that, while sharing an empirical basis, ran counter to the Western scientific tradition. But the story told here is neither one of Black people rejecting Western science, nor one of their fashioning their own counter-science. It is instead a story of Black people *engaging* with Western science, insisting on Black people's place within that culture, but also reshaping it for their own liberation.[9]

It is also worth emphasizing that Black people did not always use scientific arguments in the same way as white men of science. Though Black and white activists often used the same scientific arguments against slavery, Black people tended to use those arguments to push for a more radical agenda. Unlike their white counterparts, Black abolitionists made full Black citizenship essential to their antislavery agenda, and they saw no contradiction in discrediting slaveholders with scientific arguments while also calling for a host of other political actions, including, if need be, violence. By contrast, white antislavery men of science were often ambivalent about Black political equality and feared any abolitionist activity that might too abruptly challenge the status quo. Not infrequently, they offered planters scientific solutions to slavery as a wiser alternative to the kind of radical actions that enslaved people seemed to threaten, and that radical white and Black abolitionists seemed to condone. Black abolitionists were not afraid to challenge their more moderate white scientific allies, in regard either to their political agenda or to their racial science. But in most other scientific realms they generally embraced their work. It was an "age of science," Frederick Douglass said in an 1854 commencement address, and Black people could either reject the scientific narrative of their white allies because of its racist premises or the tepid political agenda it supported, or engage with their work and shape it to their own ends.[10]

Scholars have largely ignored the scientific arguments abolitionists made. If science is discussed at all, the focus tends to be narrowly centered on scientific theories about race; most scholars agree that the era's dominant racial science tended to justify racist ideas about Black inferiority. Scientific thought is thus characterized as having strengthened slavery, "justify[ing] severe restraint and rigorous exclusion," as one prominent scholar recently summarized it. But lost in these broad generalizations, even if we accept the importance of racial science to debates over slavery—and we should—is an appreciation of the full range of scientific arguments that men of science and abolitionists made against slavery—ones that included racial science arguments, but that extended far beyond them.[11]

Just as historians of abolition have neglected to study science's role in the movement, historians of science have largely neglected the study of abolitionism. To be sure, scholars have recently begun to explore the relationship between science and slavery, but not *anti*slavery. Some are showing how European naturalists employed slave ship captains to collect natural objects from Africa's shores; others have detailed how European explorers relied on the knowledge of enslaved Africans to map the African continent. Scholars in the history of medicine and the history of technology have been even more active. Like historians of science, they are demonstrating how Western medicine and technology sustained slave regimes, or how enslaved people contributed to putatively "Western" medicine and technology, either by sharing indigenous knowledge or by working as aides to white scientific practitioners. This book is deeply indebted to this scholarship, but it draws attention to the other side of the coin: science's relationship to antislavery, not slavery itself.[12]

This book also hopes to open a new line of inquiry into the broader question of slavery's relationship to modernity. Scholars often treat capitalism as a synonym for modernity, and many scholars now agree with Eric Williams, who, in *Capitalism and Slavery* (1944), argued that slavery helped fuel capitalism's rise. What has fared less well is Williams's second claim: that the transition to nineteenth-century industrial capitalism undermined the profitability of slavery, which paved the way for emancipation. Scholars have recently shown slavery to be a highly lucrative and dynamic institution that continued to develop alongside capitalism. Slavery was not undone by capitalism's maturation, but was deeply entwined with it, they argue—slavery was, in a word, modern.[13]

Much of the recent scholarship is trying to overturn the notion that slavery was a relic of a premodern era, that slavery was a check on progress. But few scholars have seriously explored how that myth came into being in the first place. This book poses that question and points toward one part of the answer. To be sure, historians are well aware that political economists, starting in the latter eighteenth century, began to portray slavery as an increasingly archaic institution, in the sense that it stalled economic growth. Scottish Enlightenment thinkers such as John Millar and Adam Smith depicted slavery as an intermediate stage in mankind's broader socioeconomic development, somewhere between an earlier, primitive, hunter-gatherer stage and a final stage defined by global commerce, universal freedom, and widespread prosperity. By the mid-nineteenth century, the idea that industrial, wage-based economies—the essence of nineteenth-century capitalism—would replace slave labor was popular enough that even Karl Marx subscribed to the general principle. It was also true that, by the 1850s, southern slaveholders themselves portrayed their "peculiar institution" as more humane than wage-based labor, thus contributing to the image of slavery as being at odds with capitalism. But the proslavery critique of wage labor was fleeting, and for decades before the 1850s, southern slaveholders routinely portrayed slavery as an engine of social progress, the vanguard of modernity. In any event, few historians have asked whether Enlightenment-era political economists, or even midcentury American slaveholders, were the only, or even the main, intellectuals to depict slavery as fundamentally at odds with modernity.[14]

They were not; men of science and scientific ideas in general were no less essential. Science, medicine, and technology symbolized progress and modernity as much as capitalism, and though these fields were tethered to capitalism and sometimes enmeshed in economic discourse, they inhabited distinct intellectual worlds. By the mid-nineteenth century, many groups of people, from white farmers and factory workers in the American Midwest and Manchester to middle-class white women and free Blacks throughout the Atlantic world, saw scientific knowledge and technological innovations as liberating, a set of tools and ideas that could blot out ignorance, promote human progress and prosperity—and perhaps even help solve the problem of slavery. Abolitionists and their scientific allies were therefore keen to marry the image of science to the cause of antislavery. And despite their internal divisions, they ultimately succeeded. On the eve of both British emancipation and the

American Civil War, slaveholders were routinely portrayed as scientifically backward—the enemies of progress.

A significant part of this book is told through the men of science who joined or publicly supported antislavery societies in Britain and the United States. Though perhaps only a minority of men of science joined abolitionist societies, the ones who did were often prominent—and highly visible—members. This was especially the case in the movement's early decades, between the 1770s and 1820s, when the first abolitionist societies took shape and were exclusively for white men. To take just one example: between 1787 and 1818, three of the four presidents of the Pennsylvania Abolition Society, the early republic's leading antislavery society, were men of science, including Franklin and Rush.[15] The stories told in the following chapters are based in part on their published writings, whether they appeared in scientific journals or antislavery pamphlets, and this book follows their words as they made their way into the pamphlets of nonscientific abolitionists, Black and white, and into the halls of Parliament and Congress. This study also relies on the unpublished manuscripts of men of science—journals and letters, lectures and account books—to explore how their scientific work shaped their antislavery views. Reading through their manuscripts suggests that at least one reason men of science may have been reluctant to embrace a more radical abolitionist stance was that slaveholders were often their scientific patrons. These unpublished manuscripts also reveal the unacknowledged role Black women and men played in helping them conduct their work.

Before the 1830s, white women and Black people of any gender were largely barred from scientific societies. But some nonetheless received a semblance of scientific recognition, whereas others became outspoken popularizers of scientific knowledge. To understand their role—whether it is Benjamin Banneker, the free Black astronomer; Frances Wright, a white women's rights advocate and abolitionist; or Paul Cuffe, a free African American sailor—I similarly relied on published and unpublished writings. For the final chapter, which covers antebellum America, it was possible to explore even more closely how Black abolitionists, including a few credentialed Black men of science, used scientific arguments for their own ends.

The creation of an independent Black press in the late 1820s, which allowed Black authors to write without the censorship of white editors, opened a window into how Black thinkers engaged with scientific ideas bearing on slavery. As the final chapter demonstrates, Black writers were among science's strongest advocates, but they did not shy away from confronting white scientific writers when their theories about race justified oppressive ends.

It may be useful for readers to have an overview of the abolitionist movement. To begin with, it must be noted that long before white people established formal antislavery societies, Black people had been resisting their enslavement. Africans captives fought to flee their enslavers upon seizure on the African continent, West African leaders occasionally shut down the slave trade to Europeans, and enslaved people revolted and committed suicide aboard slave ships and in the New World. In addition, as early as the sixteenth century a few European clerics began to voice their dissent. But it was not until the Revolutionary era that antislavery societies took root, ones that began to systematically challenge the institution of slavery itself. These organizations drew inspiration from Patriot rhetoric, with its emphasis on universal liberty, and were goaded by Quaker activism and Black-led petitions and lawsuits. In 1771 James Somerset, enslaved to a Boston colonial official, escaped while traveling with his owner to England, which led to a landmark decision by Britain's highest court one year later that undermined slavery's legality in England. In 1775 Philadelphia Quakers established the first abolitionist society in the Atlantic world, the Pennsylvania Abolition Society (PAS). By 1787 white abolitionists in England established their own organization, the Society for Effecting the Abolition of the Slave Trade (SEAST), which focused on abolishing the transatlantic slave trade.[16]

Like scientific societies, the first abolitionist societies were open only to white men. Members were often elites—lawyers, physicians, politicians, dukes, wealthy industrialists—and several of them had once owned enslaved people. In general, these first antislavery societies favored a gradual agenda, hoping to slowly phase out the institution rather than liquidate it immediately. To that end, they emphasized ending the transatlantic slave trade first, reasoning that cutting off the supply of enslaved Africans would encourage planters to take better care of the enslaved laborers they already owned. They favored emancipation laws that would require enslaved people to serve into their twenties, and they offered payment to slaveholders—not enslaved people—as an added enticement. They also deferred to the interests of planters, not enslaved or free Black people, reassuring planters that they would not be financially harmed by emancipation. Whether in spite or because of their cautious agenda, the first antislavery societies had significant early successes. The United States and Britain abolished the transatlantic slave trade in 1807, and every northern state had enacted some form of emancipation law by 1804. Yet in spite of these victories, slavery continued to expand throughout the Atlantic world. From 1770 to 1815 the number of enslaved people in the United States, the Caribbean, and Brazil increased from 2,340,000 to 3,000,000.[17]

It is also worth emphasizing that slavery's nineteenth-century growth was quietly being abetted by the very scientific innovations that abolitionists and antislavery men of science claimed would defeat it. Eli Whitney's cotton gin, invented in 1793, put slave-picked cotton at the center of Britain's and the United States' economies, spurring slavery's growth in the United States. Georgia's enslaved population doubled between 1790 and 1800, to 21,000; in South Carolina, it nearly quadrupled between 1790 and 1810, to 70,000. Newly patented steamboat technology also fueled cotton's expansion. In 1817 nearly twenty of the earliest steamboats shipped hundreds of bales of cotton up the Mississippi River. Three decades later, there were seven hundred steamboats making that same journey, and further innovations enabled each boat to carry twice as much cotton. All this scientific innovation, coupled with the seizure of Native American land and the development of more sophisticated financing and business management techniques, helps account for slavery's rapid expansion. In 1860 the United States held four times as many enslaved people as in 1800, and, with four million people enslaved, the United States had become the largest slaveholding nation in the world.[18]

Yet even as slavery expanded, the first wave of abolitionist leaders continued to embrace gradualist measures. Britain committed itself to policing the international slave trade and enacting policies that might reduce slavery's harshness in its Caribbean colonies, a process called *amelioration*. British and American abolitionists also promoted the creation of new free-labor (the opposite of slave-labor) agricultural settlements, whether in the expanding American frontier or in Britain's vast overseas empire; over time, they believed, these new settlements would outperform slave-labor plantations. The commitment of white abolitionists to racial equality also remained suspect, if they were not downright hostile to it. By the 1820s many white American abolitionists came to believe that gradual emancipation would occur only if freed Black people were voluntarily resettled outside the United States, an idea called *colonization*. To many, the curtailment of free Black rights in the North and Midwest seemed to leave no other option. In 1821 most free Blacks in New York lost the right to vote. Between 1803 and 1818, Ohio, Indiana, and Illinois abolished slavery in their new state constitutions, but at the same time they barred Blacks from voting and drafted laws discouraging free Black immigration. To address these realities, abolitionists on both sides of the Atlantic continued to encourage free Blacks to resettle in Sierra Leone, hoping that it would provide a model for other colonization projects.[19]

Antislavery activism changed dramatically in the 1830s. During this second wave of the movement—the more familiar story to the general public today—abolitionist societies became more radical. They became racially integrated, many accepted female members, and most demanded the immediate, rather than gradual, end to slavery. Most abolitionists also now rejected slave owner compensation and colonization, and many pressed for full Black citizenship alongside emancipation. Yet unlike proponents of the earlier gradualist movement, second-wave abolitionists were considered a radical fringe, and in general northern men of science opposed to slavery kept their distance. But if northern men of science were less likely to join abolitionist societies, many of them still publicly voiced antislavery views. Like most white Northerners, they tended to embrace the more cautious antislavery agenda of the Republican Party, an agenda that, before the Civil War, called only for slavery's non-extension and stayed silent on the question of Black citizenship. Meanwhile, in Britain, the earlier gradualist leadership held on to the movement long enough to dictate the terms of British emancipation. In 1833 Parliament emancipated all of Britain's 800,000 enslaved laborers, but on the condition that they serve as apprentices for several years, and that planters receive compensation. American emancipation took a different course, coming in the midst of war. Abolitionist agitation and fugitive escapes pushed the nation to the brink of emancipation, but Lincoln's 1863 Emancipation Proclamation was less a product of radical abolitionist conviction than of wartime necessity. Emancipation, formally enacted through the Thirteenth Amendment in 1865, was an undeniable achievement, and an exhilarating one. But few abolitionists, and even fewer Black people, doubted that the fight for full equality had only just begun.[20]

The following chapters uncover the wide range of scientific arguments abolitionists and men of science made against slavery. It tells this story through the men of science who either joined or supported an antislavery agenda, and by showing how nonscientific abolitionists made use of their work. The first chapter is set in Revolutionary America, when the early abolitionist movement's leadership began to seek out prominent men of science, focusing on Benjamin Franklin, Benjamin Rush, and the Black astronomer Benjamin Banneker. The movement's early Quaker leaders believed men of science could lend the movement legitimacy and could make abolitionism appear not just moral, but scientifically respectable. But as they sought out figures with different degrees of commitment to activism and the movement's particular agenda, internal tensions emerged that were never fully resolved: namely, the place of freed Black people within American society.

The second chapter moves to late eighteenth-century Britain, demonstrating how key scientific thinkers behind the Industrial Revolution helped align the image of science with antislavery. Erasmus Darwin, a physician and botanist, and Josiah Wedgwood, a potter, inventor, and chemist (both grandfathers of Charles Darwin), played important roles in the Industrial Revolution, and as abolitionists they began to argue that the adoption of industrial technologies and scientific know-how to slave plantations would reduce the need for slave labor. Meanwhile, a close comparison of the scientific and antislavery writings of Joseph Priestley shows how the values of science could be used to echo and reinforce the ideology of mainstream abolitionism. Yet the Haitian Revolution, which rapidly and violently destroyed slavery in French Saint Domingue, posed a challenge to these abolitionists. The Black-led revolution did not fit neatly with their scientific worldview, one that understood progress to be gradual, and social change to be the result of what they considered to be rational behavior.

The third chapter moves to Sierra Leone. From the 1770s to the early 1800s, naturalists, physicians, and explorers traveled to the West African territory on behalf of abolitionists, depicting its natural environment as idyllic. They presented its climate, soil, and indigenous inhabitants as perfectly suited to free-labor plantation agriculture, bolstering abolitionists' claim that free-labor agricultural colonies in West Africa could gradually replace slave-labor plantations in the Americas. Though they received little recognition for it, Sierra Leone's indigenous inhabitants and Black British Loyalists—formerly enslaved Black Americans who won their freedom fighting for the British during the Revolutionary War—were vital to collecting scientific evidence. This chapter also argues that the deep ties between men of science and slave owners, rather than undermining their credibility, helped insulate antislavery men of science from accusations of bias.

Chapter 4 focuses on the antislavery movement in the United States during the early republic period. White abolitionist leaders in the 1810s and 1820s promoted various colonization projects, often with avid scientific support. In the 1820s Benjamin Silliman, Yale's first chemistry professor and a prominent antislavery voice in New Haven, wrote a report for the federal government that depicted southern slaveholders as dangerously averse to new labor-saving technologies. William Maclure, a prominent geologist and supporter of colonization, wrote articles in the popular press that portrayed the expanding western frontier as ideal for free-labor agriculture. In the late 1820s Maclure joined Frances Wright, the women's rights advocate, to run an experimental plantation in western Tennessee, called Nashoba. Relying in part on Maclure's educational theories, Wright tried to prove that, if plantations were managed scientifically, planters could recoup the price of their enslaved laborers, free them, and send them wherever they chose.

Chapter 5 returns to Britain, focusing on the final three decades leading up to British emancipation. It highlights an underappreciated aspect of amelioration—the encouragement of planters to adopt labor-saving technologies—and shows antislavery men of science insisting that these new tools would ease the burdens on enslaved laborers. During this period, British abolitionists also continued to promote free-labor colonies in Britain's overseas colonies, hoping that their financial success would reassure policy makers that ending Caribbean slavery would not bring economic ruin. Explorers and naturalists—again, often with the unacknowledged support of Black aides—provided essential evidence about the topography and climate of these colonies, from Sierra Leone to Java. By the early 1830s, however, the gradualist approach was losing favor. A spate of slave rebellions in the Caribbean, in addition to the infusion of younger, more radical abolitionists, forced the abolitionist movement's aging elites to take a more confrontational stance. Increasingly, abolitionist leaders stopped pleading with slaveholders to adopt scientific solutions and instead presented them to the public as scientifically inept.

Chapter 6 moves to antebellum America—the 1830s to the Civil War—where a wide range of antislavery views prevailed, from radical new ideas supported by Black and white abolitionists to more cautious agendas embraced by the majority of white antislavery Northerners. As abolitionist societies became more radical, fewer antislavery men of science joined these groups, but men of science still contributed to the broader antislavery discourse. In the face of a virulent new form of scientific racism, many northern men of science opposed to slavery insisted on Black people's essential humanity. Yet some devised new racial theories that suggested Black people were best suited to tropical climates, a theory that conveniently allowed them to attack slavery while also rejecting Black citizenship. No less important were the arguments antislavery men of science made about technology, particularly ones pointing to new agricultural technologies as harbingers of emancipation. While all these scientific ideas, whether racial or technological, helped broaden antislavery's appeal, they were frequently wedded to a cautious antislavery agenda that few radical abolitionists embraced. Meanwhile, even though Black and white abolitionists wed similar scientific arguments to a more radical political agenda, they still contributed to the same basic narrative: slaveholders were the enemies of progress. Whether it was true in reality was irrelevant. What mattered was that scientific ideas gave the broader antislavery movement legitimacy, and with them, the movement would help set the enslaved free.

ONE

Stars and Stripes

On July 14, 1792, Benjamin Rush delivered a remarkable address before the American Philosophical Society in Philadelphia. Rush, a signer of the Declaration of Independence, an abolitionist, and one of the nation's leading physicians, declared that leprosy caused the darkening of Black people's skin. Once physicians found a cure, Black Americans would turn white, he argued; this would "add greatly to *their* happiness," for however much Black Americans "appear to be satisfied with their color, there are many proofs of their preferring that of white people." The implications were obvious. His theory would prove, once and for all, that "the whole human race . . . descended from one pair," countering the proslavery claim that Africans were a separate and inferior species, predisposed to slave labor.[1] Equally important, it would resolve a dilemma that dogged the early abolitionist movement: What to do with enslaved people after they were free? Resistance to free Black citizenship posed a serious threat to abolitionism, in northern states as much as in southern ones, and Rush hoped his theory would ease those concerns, suggesting that physicians would soon find a cure for Blackness. In the meantime, the nation should emancipate its enslaved laborers, and with the aid of medical science, Black Americans could eventually be integrated into the nation as full, equal—and white— American citizens.

Rush's belief that Black people could turn white was neither unique nor short-lived. Throughout the 1790s and into the early 1800s, leading men of science on both sides of the Atlantic frequently inspected the bodies of Black women and men whose skin appeared to be turning white, convinced it was evidence of the body's natural whiteness.[2] What makes Rush's theory noteworthy is that he offered an explanation for Blackness rooted in his own expertise: medicine. More important, by tying his theory to antislavery, Rush gave abolitionism the backing of medical science. In fact, Rush spent years writing medical lectures against slavery, even coining names for new diseases that he claimed Blacks contracted only "after they enter upon the miseries of slavery."[3]

When the first organized antislavery societies took shape during the Revolutionary era, abolitionist leaders eagerly sought out men of science. Not only did they elect three men of science to be president of the Pennsylvania Abolitionist Society (PAS), the nation's leading antislavery organization, between 1787 and 1818, but they also helped Benjamin Banneker, a free Black astronomer, publish antislavery almanacs in the movement's behalf. Rush, Franklin, and Banneker reflected the wide range of men of science—astronomers, physicians, electricians—that abolitionists would seek out in the decades to come, and they also illustrate what the movement stood to gain by attracting scientific thinkers. Before the 1770s, the movement's original Quaker leadership critiqued slavery in religious and moral terms.[4] But men of science spoke about the problem of slavery in *scientific* terms. Science had the veneer of being universal and disinterested, above the interests of any one religious sect. Men of science could thus make antislavery appear less like a religious crusade and more like a rational decision backed by science. But in cultivating these men, the movement's leaders exposed profound disagreements about the movement's precise goals and objectives— and nowhere more significant than in regard to Black citizenship.

There was no shortage of antislavery sentiment before the American Revolution. Yet the American Revolution transformed what had been disparate and uncoordinated attacks on slavery into an organized political movement, and Quakers in particular were at the forefront. Philadelphia's Quaker community banned all their members from buying enslaved people in 1755, then outlawed slaveholding entirely among Quakers in 1774.[5] One year later, they established the PAS, seizing on the politicization of slavery amid the colonies' conflict with Britain. Meanwhile, enslaved Black Americans saw in the Patriots' call for liberty an argument for their own emancipation. On July 13, 1777, enslaved New Englanders petitioned the Massachusetts legislature for an emancipation law, rooting their claim in "the Natural right of all men . . . in this Land of Liberty"; they also pointed out the "inconsistency of [Patriots] acting themselves the part which they condemn and oppose in others."[6] Patriots had made a habit of likening their political standing in the British Empire to slavery, and British Royalists were all too eager to point out the irony: how strange it was for Patriots to compare themselves to slaves when roughly one in five colonists was actually enslaved.

Patriot leaders responded that they cared little for slavery and would rather see it disappear if only Parliament would let them. They were not being entirely disingenuous, either. In 1772 the Virginia colonial legislature passed a bill to curtail the slave trade, only to see Parliament reject it; two years later, the Continental Congress adopted a resolution in favor of banning slave importations.[7] But the rationale behind these moves differed from Quakers'. Quakers opposed slavery largely for religious reasons, while Patriots did so mainly for political ones. For many Patriot leaders, the main problem with slavery was that it threatened the white population. They feared that it would squeeze whites out of the labor market, making the project of nation building more difficult and creating a permanent subversive class, an entrenched "internal enemy."[8] The PAS's Quaker leaders understood these concerns, and some probably shared them. But they needed influential Patriot allies, ones who, regardless of their religious affiliation, could argue on the movement's behalf. In this enlightened age, better still were men who could frame antislavery arguments in the language of reason. What they needed, in short, were men of science.

No one fit that bill better than Benjamin Franklin. By the 1750s, Franklin had emerged as the colonies' most distinguished man of science. The Royal Society in London, Britain's most prestigious scientific society, awarded him its highest award, the Copley Medal, for his work on electrical currents in 1753; three years later it offered him membership, making him the first American inducted into the society. His scientific reputation enhanced his political one. For eighteen years, between 1757 and 1775, he represented the colonies in Whitehall, and during those long years in London he accrued a deep network of political and scientific allies, ones who would prove useful to abolitionist leaders. But there was one problem: Franklin's commitment to antislavery was suspect. For one thing, Franklin owned enslaved people, several of them. In 1735 Franklin, then twenty-nine, bought his first enslaved servant, Joseph. Over the course of his life, he would acquire at least seven more, most of whom lived in his Philadelphia home, and two of whom, Peter and King, traveled with him to London in 1757. He freed none of them during his lifetime. Some died under his family's control; King escaped to freedom three years after arriving in London; and only one, Bob, was manumitted "after my decease," as Franklin wrote in his 1788 will.[9] A good portion of Franklin's wealth indirectly derived from the sale of enslaved people: at least 20 percent of the advertisements published in Franklin's newspaper, the *Pennsylvania Gazette*, between 1729 and 1748 concerned enslaved or indentured laborers.[10]

Franklin's investment in slavery did not necessarily make him a hypocrite. Slavery touched nearly every facet of eighteenth-century colonial life, and many early abolitionists had once owned slaves or profited from slavery. But Franklin's late-in-life endorsement of the PAS was hardly a tale of a compromised man who finally came to his senses. Near the end of his life, Franklin became the public face of the PAS—its leaders elected him president from 1787 to 1790—but he was, in fact, a "rather strange friend of antislavery," as one scholar has put it.[11] In contrast to the Quaker community, Franklin emphasized the racial threat that slavery posed to an imagined all-white America, and, like many Patriot elites, he blamed Britain for enacting the slave trade in the first place.

But what distinguished Franklin from other antislavery Patriots was the role science played in his activism. Political arithmetic, a form of scientific discourse, gave him the language to denounce slavery in some of his earliest critiques of the institution. The friendships Franklin forged over science with political figures in London, several of them abolitionists, also offered Franklin a correspondence network through which he could privately experiment with antislavery ideas. By the 1780s the scientific work he published on the Gulf Stream began to reflect a man pondering what the nation might look like without transatlantic ties to slavery. Franklin was hardly a stalwart abolitionist. But by using scientific discourse and scientific networks to think through antislavery ideas, he helped the movement shed its pious image and made it look rational, even backed by science.

Near the end of his life, Franklin liked to boast that he published some of the earliest antislavery writings in the thirteen colonies. "About the year 1728 or 29 I myself printed a book for Ralph Sandiford," and "another book on the same subject by Benjamin Lay" nine years later, he wrote to John Wright, a British antislavery sympathizer, on November 4, 1789.[12] Franklin did in fact publish antislavery pamphlets by Sandiford and Lay, two of the first Quaker critics of slavery, but that did not mean he necessarily agreed with them. Franklin was a relatively new printer in town and, in pursuit of business, he printed their antislavery writings—so long as his name was kept off their pamphlets, and so long as they distributed their works themselves. The antislavery arguments of Sandiford and Lay, like those of later Quaker abolitionists, stood in stark contrast to the ones Franklin would eventually make. Early eighteenth-century Quakers focused on slavery's violation of religious doctrine, emphasizing its sinfulness, and concentrated on purifying their religious community from this ungodly act; they were not yet committed to making a political case beyond the Quaker community. In the 1737 essay Franklin printed, Lay reflected those priorities, calling slavery a "filthy Sin" that violated the Quaker principles of "preaching up Perfection" and "the universal Love of God to all people, of all Colours and Countries."[13]

Franklin, raised by Calvinists and sympathetic to deism, admired the Quakers' egalitarianism, but he had little interest in their professions of religious purity. He saw their strident religious moralism—their slavery ban, their pacifism—as a kind of fundamentalism, a stark way of seeing the world that flew in the face of what it meant to be an enlightened, scientific thinker. Establishing probability, not certainty, was the ideal of Enlightenment science. The enlightened man of science established truths not by adhering to religious dogma, but by reasoning from worldly particulars. In his *Autobiography*, written in 1771 and published posthumously, Franklin likened what he took to be Quakers' biblical literalism as akin to being "wrapped in the Fog": they saw clearly only what was right in front of them, but they could not comprehend the complexities of the world beyond. He preferred to reduce moral decisions to a calculus, not a catechism. In 1773 he wrote to Joseph Priestley, a prominent British chemist and antislavery supporter, that he had long been in the habit of penning a chart, one column headed "Pro," the other "Con," when he needed to make an important decision. That way, "I can judge better, and am less likely to make a rash Step," he wrote, calling this method *"Moral* or *Prudential Algebra."*[14]

Franklin made his first public comments against slavery in 1755, in an essay titled "Observations concerning the Increase of Mankind." His critique showed none of the Quaker emphasis on sin and instead focused entirely on the threat slavery posed to Britain's economy and the colonies' racial homogeneity. Slavery was discussed as one of several ill-conceived policies in an essay written in response to a recent British law that prohibited building iron furnaces and mills in the American colonies. Franklin argued that the policy was cutting off employment for potential English immigrants and in turn threatened the colonies' white majority: the policy "greatly diminish'd the Whites" by "depriv[ing] of Employment" recent British immigrants, he wrote. In any event, he asked, "Why increase the Sons of Africa, by Planting them in America where we have so fair an Opportunity, by excluding all Blacks and Tawneys, of increasing the lovely White?" Franklin's racial favoritism stood in contrast to Lay's insistence that Quakers must love people of "all Colours." In addition, Franklin, unlike Lay, made no mention of the harm, physical and otherwise, that slavery did to the enslaved, instead focusing only on the harm it did to white colonists. "Whites who have Slaves" become "enfeebled" and lazy, he wrote, eventually becoming "proud, [and] disgusted with Labour."[15]

The publication of "Observations" may have put Franklin on the abolitionists' radar. But his wife, Deborah Read Franklin, probably secured his antislavery reputation. Deborah Franklin became involved with Philadelphia's antislavery activists in the late 1750s, and at least from 1758, she kept a portrait of Lay in their Philadelphia home. Her avid support for a new school for enslaved children in Philadelphia, established in 1758, also may have bolstered Franklin's antislavery image. On August 9, 1759, she enthusiastically described the school in a letter to her husband, who was back in London: she had "a great deal of Pleasure" seeing "the Negro Children catechized," and planned to enroll their own newly acquired enslaved servant, Othello.[16] One year later, perhaps to please his wife, Franklin accepted the role of board president. Deborah Franklin also provided an important familial link to Anthony Benezet, the influential Quaker abolitionist who would play a critical role recruiting Rush and Franklin to the movement: Benezet's brother had married a relative of hers.

Franklin's involvement with the school for enslaved children should not be misread. Established by a philanthropic arm of the Anglican Church called the Associates of Dr. Bray, the school intended only to educate enslaved children, not to free them. Planters may have viewed the education of enslaved people as a Trojan horse for abolition, but the Associates of Dr. Bray was decidedly not an antislavery organization. Privately, Franklin also showed less moral growth than historians sometimes suggest. Some scholars have cited a letter he wrote to the school's founder, John Waring, after a visit in 1763, as evidence of the beginning of a moral transformation. Franklin told Waring that he "conceiv'd a higher Opinion of the natural Capacities of the black Race, than I had ever before entertained" after seeing his school. But in the same letter he also admitted that he still had doubts: "You will wonder perhaps that I should ever doubt it," he wrote, referring to Black intelligence, adding, "I will not undertake to justify all my Prejudices, nor to account for them."[17] Even more important than Franklin's personal racial views was what he did politically, in behalf of slaveholders. In 1768, as the newly appointed colonial agent for Georgia in London, he petitioned Whitehall to approve the colony's proslavery laws.[18]

Franklin nevertheless increasingly critiqued slavery in print during the 1770s. All these writings took place in the Revolutionary context, however, and were concerned more with protecting the colonists against charges of hypocrisy than with protecting the interests of enslaved people. In his first full essay against slavery, published anonymously in London's *Public Advertiser* on January 30, 1770, Franklin admitted that slavery was a "Crime." Yet he dwelled almost entirely on discrediting the notion that Americans were disingenuous. If Royalists accused Patriots of being insincere, then how ironic it was for Britain to now pose as enslaved people's sudden friend: "Remember," Franklin wrote, England "began the Slave trade."[19] Two years later, Franklin was forced to take up the issue again. In 1772 England's chief justice ruled that James Somerset, an enslaved Bostonian brought to England, could not be considered a slave when he arrived in England because slavery was never legal there. Enslaved communities and abolitionists throughout the Atlantic world interpreted the ruling as effectively freeing all enslaved laborers in England—it did not. But the misinterpretation bolstered the hopes of enslaved communities and their allies, regardless of which nation issued the ruling.[20]

Franklin's June 18, 1772, letter on the Somerset case, again published anonymously in the *London Chronicle*, walked a fine line, on the one hand endorsing the gradualist abolitionist agenda, but on the other puncturing Britain's self-regard. Franklin gave his support to what had become the basic abolitionist strategy—ending the slave trade first, then enacting laws "declaring the children of present Slaves free after they become of age." But he then deflected attention from slavery in North America and focused almost exclusively on British hypocrisy. "*Pharisaical Britain!*" he wrote: "To pride thyself in setting free a single Slave that happens to land on thy coasts while thy Merchants . . . continue a commerce whereby so many hundreds of thousands are dragged into a slavery!"[21]

Franklin's Somerset article was noteworthy for his use of political arithmetic. Political arithmetic emerged in the late seventeenth century to help imperial officials calculate the potential for population growth, then seen as an important measure of political strength. Franklin used political arithmetic in his earlier 1755 essay, "Observations concerning the Increase of Mankind," to show how slavery slowed the growth of the colonies' white population, concluding that "slaves can never be so cheap here as the Labour of working Men is in Britain. . . . Any one may compute it." In his Somerset essay, Franklin deployed political arithmetic even more extensively. Nearly one-third of the letter described, in strictly numerical terms, the extent of slavery both in the colonies and through the Atlantic slave trade. "By a late computation made in America," he wrote, nearly "eight hundred and fifty thousand Negroes" lived in all the British colonies, while "about one hundred thousand" more were imported annually, "of which number about one third perish" from disease during the Middle Passage. Franklin took these numbers directly from Benezet, who wrote to Franklin on April 27, 1772, asking for help combating this "terrible evil."[22] Shrewdly, Franklin did not publish his own name on the article—to do so risked alienating the southern colonies whose support the Patriots needed to win the war. But Benezet's solicitation and sharing of data to make the case indicate how important scientific arguments were becoming to the nascent antislavery movement. Benezet did not even need Franklin's name and the scientific stature it conferred on these articles: the scientific arguments themselves were good enough.

In April 1782 Patriot leaders selected Franklin to be one of the lead American negotiators in peace talks with Britain. Abolitionists had good reason to believe that Franklin might include some kind of antislavery legislation in the final treaty. After all, slavery had received a serious blow during the war years. Tens of thousands of enslaved Americans escaped to British lines during the war; in turn, northern abolitionists capitalized on those gains by enacting gradual emancipation laws, state by state. Franklin's Pennsylvania epitomized the way slave-led resistance propelled antislavery legislation. In 1780 PAS abolitionists lobbied Pennsylvania's new state government to enact the nation's first gradual emancipation law, which guaranteed freedom to all enslaved people born after March 1, 1780, after twenty-eight years of service. But by then, slavery had already sharply declined. In Philadelphia, which had the largest concentration of slaves anywhere in the state, the enslaved population dropped by roughly two-thirds between 1767 and 1780—that is, *before* the gradual emancipation law even passed. The decline resulted in part from individual manumissions of Quaker slave owners, for whom slaveholding was increasingly unprofitable, but also from the sharp rise in numbers of enslaved people who had escaped during the war. The momentum created by enslaved peoples' resistance and Quaker manumissions made the 1780 law possible, which in turn hastened emancipation everywhere else in the state.[23]

As an American negotiator at the Paris peace talks, Franklin had an opportunity to capitalize on these gains. He even showed some initiative, writing down various antislavery proposals that might be included in the peace treaty with Britain. In an unpublished essay, titled "A Thought concerning the Sugar Islands," written shortly after July 10, 1782, he argued that Britain should consider eradicating slavery in its Caribbean islands. Too many wars had been fought, too many lives lost, over empires trying to seize control of the Caribbean islands, he wrote. But what was notable was that Franklin said nothing about ending American slavery, and he again emphasized the toll slavery exacted on "white Nations," who he claimed disproportionately died "fighting for those Islands."[24] In addition, the essay illustrated how Franklin began to use antislavery rhetoric for the political leverage it would give the newly independent American states over British officials. Attitudes toward slavery may have been changing, but the reality was that Caribbean slavery still remained profitable. Therefore, if Britain dismantled slavery in the West Indies, the real beneficiary would be America.[25]

American abolitionists were not involved with Franklin's "Sugar Islands" essay. But to the extent that they would have supported its basic proposal—Caribbean emancipation—they would have been glad to know that Franklin's scientific networks helped facilitate its circulation. On May 8, 1783, Franklin sent the "Sugar Islands" essay to David Hartley, an inventor and chemist with whom Franklin had forged a friendship over science, and who was now the lead British negotiator in the Paris peace talks—an appointment he received in part because of their personal rapport. Hartley was a well-known abolitionist, having put forward the first resolution in the House of Commons to denounce the slave trade, in 1776. And Franklin played to his interests in science and antislavery to ease the tense political talks. "I speak to you as a Philosopher and a Philanthropist," Franklin wrote to Hartley in 1779, when they began unofficial peace talks, using common terms for a man of science and humanitarian, respectively. "The subject being peace," Franklin thought they should take measure of "the latitude & longitude of the Politicks of nations" so as to bring an end to the war as soon as possible.[26] Regardless of Franklin's tepid embrace of abolitionism, the fact that several of his scientific friends took up the cause forced him to remain engaged. To retain their respect, he would continue to play the role of abolitionist.

Indeed, science and antislavery often mixed in Franklin's correspondence with men of science. In December 1773, for instance, the Marquis de Condorcet, a renowned man of science and abolitionist in France, asked Franklin about the condition of free Blacks in Philadelphia in relation to a work of natural history he was researching. "Generally improvident and poor," Franklin responded, though he added that they were "not deficient in natural Understanding, but they have not the Advantage of Education." Philadelphia's abolitionist leaders seem to have understood that abolitionism was gaining support among scientific elites in Europe, and they used that fact to bring Franklin closer to the movement. In Benezet's first letter to Franklin, written in 1772, he mentioned that other "most weighty" men of science had also been contacted, such as the physician and botanist John Fothergill, who would play a critical role in establishing Sierra Leone.[27] Benezet seems to have believed that he could convince Franklin to join the crusade by noting that abolitionism was catching the imagination of many of the scientific colleagues he admired.

Whatever hopes abolitionist leaders had regarding antislavery measures at the peace talks, they must have been disappointed. The Treaty of Paris, which officially ended the war on September 3, 1783, and which Franklin and Hartley helped draft, sacrificed abolitionist goals to the demands of American slave owners. Most important, the treaty allowed American states to pursue their "confiscated" property—that is, their enslaved laborers—that had escaped to "his Majesty's arms" during the war years. Yet the PAS would not leave empty-handed. When Franklin returned to the United States, he incorporated the antislavery ideas in "A Thought concerning the Sugar Islands" in the final scientific essay he ever wrote. Titled "Sundry Maritime Observations" and printed in the American Philosophical Society's *Transactions* in 1786, the essay served as a clearinghouse for all the sea-related knowledge he had gathered over the previous three decades, from how to build a ship that balanced properly to what to pack when braving a transatlantic voyage. He also included, almost verbatim, passages from "A Thought concerning the Sugar Islands." When considering "the wars we make to take and retake the sugar islands from one another," he wrote, virtually repeating his earlier lines, sugar ought to be seen as "stained with spots of human blood!"[28]

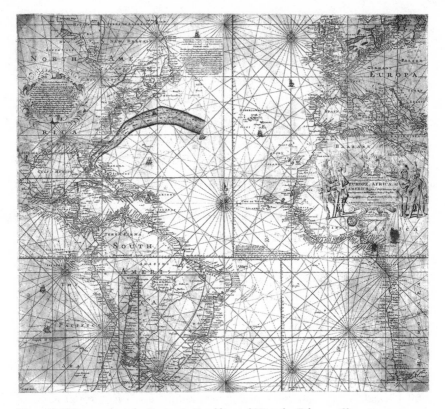

Fig. 1. Gulf Stream chart, by Benjamin Franklin and Timothy Folger, 1768.

Importantly for abolitionists, Franklin enlisted scientific knowledge—a discussion of ocean currents in particular—to present his hopes for a slave-free future. "Sundry Maritime Observations" included his fullest description yet of the Gulf Stream, a fast-flowing current that starts on the eastern coast of Florida, sweeps up the Atlantic Seaboard, and trails off deep into the northern Atlantic Ocean. If sailors heading west from England accidently steer into it, they effectively start sailing against the tide. In 1768, before the Revolution, Franklin wrote a scientific treatise on the Gulf Stream in order to expedite shipping between the colonies and Britain. But the final 1786 version came with a distinct American imprint. He no longer mentioned the Earth's rotation as a cause of the stream, and instead focused on the massive amount of water that accumulated on the Eastern Seaboard of the

Fig. 2. A chart of the Gulf Stream, by Benjamin Franklin and James Poupard, 1786.

Unlike Benjamin Franklin's 1768 Gulf Stream map (fig. 1), this 1786 map focuses only on the Eastern Seaboard of the newly independent American colonies. By cropping out Africa and the Caribbean, this later map encouraged viewers to imagine a nation cut off from slavery. Maps courtesy of the Library of Congress, Geography and Map Division.

American states. In addition, the 1786 map included with the essay showed only the Gulf Stream's connection to the North American seaboard, cutting out Britain and its Atlantic empire entirely. As one Franklin biographer has argued, the change in focus—from the global to the local—mirrored the shift in Franklin's geopolitical thinking. He was no longer thinking like a British imperial agent, one for whom the world was one's field of inquiry. He was thinking like an American, one whose own borders came first.[29]

But "Sundry Maritime Observations" also showed Franklin thinking like an antislavery American. In depicting the Gulf Stream as being caused by natural phenomena found only on the Eastern Seaboard, and by cutting out Africa and the Caribbean from the 1786 map, Franklin was choosing to ignore the Atlantic slave system. He was allowing his readers to imagine a future without any ties to the Caribbean or Africa, as if the problem of slavery might someday disappear. In addition, Franklin's essay suggested that Africa, and people of African descent, had nothing useful to offer America. Part of "Sundry Maritime Observations" argued that white Americans could learn from the boat-making technological skills of the Chinese, Pacific Islanders, and Native Americans, yet he said nothing about African or African American seafaring knowledge, despite the ubiquity of Black sailors throughout the Atlantic world.[30] "Sundry Maritime Observations" ultimately embodied all the contradictions in Franklin's antislavery views. Though he increasingly expressed them in public, the ideas themselves would focus primarily on antislavery's benefits to an implicitly white America. Franklin's Gulf Stream map provided a perfect visual analogue. He hoped to rid America of slavery not so much by attacking the institution directly, but by pretending that it barely existed in the new nation, and by somehow removing the nation from the wider Atlantic world in which it did.

Franklin's antislavery politics were at odds with many PAS leaders' views, but that did not seem to bother them. His willingness to endorse their cause in the pages of the nation's premier scientific journal, *Transactions*, was too great an asset to quibble with the particulars. In 1787, within a year of the publication of "Sundry Maritime Observations," the PAS elected Franklin president. They did so in part because of his political reputation and power: after all, he would soon be participating in the Constitutional Convention, and abolitionists hoped he would advocate for antislavery measures on their behalf. But his political authority in no small part stemmed from his scientific renown. In addition, by packaging antislavery arguments within the cool, dispassionate language of science, Franklin offered a way to speak about slavery that appeared more rational than the emotional debates that abolitionism invariably provoked.[31]

Until Franklin's dying day, abolitionists found his reputation, scientific and otherwise, useful. On February 5, 1790, as Franklin's health was rapidly declining, James Pemberton, a Quaker, the PAS's vice president, and the society's shrewdest strategist, pressed Franklin to sign a petition to Congress demanding an end to the slave trade. One year later, Pemberton would play an equally important part in soliciting Benjamin Banneker to write a scientific almanac in the movement's behalf. The PAS may have even hoped that Franklin's signature would counter claims that the petition was merely another protest by pious Quaker fanatics. To an extent, the PAS's strategy worked: the 1790 petition sparked one of the first debates in Congress about slavery's place in the nation's future. But the debate's resolution was hardly worth celebrating. On March 5, 1790, Congress issued a report that concluded that what the Constitution said "relating to the abolition of slavery" was clear: though Congress had some authority to regulate the importation of African slaves, it had no power to interfere "in the emancipation of slaves" or with those who may be "imported into or born within any of the said States." The report did, however, express hope that individual states might take it on themselves to revise their laws "from time to time" and make every attempt to "tend to the happiness of slaves."[32] Franklin could not have said it better. Compromised, and calculated, the report allowed the nation, represented by Congress, to present itself as a defender of liberty while excusing slavery's continuation in a few aberrant states.

Franklin died on April 17, 1790, roughly five weeks after Congress issued its report, and just as the organized antislavery movement was getting on its feet. But Franklin knew exceedingly well the person who would help steer the movement forward in the following two decades: Benjamin Rush. A member of the PAS since 1784, Rush became its president in 1803 and served in that position until his death ten years later. Born forty years after Franklin, in 1746, Rush met Franklin while studying for his medical degree at the University of Edinburgh, in the late 1760s, and corresponded with Franklin for the remainder of his life. Rush worked hard to secure Franklin's antislavery reputation. After privately visiting Franklin's barely expired corpse, Rush wrote in his diary: "To record all the exploits of his benevolence . . . would employ a volume," alluding to Franklin's antislavery work. As Franklin's body lay cold and lifeless, Rush plucked a "lock of hair" from the body, sending it to a mutual scientific colleague in England.[33] It was as if Franklin's physical being embodied science and antislavery, twin pillars of a transatlantic enlightened culture.

Like Franklin, Rush believed slavery threatened the nation's future as a white republic. And like his mentor, Rush was avidly pursued by the PAS's leaders. But the organization's leaders got far more out of Rush than political stature alone—they got one of the most sustained scientific critiques of slavery the movement had yet seen. In one of Rush's earliest antislavery pamphlets, he relied on environmentalism, the reigning scientific theory of race, to prove Black people's essential equality. But over the subsequent decades, he went far beyond environmentalism, leaning heavily on his own expertise: medicine. Whether from his lectern at the University of Pennsylvania's medical school or from the dais of the American Philosophical Society, Rush developed a litany of medical theories against slavery, all of which contended that slavery was as detrimental to the health of the enslaved as it was to the slaveholder. Indeed, in Rush's medical mind, slavery imperiled the physical well-being of the entire nation and put his hope for a slave-free white republic at risk.[34]

In 1766, a few years after graduating from the College of New Jersey (now Princeton University), Rush traveled to the University of Edinburgh, one of Europe's leading medical schools, for his medical degree. Rush had not yet met Franklin, but he dedicated his dissertation to him and sent him a copy. Flattered, Franklin invited the twenty-three-year-old to London and introduced him to his prestigious network of scientific friends, men like Sir John Pringle, "then the favorite physician of the Queen," as Rush recounted in his diary, and Dr. John Fothergill, the botanist and an architect of Sierra Leone. Franklin offered to pay for Rush's subsequent trip to Paris, providing him letters of introduction to similarly renowned scientific figures in France, many of them known antislavery sympathizers, such as the chemist Antoine Baumé and the Versailles botanist Bernard de Jussieu. Upon meeting Jacques Barbeu-Dubourg, a physician and Franklin's French translator, Rush noted in his memoir that Dubourg greeted him "in the following words: "Voila! Un ami de Mons. Franklin." Franklin opened doors, and in the years to come, Rush would rely on these scientific networks to circulate his antislavery writings and tighten the links between abolitionists and men of science.[35]

Upon returning to Philadelphia in 1769, Rush was appointed professor of chemistry at the medical school of the future University of Pennsylvania. Three years later, Benezet asked Rush to write an essay in the movement's behalf. It is unclear how Benezet knew Rush sympathized with abolitionism, or exactly why he contacted him, but Rush's scientific reputation probably played a role. With the help of Franklin and the university appointment, Rush's reputation as a man of science had been firmly established. Benezet may have also assumed, given Rush's scientific relationships, that he, like his friends, sympathized with the slave's cause. Whatever Benezet's precise logic, Rush eagerly accepted his offer, and the resulting essay, *An Address to the Inhabitants of the British Settlements, on the Slavery of the Negroes in America*, published anonymously in 1773, was the first antislavery act—indeed, one of the first political acts—Rush ever took.[36]

In *An Address*, Rush drew heavily on environmentalism, the dominant scientific explanation for racial difference, to refute planters' claim that Africans were innately inferior and predisposed to slave labor. Environmentalism was rooted in natural history, though one of its core assumptions—that all races derived from a single human pair, or what historians call monogenism—came from the Bible. According to environmentalism, racial differences emerged as humans migrated across the globe, their bodies, customs, and mental habits changing in accordance with their new natural environments, or "climates," as well as changes in the social environment. Yet while environmentalists assumed that all races were fundamentally equal, all equally human, they had no problem assigning certain physical and intellectual traits to individual "races." The key difference between environmentalists and later nineteenth-century racial theorists was that environmentalists believed race was malleable, changing as people moved to new environments. Over time, Blacks could turn white, and whites Black, the theory supposed—or, in a word, race was fluid, not fixed.[37]

In *An Address* Rush argued that, whatever inferior traits enslaved Africans allegedly exhibited, slavery itself—not anything innate to Black people—could explain them. Rush's claim fit neatly within the environmentalist framework, since slavery could be understood as a social environment that "degraded" the more positive traits of indigenous Africans. To make his case,

Rush cited the work of scientific explorers in Africa to demonstrate that, on their own continent, Africans were in every way "equal to the Europeans." Only when Africans were enslaved and brought to the New World did they begin to show signs of moral and intellectual degradation: "All the vices which are charged upon the Negroes in the southern colonies and the West-Indies," he wrote, are "the genuine offspring of slavery." Rush immediately sent copies of *An Address* to Granville Sharp, Britain's leading abolitionist, making sure he knew Rush's scientific credentials: he was in "the profession of physic," as medicine was often called. He also sent copies to his scientific friends, including Franklin and "our good Friend Mons. Dubourg of Paris." Though Franklin offered polite encouragement, not everyone was so kind. Richard Nisbet, a Caribbean slave owner, published a rebuttal dismissing the environmentalist explanation altogether: "The stupidity of the natives [of West Africa] cannot be attributed to *climate*," he wrote, concluding that "on the whole, it seems probable, that they are a much inferior race."[38]

Rush soon published a response, extending his environmentalist argument. While again stressing the essential equality of Europeans and Africans, and underscoring how slavery itself degraded Africans, he now naturalized racial stereotypes. He began from the premises that "Human Nature is the same in all Ages and Countries" and that "all the difference we perceive . . . may be accounted for from Climate, Country, Degrees of Civilization, form of Government, or other accidental causes." But to explain Africans' particular traits, none of them exactly favorable, he pointed to West Africa's climate and geography. The region's vast terrain created insurmountable distances between people, making it difficult to establish organized governments; without proper government, he argued, civilization could not take root. Meanwhile, the continent's extreme heat brought about "Indolence of Mind, and Body," making Africans vulnerable to enslavement—in other words, the climate made native Africans too lazy, too unintelligent to defend themselves. Here Rush recycled what had become a truism among eighteenth-century Euro-American naturalists, regardless of their views on slavery: that Africa's tropical environment, its "tropical exuberance," created a surplus of food that grew without effort, making its inhabitants unused to work. He then used this theory to explain, scientifically, why slaveholders often resorted to violence, and why, by extension, slavery was unnatural: "the Love of Ease which is peculiar to the Inhabitants of Warm Climates" necessitated "severe Laws and Punishments," he wrote. In Rush's environmentalist view, the violence of slavery did not simply emanate from planters' inhumanity—it was an inevitable consequence of the natural environment, of the "laws of nature."[39]

Having deployed environmentalism, Rush now turned to his true expertise—medicine—beginning a line of argument that he would develop over the next several decades. Rush countered Nisbet's argument that enslaved Africans' "carelessness in preserving their health" accounted for their high mortality rate, and he instead pointed to the conditions in which enslaved people lived. He attributed diseases associated with enslaved children, such as "jaw disorder," in which a child's jaw remained tightly shut, to poorly ventilated slave cabins. Relying on the medical testimony of Caribbean physicians, he argued that rape, euphemistically called "debauchery," led enslaved women to "procure repeated Abortions, which incapacitates them for Child bearing." Relatedly, he contended that their enslaved husbands lacked confidence in "the Fidelity of their Wives"—a direct reaction to slave owners' raping their wives—which made them less likely to care for their children and increased the likelihood of childhood illnesses. Fifteen years later, Rush expanded on these ideas and included them in a medical school lecture, sending it to a fellow physician and abolitionist in London, John Coakley Lettsom, who published parts of it "in the Morning Chronicle." In the unpublished medical school version, titled "Diseases of Negroes," Rush added new details. For instance, he redefined a disease in which enslaved people were found eating dirt—probably the symptom of a poor diet, perhaps a vitamin-B deficiency—and attributed it to slavery, not to Africans' innate savagery, as slave society physicians often argued. "Dirteatis," as Rush renamed it, occurred only "after they enter upon the miseries of their slavery."[40]

An Address also showed Rush's ambivalence about Black people's place within American society. Rush supported the standard gradualist approach, calling for an end to the slave trade first, but he also argued that the colonies needed to increase the number of "White People," so as to dilute what he implied was an unwanted Black demographic. When he addressed the more nettlesome question—what to do with those already enslaved in the thirteen colonies?—he offered what would become his trademark equivocation. "I would propose," he wrote, that enslaved people who were too old or had "acquired all the low vices of slavery" should remain enslaved, whereas younger captives should be educated and taught Christianity, so that they could support themselves once free. Provocatively, he suggested that the emancipated should be given "all the privileges of free-born British subjects." But then he hedged. He conceded that the issue of free Black equality was "a most difficult question," and concluded, "Let every man contrive to answer it for himself."[41]

As the conflict with Britain intensified, Rush began to use medical theory to link the Patriots' cause to the fight against slavery. He relied on the ideas of his Edinburgh professor, the physician William Cullen, who helped establish a new paradigm for understanding disease. Cullen broke away from the Galenic view, which held that diseases resulted from an imbalance of the four bodily humors, and instead argued that diseases were caused by either over- or understimulation of the nervous system, which in turn created a bodily imbalance. Cullen also argued that physical and psychological factors could contribute to the same disease. Because all diseases shared a common cause—a nervous system gone haywire—diseases were named according to the symptoms being shown, rather than the root cause. In the 1780s Rush would slightly amend Cullen's system, placing the vascular system at the center of his theory, rather than the nervous system. But the cures Rush prescribed followed the same basic logic: provide stimulants to enervating diseases, and depressants to exciting ones. In either case, the key was to restore balance.[42]

In the 1770s Rush used Cullen's framework to argue that any excess of power, any excess of wealth—whether wielded by British politicians or plantation owners—predisposed both the mind and body to disease. In a 1774 lecture delivered before the American Philosophical Society, titled "Inquiry into the Natural History of Medicine among the Indians," Rush argued that Britain's ruling elites became more vulnerable to diseases as they became overly dependent on commerce and the labor of their colonial subjects. Divorced from the benefits of moderate physical labor and lacking any purpose, they became prone to vices like drinking, which in turn predisposed them to disease. "How fatal are the effects of idleness and intemperance among the rich[!]" he exclaimed. Native American society provided Rush with an ideal contrast, a model for modern living. Living a simpler, independent life, neither dependent on nor oppressed by others, Native Americans had none of the diseases that characterized Britain's urban elites—venereal diseases, hysteria, "the NERVOUS FEVER." In addition, they experienced none of the emotional toil that resulted from the loss of liberty—or, as he put it, "political slavery"—which predisposed common British subjects to diseases that "most deform and debase the human body." To stave off a public health disaster in the colonies, he advocated for a society of independent yeoman farmers, imagining a mythic European past in which this once existed: "The abolition of the feudal system in Europe," he wrote, "by introducing freedom, introduced at the same time agriculture." Only a return to such a past—an agricultural society where freedom flourished and slavery was unknown—would "put a stop to these disorders."[43]

Rush's fame spread quickly after the 1774 lecture, and once the war broke out, he began to participate in politics. In September 1774 John Adams and Samuel Adams stayed with him when the first Continental Congress met in Philadelphia; Patrick Henry became his patient (Rush gave him an "inoculation for small pox," he later recalled); and Thomas Paine, newly arrived from Britain, asked Rush to comment on early drafts of *Common Sense* (1776), perhaps the most influential piece of Patriot propaganda from the period. In the summer of 1776, Rush became the only physician to sign the Declaration of Independence, and a year later he served as a surgeon general in the Continental Army. By the war's end, Rush emerged as not only one of the nation's most respected physicians, but also one of its most recognizable Patriots and abolitionists. Yet Rush was not above slaveholding. In 1784 he purchased William Grubber, his only enslaved servant; he freed him ten years later. Whatever reservations the PAS may have had about Rush's status as a slaveholder, they ignored it; among other attributes, his ability to wed science to antislavery made him too valuable. In 1787, the moment the society expanded membership beyond Quakers, it invited Rush to join. And like Franklin, Rush gladly accepted.[44]

The PAS got more than they could have wished. Throughout the 1780s, Rush developed even more medical theories that could explain the harmful effects of both political and chattel slavery. In a medical essay from 1789, Rush pathologized certain political behaviors, drawn from his experience in the War of Independence. He claimed that the deaths of four Loyalists were attributable to the "loss of former power or influence in government," which psychologically devastated them; he called this disease *"Revolutiana."* He contended that Patriots who became too zealous for freedom, too committed to radical democracy, suffered from "a species of insanity," a mental disorder that resulted from an "excess of the passion for liberty," which he called *"Anarchia."* In an unpublished lecture, titled "Diseases Caused by Government," he elaborated, writing that two opposing political systems—"despotic," or popular democracies, at one extreme, and monarchies at the other—led to emotional imbalances in both the ruler and the ruled that could trigger disease. In monarchies, "the alternate influence of liberty and slavery on the mind [acts] like a variable climate" and "produces a succession of extremes of excitement and debility, which have an unfriendly influence on the body." Republics offered the ideal middle ground, naturally balancing the mind's innate desire for liberty. A limited franchise for property-owning men furnished "an easy and certain channel thro' which those passions vent themselves."[45]

Rush's medical argument against tyranny easily transferred to slavery. At bottom, both tyranny and slavery stifled the mind's natural desire for freedom, causing an imbalance in the body, which induced disease. In "Diseases Caused by Government," Rush explicitly linked political and chattel slavery, writing, "There is in all slaves the absence of the stimulus of the love of liberty . . . and hence animal life exists in them in a feeble state," which predisposed them to diseases. Humanitarian causes, like abolitionism, functioned as a kind of medicine: they "have a friendly influence upon health," he wrote, presumably meaning that they buoyed the spirits and rebalanced the body, and added that such benevolent acts are "exerted most in a Republic."[46] In a related essay, titled "On the Different Species of Mania," Rush, somewhat playfully, suggested that anyone who showed too much enthusiasm for an "extreme" cause, political or otherwise, was simply suffering from insanity. He gave various examples. Defenders of slavery suffered from "Negro Mania," he wrote, since it must be some kind of insanity to employ enslaved people when there was "no reason why rice and indigo may not be cultivated by white men." Similarly, monarchists suffered from "monarchial mania," mistakenly believing that it was "criminal to depose tyrants."[47] Unlike other Patriot leaders, Rush did not simply make a rhetorical link between political and chattel slavery: he made the connection literal. Using medicine, he invited Patriots to see their own cause, of freedom and independence, as scientifically inseparable from abolition.

In the 1790s the antislavery movement faced increasing headwinds. The Haitian Revolution made antislavery societies easy scapegoats, as proslavery advocates argued that abolitionists would encourage a similar slave revolt if they continued their agitation. Meanwhile, northern white resistance to free Black integration served as another check on abolitionists' agenda.[48] To make matters worse, Rush increasingly aligned himself with a political party that drew considerable support from southern slaveholders. All these factors put immense pressure on Rush. Like many white abolitionists, Rush had to come up with answers for a nation increasingly skeptical of emancipation and bitterly divided over the nation's political direction. The medical answers Rush provided seem strange in hindsight, but they reflected common arguments white abolitionists found themselves making as they ran into a wall of resistance—the resistance of a nation that had never grappled with the consequences of emancipation for the nation's political future.

Not long after the Constitution was ratified, in 1789, a raucous political debate broke out over the direction of the new republic. On one side were Federalists, who supported a strong central government and a manufacturing-based economy; on the other side were Democratic-Republicans, who favored decentralized political power and an agricultural economy, preferably worked by independent white yeoman farmers. Rush's views aligned closely with Democratic-Republicans', and though antislavery was never part of the Democratic-Republican platform—indeed, the party's intellectual architect, Thomas Jefferson, captured all the party's contradictions about slavery—Rush often combined Jefferson's white yeoman ideal with his own more committed antislavery views.[49]

Science offered Rush proof that a nation of independent yeoman farmers, rather than slaveholders, manufacturers, traders, and bankers, provided the best foundation for a healthy republic. In a 1798 essay, Rush argued that Pennsylvania's state government, in his view a shining example of yeoman republicanism, grew organically from the state's natural environment. The colony's first settlers entered a vast, disease-ridden forest, he wrote, but over time, they tamed the land and planted modest farms; eventually, the average Pennsylvania yeoman worked his own farm and became "a man of property and good character," who "values the protection of laws" and "punctually pays his taxes towards the support of the government." If farming the land oneself led naturally to responsible citizenship, then slave owning did just the opposite. As he wrote, Pennsylvanian farmers may "possess less refinement than their southern neighbours, who cultivate their land with slaves," but "they possess more republican virtue."[50]

In another unpublished essay, titled "Diseases from the Different States of Society," Rush suggested that the more commercial and urban a society became, the more unequal and unhealthy it grew, referencing his 1774 essay on British and Native American diseases. In the margins of the unpublished essay he also noted that "one in six slaves die annually in the West Indies," suggesting that he may have wanted to draw a parallel between the unequal conditions in commercial societies and the unequal conditions in slave societies.[51] Taken together, Rush's medical theories implied that decadence in London was no different from decadence in Savannah. Extreme inequality—whether it was wealthy commercial elites relying on the labor of urban artisans, or slave owners relying on the labor of their slaves—created bodily imbalances that made both parties vulnerable to disease. By contrast, yeomanry encouraged independence, purpose, and equally shared if not ostentatious wealth, which in turn secured bodily balance and the health of the larger republic.

As the northern free Black population grew, Rush also had to address the question of Black citizenship. In his 1773 antislavery essay, he said little about skin color, writing only that Black skin was "far from being a curse" and in fact "qualifies them for that part of the Globe in which providence has placed them," meaning Africa.[52] But as the free Black population grew, race, with skin color as its shorthand, emerged as a fraught marker of citizenship. The nation's free Black population expanded significantly at the century's close, nowhere more than in Rush's Philadelphia. In 1790 the city's free Black population grew to 1,849, a sixteen-fold increase from 1775, when it stood at 114. Nationwide, the free Black population totaled 59,466, about 8 percent of the entire Black population, in a nation numbering nearly four million. When nearly all Africans had been enslaved, Black citizenship was a moot point. But as the free Black population grew, the nation increasingly had to consider whether they deserved equal political standing.[53]

The exact requirements for citizenship were not defined at the nation's founding. The Constitution made no mention of racial qualifications, and in certain northern states free Blacks could technically vote. But in practical terms, de facto discrimination made a mockery of the Black franchise and Black citizenship. Laws from the colonial era barring intermarriage remained intact; free Blacks were routinely stuck in the least desirable jobs, often teetering on the brink of poverty. Because suffrage was usually based on property ownership (and of course, gender), free Blacks rarely could vote, regardless of whether they had the right. Of more immediate concern was their safety. The 1793 federal Fugitive Slave Law allowed slaveholders to cross state lines to re-enslave runaways, which led to a wave of free Blacks being captured in northern states and sold back into the South. Meanwhile, in the 1790s "a climate of postrevolutionary racial retrenchment" undermined Black attempts at integration. White churches often barred Blacks from joining or forced them to sit in segregated pews; mob violence became routine. Northern states began to restrict the number of free Black residents, especially as thousands of Black Haitian refugees arrived in the United States in the 1790s. In 1800 Boston lawmakers even expelled 239 Blacks from out of state.[54]

It was in this climate that Rush delivered his 1792 lecture on the causes of Black skin. Rush's theory was one of several like it, his own essay framed as a response to a famed 1787 lecture by Samuel Stanhope Smith, a Princeton professor and minister. Smith's address, "Essay on the Causes of Variety of Complexion and Figure in the Human Species," attacked polygenism, the theory that Black and white people might be entirely separate species, which slaveholders like Thomas Jefferson increasingly aired in public. Smith argued that polygenism defied both religion and science, drawing on environmentalism to prove it. Every respectable naturalist knew that racial groups derived from a single human pair, Smith wrote, adding that dark skin simply resulted from centuries of exposure to the scorching sun. According to Smith, constant exposure to the sun's heat increased the amount of bile in the body, which in turn darkened the skin. Rush fully endorsed Smith's monogenist premise. But what made Rush's essay unique was that he tried to move beyond the standard environmentalist narrative and base his explanation in distinct medical terms, not environmentalist generalities. In the unpublished draft of his lecture, Rush underscored this distinction, writing that while environmental factors such as diet and climate were important, disease was central: Blackness, he wrote, "must be the effect of *Disease*."[55]

For Rush, leprosy lay at the root of Black physical features, a disease he found useful for several reasons. For one, Europeans had long associated leprosy with uncivilized, unclean ethnic groups—a common stereotype of Africans. The disease was also believed to be hereditable, and Rush could cite a lengthy record of leprosy in both modern and ancient Africa, the latter where he claimed Black Americans' ancestors first contracted the disease, along "the ancient miry canals of Egypt." Though white skin lesions were then considered a common symptom of leprosy, Rush could nonetheless cite numerous case studies where Black lesions seemed equally symptomatic. Perhaps more important, other physical features white people then ascribed to the Black body—"thick lips, woolly hair . . . short, ugly, ill proportioned"—could be found in leprosy cases. What made leprosy additionally useful was that it was thought to be potentially contagious, a fact that allowed Rush to temporarily endorse segregation, or, as he put it, "the necessity of keeping up that prejudice against such connections with them." In a letter Rush wrote to Jefferson in 1797 about the leprosy lecture, Rush underscored the point: "The inferences from" the lecture, Rush explained, were that whites should be "keeping up the existing prejudices against matrimonial connections with them"—meaning former enslaved laborers.[56]

But more important than the specific disease was the fact that a disease could explain Blackness at all. By contending that Black physical features were the symptom of a disease, rather than being caused by the natural environment, Rush could suggest, however awkwardly, that Black people could one day be assimilated into white society; after all, if Blackness emanated from illness, then it could be cured. Portraying Blackness as the symptom of a disease also allowed Rush to reassert Black peoples' humanity, since, according to his logic, beneath their diseased symptoms, they were just as white, just as human, as everyone else. In addition, Rush could use the idea that Black people were suffering from a disease to induce sympathy in his readers: Black people deserved "a double portion of our humanity," not disdain, he wrote. Last, by pathologizing Blackness Rush could summon the heroic narrative of medicine and implicitly encourage more men of science to join the antislavery crusade: "Let science and humanity combine their efforts, and endeavor to discover a remedy for it," he wrote. In the process, "we shall in the first place destroy one of the arguments in favor of enslaving the negroes."[57]

Rush ultimately capitulated to white fears of Black citizenship, but he also subtly challenged them. At a time when most white Americans believed that the only alternative to perpetual enslavement was either freedom and second-class citizenship or the forced removal of Black Americans, Rush was suggesting that Black Americans could become full and equal citizens—once they were cured of their color. Though this theory betrayed the same deep-seated racism that suffused much white antislavery thought at the time, it also showed Rush struggling to find an answer to the question of Black citizenship, and one that did not necessitate their removal.[58]

Rush's theory is easy to write off as a curiosity. But the essential assumption it upheld—that Blackness was somehow unnatural—had many scientific adherents. Throughout the eighteenth century, European naturalists assumed that, if Africans were transported to a cooler northern climate, they would gradually turn white. Travel narratives to Africa pointed to African albinos as proof of the body's natural whiteness, a fact Rush's lecture mentioned. Then there was the peculiar fascination with "white negroes," or Black people who appeared to be turning white, which captured the attention of many men of science at century's end. Rush cited several examples, including the most famous case at the time, a free Black Virginian named Henry Moss. Moss had what modern physicians call *vitiligo*, or the gradual loss of skin pigmentation. But to Rush and many other scientific figures, the condition suggested that white was the natural, healthiest skin tone. Rush mentioned numerous experiments performed on Moss and other vitiligo patients, from splashing acid on Black people's bodies to bleeding and purging them, all of which seemed to "lessen the Black color in negroes." As late as 1803, Rush was still experimenting on Moss, who reported that Rush had "blistered him &c, but to no purpose."[59] Rush's specific theory about leprosy may not have gained many adherents, but it clearly echoed a widespread belief that Blackness was somehow deviant from the white norm, and perhaps, one day, science might help it disappear.

Rush's attacks on slavery were not unanswered. Proslavery physicians responded vigorously, if not to Rush's specific ideas, then to abolitionist critiques in general. When the context was right, as it was in the 1790s, they could hitch their proslavery medical theories to the public mood just as effectively as Rush did himself. Proslavery men of science mounted a particularly threatening assault on abolitionists in the 1790s and early 1800s, amid the yellow fever outbreaks that struck numerous Atlantic port cities. Philadelphia experienced one of the worst outbreaks in the summer of 1793: the death toll reached four thousand, nearly one-tenth of the city's population. The inability to stop the epidemics created a crisis within the transatlantic medical community. Because so many physicians were offering opposing theories, political insinuations were often leveraged to make an argument stick. One core disagreement revolved around whether yellow fever was an imported, contagious disease passed from human to human, or whether it was a noncontagious, domestic disease caused by distinct local conditions. Many factors influenced whether a physician fell within the contagionist or localist camp. But the politics of antislavery provided an important subtext, providing Rush with a challenge that perhaps only he knew how to answer.[60]

Prominent proslavery British physicians tried to bolster the contagionist theory by associating localism with abolitionists. In 1795 Colin Chisholm, a Royal Navy physician and Grenada plantation owner, published a widely circulated essay that argued that the recent epidemics had originated from an abolitionist ship, the *Hanky*, that docked in Grenada in 1793. The *Hanky* was "chartered by the Sierra Leona [*sic*] Company," he wrote, referring to Sierra Leone, the colony for freed Black people established by British abolitionists in 1787. The Sierra Leone Company's purpose, he reminded readers, was "the fanatical enthusiasm for the Abolition of the Slave Trade." Chisholm was probably motivated not only by his personal disdain for abolitionists, but also by Britain's failing attempt to capture Haiti and turn it into a British slave colony. The Royal Navy had been trying to take over Haiti since 1793, but yellow fever and malaria outbreaks were undermining their efforts.[61] By arguing that yellow fever did not originate in Haiti, or anywhere in the Caribbean—but in West Africa, and from an antislavery colony no less— Chisholm could deflect attention away from both diseases' decimation of British sailors and encourage British officials to continue fighting for Haiti.

Many of Rush's former students called him out for trying to smear abolitionists. In 1798 Elihu Smith, a New York physician and antislavery sympathizer, wrote that Chisholm, whether "from inattention, or ignorance, or design," was tainting "philanthropic gentlemen" whose only crime was trying to stop "the iniquitous traffic in human flesh." One year later, Noah Webster, a staunch abolitionist, highlighted Chisholm's proslavery agenda, calling it "a labored attempt." Webster cited Rush generously in his text, then wrote to Rush directly, asking if he would publicly support the localist theory: "I believe *you* to be the advocate of *truth*," he wrote to Rush in December 1797.[62]

By the turn of the century, the association between the contagionism and proslavery had become clear. Only after that point did Rush side with the localists. In 1803 Rush published an essay in the *Medical Repository* announcing that he no longer believed the fever "was a highly contagious disease," and that he was "indebted" to "Mr. Webster's publications," as well as his medical students, for changing his view. Rush also apologized for having previously supported contagionism, noting that he had in part been misled by "West-Indian writers," a subtle jab at slaveholders. By siding with the localists, however, Rush risked implying that the local American environment caused the disease, which in part explained why so many American physicians avoided it.[63] Rush skirted the appearance of being unpatriotic, however, by associating the outbreaks not with the nation writ large, but only with the aspects of the nation he found disreputable: namely, slavery and commercialization.

In Rush's account of the 1793 Philadelphia outbreak, he named the unsanitary urban environment—not the healthful, airy "hilly country"—as a key contributing factor. Only moist, hot, unventilated urban spaces, he argued, provided the necessary conditions in which the disease could spread. Federalists accused Rush of being "very mischievous," scheming to undermine the nation's commercial interests by implying that cities were in part to blame for the disease. But slaveholders may have been equally annoyed, for Rush's theory also impugned slavery. He tied the Philadelphia outbreak not only to the urban environment, but also to decaying coffee—widely known as a slave-grown cash crop—that had been dropped off on the city's docks. Left to fester in the humid ports, he reasoned, the rotting coffee released noxious fumes that, when inhaled, triggered the disease.[64] Rush's theory ultimately advanced on all the medical arguments he had been making in favor of a slave-free yeoman republic for decades. He rejected the contagionist theory partly because of its proslavery associations, then shifted the blame to Philadelphia's urban environment and the slave-grown goods its merchant class imported. In doing so, he implied that there was only one way to prevent future yellow fever outbreaks: create a yeoman society, one in which Americans no longer needed to depend on commerce and slavery.

Rush spent his final years defending his scientific and political reputation. The two had become so intertwined that defending one almost always meant defending the other. When Jefferson won the presidency in 1800, William Cobbett, the British-born Federalist, journalist, and slavery apologist, highlighted Rush's antislavery views to ridicule his medical theories. He emphasized Rush's work alongside "two Revered Negroes"—Absalom Jones and Richard Allen, leaders of the free Black community in Philadelphia—during the 1793 outbreak to impugn Rush's use of bloodletting. Jones and Allen, Cobbett wrote, had the least bit of training, yet they ran about the city yelling, "Purge and bleed! Purge and bleed!" Rush's remedies, he sneered, probably received only the approval of "King Touissant," a clever allusion that linked the blood spilled in Haiti to the blood spilled in bloodletting. Cobbett even singled out Rush's essay on Blackness, mocking the notion that "the colour of the Negroes proceeds solely from the *Leprosy*, and that, when the race shall be *purged* of that disease, they will all turn white!!!"[65]

Rush defended himself in print throughout the following decade. He attacked his critics for using politics, not science, to delegitimize his medical work. In an essay titled "A Defence of Blood-Letting," reprinted in 1809, he lamented how political rivalries often led British physicians to disassociate themselves from any practice deemed too American. "In contemplating the prejudices against blood-letting," he wrote, "I have been led to ascribe them to a cause wholly political." If a political enemy of the British ate soup, he wrote, the British would eat meat, and likewise, if physicians of a political rival "advise bleeding," then "English physicians forbid it." "Here then," he concluded, "we discover the source of the prejudices and error . . . upon the subject of blood-letting. They are of British origin."[66] But it was a futile line of argument. Rush had long embedded political agendas in his medical work. To chastise his critics for doing the same smacked of hypocrisy.

While defending his reputation in print, Rush continued to work quietly in the PAS's behalf. After becoming president in 1803, he supported the organization in their lobbying of Congress to enforce the African slave-trade ban, passed in 1807. On the state level, he oversaw the PAS as it sued for the freedom of enslaved people who ran to freedom in Pennsylvania. Yet all the while, Rush seemed ambivalent about Black citizenship. On the one hand, he raised money for the first free Black institutions in Philadelphia, which suggests that he may have hoped Blacks and whites could live equally, side by side. "Our African friends continue to flourish in Philadelphia," he wrote to Granville Sharp on June 20, 1809, hopeful about Black people's future. On the other, evidence indicates that he might have continued to support segregation: in 1804 he purchased land in western Pennsylvania for a free Black settlement, which he had described years earlier as a project that might cure Blacks of the "vices" they "contracted in slavery."[67] Perhaps he believed that segregation was only a temporary measure, something that could end as soon as Black people became more like the rest of Americans—in a word, more white. Whatever the intentions of the Black settlement, no one took him up on his offer.

By the end of his life, Rush increasingly worried about the future of both the antislavery movement and America's experiment with republican government. Every day wars throughout the Atlantic world seemed to swing nations from one political extreme to the other, he told his close friend and former U.S. president John Adams in June 1806: "If we fly from the lyon of despotism, the bear of Anarchy meets us." He began to doubt whether advances in science and the spread of revolution might ever bring about the kind of slave-free, yeoman republics he had hoped for, whether at home or abroad. Science and "even liberty," he wrote, "long ago failed in their Attempts to improve the condition of mankind." Nonetheless, he would go on using science to promote his political views. "Tomorrow I expect to close my lectures," he told Adams three years later. "The Subject of my last lecture will be 'the diseases of the *eyes & ears*,' " at last showing a bit of humor. The nation had become blind to the dangers of centralized banks, he said, and deaf to how Britain had been impressing American sailors. Difficult as these diseases were "to cure in the human body," he wrote, "they are far less so, than when they affect those two Senses in public bodies."[68] Abolitionist leaders may have been disappointed by his pessimism, but they would have appreciated the scientific analogy.

In addition to devising medical arguments against slavery, Rush scoured the country looking for Black men of science. Nothing symbolized intelligence more than scientific achievement, and in seeking out Black men with clear scientific skills, Rush felt he could help abolitionists prove that Black people were, apart from their outward appearance, equal to whites. In 1788 he wrote about the "extraordinary Powers of Calculation" of a "Negro Slave, in Maryland," named Thomas Fuller, who, if given a person's age, could calculate the number of seconds that person had lived in no more than "a minute and a half."[69] Had Rush spent more time in Maryland, he would have found yet another example: Benjamin Banneker, a self-taught astronomer and one of the PAS's most beloved "Black exhibits"—a term one scholar has used to describe the circuslike specimens white society turned Black intellectuals into, to "showcase [the] Black capacity for Whiteness." Born free in 1731, Banneker spent most of his adult life as a tobacco farmer in rural Maryland, in a small town about ten miles west of Baltimore. At night, however, he turned to his true passion: science. In 1753 he taught himself how to construct a wooden clock, a remarkable feat that turned him into a local celebrity. His unpublished astronomical journal reveals an acute observer of the natural world: in 1749 he noticed locusts that appeared one summer "but they like Comets make but a short stay with us." Every seventeen years, like clockwork, they reappeared, and Banneker soon realized he was witnessing the unique mating pattern of North American cicadas.[70]

But it was the night sky that fascinated him most. His journal is filled with the painstaking calculations required to construct an ephemeris, or astronomical table, which formed the backbone of the almanacs he would eventually publish, and which would lead historians to count him among the first Black abolitionists. Yet those who knew him personally remembered him first and foremost as an astronomer, a man devoted to his scientific work, not an antislavery activist. Martha Tyson, the daughter of George Ellicott, a close friend who lent him his scientific books and instruments, recalled that after he started to learn astronomy in the early 1780s, it "became *the one* study to which all others yielded precedence." In 1796 Susanna Mason, an abolitionist who had worked with Benezet, traveled to meet the famous "sable son of science." She found him still living in a "lowly dwelling" made of logs and eager to show off his scientific work. He made a point of showing Mason his "wooden clock of his own constructing," as well as his almanac. Jacob Hall, one of the few free Black men who had known Banneker since childhood, told Tyson: "All his delight was to dive into books." Even a local Maryland slaveholder, Charles Dorsey, remembered him as being engrossed in mathematical equations. Banneker would bring "abstruse questions in arithmetic" he composed himself to the local town store, Dorsey recalled, a sight that "made so deep an impression on my mind" that he never forgot it.[71]

Banneker's commitment to science made his antislavery reputation possible, yet scholars tend to take his scientific work for granted, casting it as a mere vehicle for his antislavery views.[72] But for Banneker, it was the opposite way around: science was at the core of his identity, and while he cared deeply about ending slavery, he was neither a radical nor a reliable abolitionist—he was a reluctant abolitionist. It is easy to romanticize Banneker's engagement with abolitionism, but to do so obscures the fact that activism, especially for a Black man, was dangerous: his home was shot at three times and burnt to the ground two days after his death.[73] Joining the movement when he did, in the 1790s, also required brooking the routine condescension of the white abolitionist elites and supporting an agenda that conflicted with his own. The movement's leaders valued Banneker not for his ideas about antislavery, Black citizenship, or even science, but for what he embodied, an exhibit of Black intelligence. Banneker understood his race made him an asset to the movement, and he even shrewdly played to the era's racial expectations. But he did so as much for what he could get out of the movement as for what they could get out of him.

Banneker's decision to aid abolitionists was in part personal. As a Black man, he had little chance of being published in an elite scientific journal. Almanacs, however, were the only print medium where scientifically skilled men, and even women, could find publishers. In the seventeenth century, almanacs' ephemerides were calculated mostly by Harvard students, but by the eighteenth century they were increasingly written by less elite figures — surveyors, navigators, statesmen, even a few women, and, in Banneker's case, a Black man. Though almanacs lacked the prestige of elite journals, they were scientific nonetheless. Perhaps more important, they were popular: by the American Revolution, no other print medium had a wider circulation, not even the Bible.[74]

Few did more to open the genre to less well-trained intellectuals than Benjamin Franklin. With the appearance of his own almanac, *Poor Richard's Almanac*, in 1732, Franklin transformed the medium from a purely practical publication into a political and educational one. Like all almanacs, the core of *Poor Richard's* was the ephemeris. These astronomical tables predicted the rising and setting of the sun, the phases of the moon, the times and heights of the tides, and the weather, all of which made them immensely popular among farmers and sailors. In addition, *Poor Richard's*, like most other almanacs, included information on court dates, interest rates, religious and political holidays, and basic medical remedies. Franklin's novelty was to embed the almanac with literary essays, scientific knowledge, history, and political writings.[75] By making almanacs intellectually respectable, even political, Franklin paved the way for their later use by Banneker and abolitionist leaders. But if abolitionist leaders saw almanacs mainly as a medium to prove Black intelligence, Banneker had a more expansive view in mind: he would use them to showcase his own scientific skills and to advocate for the kind of multiracial republic few white Americans were willing to accept.

The idea to publish an almanac originated with Banneker, but there is little evidence that he initially planned to publish one for abolitionist ends. In 1788 Banneker began to teach himself astronomy with the help of George Ellicott, the son of a white Quaker family that owned the town's general store. Ellicott, nearly thirty years younger than Banneker, lent him Ferguson's *Astronomy* and Leadbetter's *Lunar Tables,* both popular textbooks, as well as a telescope, and probably let him use his celestial globe. In 1789 Banneker told Ellicott that if his skills improved, "I Doubt not being able to Calculate a Common Almanack." It was the first indication that Banneker might want to write an almanac, but he made no mention of wanting to use it for a political purpose. In any case, the first two printers he asked to publish his ephemeris, both of whom rejected him, were not the best-known publishers of abolitionist works. Only the third publisher he approached, John Hayes of Baltimore, stood out as an obvious abolitionist printer.[76]

Hayes was a founding member of Maryland's short-lived antislavery society, established in 1789. After the first two publishers rejected Banneker, Hayes told him "he would gladly" print the ephemeris—"provided the Calculations Came any ways near the truth," as Banneker recalled in a letter to George Ellicott's older cousin, the surveyor Andrew Ellicott, on May 6, 1790. Banneker had written to Ellicott because Hayes told Banneker that he was going to send Ellicott the ephemeris, to check its accuracy. Only at that point, already twice rejected, did Banneker mention his race: "I hope that you will be kind enough to view with an eye of pitty [my ephemeris]," Banneker wrote to Ellicott, "as I suppose it to be the first attempt of the kind that ever was made in America by a person of my Complection."[77] Yet Banneker still made no mention of the almanac carrying an antislavery message, and he seems to have been playing to anti-Black stereotypes only to increase the chances of being published. After all, Banneker was not saying that his ephemeris *proved* Black intelligence, the reason abolitionists would find it useful; he was saying just the opposite—that Ellicott should forgive any mistakes he may have made *because* he was Black.

Hayes's decision to send the almanac to Ellicott backfired. Ellicott was the author of the almanacs Hayes already printed, so Hayes was in effect asking Banneker's competitor to approve his work. Ellicott's response has not been found, but Hayes ultimately turned Banneker down.[78] Yet once abolitionists got wind of Banneker's interest in publishing an almanac, they quickly strove to get his work into print—and making his race, and its relevance to the antislavery cause, the central feature. Banneker may have seen it coming. Editors had already used his scientific skills to make a point about Black potential. In 1791 Thomas Jefferson commissioned Andrew Ellicott to be the lead surveyor for the new federal city (Washington, D.C.), and Ellicott asked Banneker to be his assistant. For two months, beginning in February 1791, Banneker kept Ellicott's instruments in working order—telescopes, stopwatches, sextants, an astronomical clock—while quietly taking down calculations in his personal astronomical journal. When editors found out that Banneker, not yet widely known, had been on the survey team, they used it to discount the idea of Black inferiority: Banneker was "an Ethiopian," Philadelphia's *General Advertiser* proudly announced, emphasizing how his race undermined the notion that Black people were "void of mental endowments."[79]

If white antislavery sympathizers, like the editors at the *General Advertiser,* cared about Banneker's scientific skills only for what they said about Black potential, then Banneker's colleagues on the survey seemed not to care about his skills at all. Martha Tyson, George Ellicott's daughter, recalled years later that although the fellow engineers on the survey "disregarded all prejudice" when they realized Banneker's abilities, he still had to eat separately. Tyson, who left the only known account of the episode, tried to put a brighter gloss on the event, writing that the white engineers invited Banneker to sit with them, and that Banneker, out of "his characteristic modesty," declined the offer.[80] But it may be that he was simply never asked, or that he wanted to save them the discomfort. Either way, the experience revealed how racial prejudice belied the ideal of science as a universal enterprise, open to all.

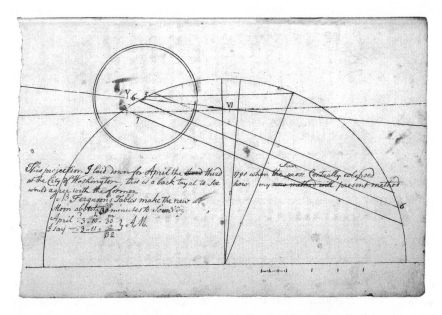

Fig. 3. Benjamin Banneker's Astronomical Journal, diagram of an eclipse, April 3, 1791.

Benjamin Banneker drew this diagram of a solar eclipse while working on the surveying team for Washington, D.C. The national attention he received on the survey enabled him to publish his first antislavery almanac by the end of 1791. Courtesy of the Maryland Historical Society, MS2700.

The publicity Banneker received for his work on the survey coincided with the PAS's own effort to get Banneker's almanac published. Franklin's successor as PAS president, James Pemberton, who had made sure Franklin signed the congressional petition just before he died, played a critical role. After Banneker's attempt to publish his own almanac in 1790 went nowhere, Pemberton asked the Maryland society president, Joseph Townsend, for help. In a November 1790 letter, Townsend told Pemberton that he would look into "the Black man's calculations" as soon as possible. George Ellicott's younger brother, Elias, elected to the PAS in 1790, also threw himself into the effort. In May, Elias told Pemberton that Banneker was now happy to make note of his race: "He thinks as it is the first performance of the kind ever done by One of his Complection that it might be a manes [*sic*] of Promoting the Cause of Humanity as many are of Opinion that the Blacks are

Void of Mental endowments."[81] By then Banneker seems to have realized that his race, if played right, could help both his personal ambitions and the cause of abolition. Meanwhile, Pemberton, the Ellicotts, and Townsend understood that Banneker offered something unique, something different from other eighteenth-century Black intellectuals. Unlike Phillis Wheatley or Olaudah Equiano, Banneker was not offering literary evidence of Black intellectual talent: he was offering *scientific* evidence.

Yet Banneker's scientific abilities were not allowed to speak for themselves. Until free Blacks established their own newspapers in the 1820s, Black authors needed to have their work publicly authenticated by white authors since many readers doubted that Blacks wrote these texts themselves. Banneker's work posed an additional challenge: it did not simply need the validation of respectable white men—it required the validation of respectable white *men of science*. Whether Pemberton realized it or not, Banneker's unique contribution to abolitionism—his scientific work—had the added benefit of seeming to enlist even more men of science into the movement; in no time, many of the nation's leading men of science publicly vouched for Banneker's work. David Rittenhouse, the nation's most renowned astronomer and president of the American Philosophical Society, sent the PAS a formal endorsement: "A very extraordinary performance, considering the Color of the Author," he wrote to Pemberton on August 6, 1791, after receiving a manuscript copy. William Waring, another astronomer and member of the American Philosophical Society, also gave his stamp of approval: "I have examined Benjamin Banneker's Almanac for the Year 1792," he wrote on August 16, 1791, and "am of the Opinion" that it deserves to be published. A former student of Benjamin Rush, Dr. James McHenry, offered an even fuller endorsement: "I consider this Negro as a fresh proof that the powers of the mind are disconnected with the colour of the skin."[82]

Pemberton solicited most of these figures, but Banneker contacted one man of science himself: Thomas Jefferson, then secretary of state. Banneker's letter to Jefferson has received inordinate scholarly attention, overshadowing the fact it was the only explicit antislavery writing Banneker ever wrote, and ignoring its original purpose. Banneker primarily meant it to be a *private* letter, in part hoping that Jefferson, privately and personally, would give him the scientific respect he knew he deserved. In the letter, Banneker made clear that he did not initially plan to write about slavery: "Originally my design," Banneker wrote, was only to send his completed almanac "as a present," nothing more. Regardless of whether he was being coy, the statement underscores that Banneker knew his most powerful antislavery weapon was not his words, but his science. His astronomical work would stand as an implicit refutation of Jefferson's notorious contention that Black people were "in reason much inferior." Moreover, Banneker offered to publish the exchange only after his Baltimore editors refused to share the McHenry letter with other almanac publishers; Jefferson's letter would function as a replacement, authenticating Banneker's work: "My letter to Mr. Jefferson and his answer to me," Banneker wrote to Pemberton on September 3, 1791, could easily substitute for McHenry's letter.[83]

Rather quickly, Banneker got what he wanted. Jefferson wrote a respectful reply, which must have given Banneker immense personal satisfaction. Better still, Jefferson forwarded Banneker's ephemeris to an even more towering man of science: "Monsieur de Condorcet, Secretary of the Academy of sciences in Paris," as Jefferson told Banneker. Jefferson seemed to get something out of the exchange too. In sending Banneker's letter to Condorcet, Jefferson could win back a bit of lost scientific respectability. Jefferson's speculation about innate Black inferiority, published in *Notes on the State of Virginia* (1787), was at odds with mainstream scientific opinion— "rather old-fashioned," as one historian has put it—despite its often being seen, inaccurately, as representative of scientific ideas about race at the time. In fact, until the early nineteenth century, environmentalism, and its assumption that all races were fundamentally equal, held sway. By sending Banneker's letter to Condorcet, Jefferson could erase his outré racial remarks and imply that he now accepted environmentalism. As Jefferson wrote to Condorcet, Banneker's skills "prove that the want of talents observed in them is merely the effect of their degraded condition."[84]

For Banneker, publishing his exchange with Jefferson would not only help him win scientific respect; it would also allow him to make public his views on slavery and Black citizenship—views that were at odds with the white republic imagined by Rush and Franklin. In the letter, Banneker told

Jefferson that slavery defied the spirit of the Declaration of Independence, asking him to remember how unnatural it felt to live under the "tyranny of the British Crown"; for the 700,000 still enslaved, Banneker implied, it felt no different. But unlike Rush and Franklin, Banneker refused to wish away the free Black population. Nor would he apologize for his Blackness: "Sir," he wrote to Jefferson, "I freely and Chearfully acknowledge, that I am of the African race, and in that colour which is natural to them of the deepest dye."[85] Banneker's confident identification with Blackness, combined with his justification for emancipation that was based on the nation's founding document, implied a desire for full Black citizenship. He offered no plan to voluntarily emigrate, an idea that would soon gain popularity among many white antislavery supporters; he offered no hope that Blackness would be cured.

For obvious reasons, a later generation of abolitionists and historians would focus on Banneker's letter to Jefferson without considering his motives for engaging with the movement. Foregrounding his scientific ambitions suggests that he was less a pioneering Black abolitionist and more an ambivalent one, a man who devoted the time he had outside farming to science, not activism, and who became involved in the movement only when he saw that his scientific abilities were suddenly marketable. Race may have bounded his life, but he was not bound by racial politics. He played no role building the free Black institutions—newspapers, churches, schools—to which men like Absalom Jones and Richard Allen dedicated their lives. And after writing his letter to Jefferson, he never wrote anything else against slavery. Instead, he allowed the publishers of his almanacs, twenty-eight editions printed over the next five years, to reprint other people's antislavery writings alongside the ephemeris he calculated each year. In a very literal sense, then, Banneker's almanacs embodied the nature of his abolitionist work: he did the science and gladly left it to others to write the antislavery words.

Banneker's own antislavery writings were minuscule, but his almanacs provided an important platform for the writings of other men of science. His editors were in charge of selecting the accompanying writings, and they seemed to relish works by or about Benjamin Rush and Benjamin Franklin, which enhanced the almanacs' scientific veneer. For a 1793 almanac, Banneker's Philadelphia editor, Joseph Crukshank, included an essay by Rush in which he called for a "Peace-Office" instead of a department of war. The same almanac featured a poem invoking Franklin, casting him as the embodiment of an enlightened age, one in which science, independence, and abolition went hand in hand: "The first spark that lit the mighty flame / From some lone hand—perhaps from Franklin came."[86] Of course, the inclusion of Rush and Franklin in Banneker's almanacs concealed their very different views on America's racial future. But neither the almanacs' editors nor abolitionist leaders were eager to point those differences out, for good reason.

Other Banneker almanacs used astronomy to make antislavery arguments. Some invoked the vastness of the cosmos to point out the absurdity of humankind dividing itself up according to small differences. If one could imagine observing human beings from the distance of other galaxies, a passage from a 1792 almanac read, they would see that all human beings looked exactly the same: "but a dim speck, hardly perceivable in the map of the universe." Another astronomical passage, taken from a 1775 essay written by Rittenhouse, took a similar tack, asking what extraterrestrial life might make of the violence humans inflicted on each other. The original essay hoped that life on other planets had not yet been corrupted "with our vices, nor injured . . . by violence." The almanac reprinted the lines that immediately followed, ones that attacked the enslavement of Africans "merely because *their* bodies may be disposed to reflect or absorb the rays of light in a way different from *ours*."[87] In noting that skin color represented nothing more than a difference in how pigment absorbed light, Rittenhouse suggested the fatuity of divvying up humankind by what were, at root, insignificant differences.

The history of astronomy in Africa was also invoked to prove Black intellectual potential. The 1793 almanac noted that astronomy was "first studied by the Moors, and brought to Europe" in 1201 CE, then later "taken up by Copernicus." While Banneker's editors no doubt hoped this history would remind white readers of Black people's essential humanity, it also reinforced the notion that civilization had long since collapsed in Africa. The implication was that people of African descent, whether in Africa or enslaved abroad, would again need to be civilized, retaught the sciences that Europeans had taken from them and developed. Whatever the implications, Banneker may

have approved of these passages, even though he was not responsible for selecting them. For one thing, they may have reminded Banneker of his grandfather, a West African chieftain named Bannaka, who some historians argue
may have inspired his grandson's interest in astronomy. Some time in the
late seventeenth century, slave traders captured Bannaka in Africa and sold
him to a former English indentured servant named Molly Welsh in Maryland. She soon freed him, and she married him around 1696.[88]

Banneker never met Bannaka, his maternal grandfather, but evidence
suggests that his memory left a powerful influence on the entire Banneker
family. When Martha Tyson went to interview Benjamin's remaining family
in 1836, they passed on vivid details of Bannaka: he was an "African prince,"
they told her, a "man of bright intelligence, and fine temper, with a very
agreeable presence, dignified manner, and contemplative habits." Bannaka's defiance also supports the claim of his highborn status: he refused to
convert to Christianity despite his wife's insistence; nor did he willingly
work as an enslaved field hand while under Welsh's ownership. Bannaka
may have even learned astronomy in Africa. His name suggests he was a Dogon, an ethnic group in modern-day Mali with a deep history of astronomical practice. Some scholars conjecture that, as a prominent Dogon, Bannaka
may have been taught astronomy and possibly passed that knowledge down
to Banneker through Molly.[89] Though there is no direct evidence for these
claims, it is at the very least possible that Bannaka's elite African status and
possible Dogon education provided an inspiration for Banneker to similarly
distinguish himself.

Despite the early success of Banneker's almanacs, sales plummeted after
1795. The ongoing revolution in Haiti cast a pall over the Anglo-American
abolitionist movement, cooling whatever public sympathy abolitionism had
had. Almost in an instant, his main asset—his race—had become his greatest liability. Yet Banneker continued to calculate an ephemeris in his astronomical journal, year after year, until 1805, a year before his death and long
after his editors had given up interest. Observations of the night sky also
filled his journal: he noted the appearance of a strange mist around nine
o'clock one cold evening in November 1798, the "Condensed particles of
the Atmosphere of divers colours gathering round the moon." A little more

than four years later, on February 2, 1803, he watched with wonder as the setting sun "beautified the Snow" that had fallen earlier that day.[90] That Banneker continued to calculate ephemerides even after his editors had lost interest, and that he never wrote anything else against slavery, illustrates the central role science played in his life. He sympathized deeply with abolitionism, but activism was not his life's ambition. Banneker's scientific interests also remind us that not all Black men and women devoted their lives to defending their race or fighting against slavery—and yet that, too, was its own political act. To live as one wished, not how society expected one to, was a certain kind of freedom.

In any event, Banneker had many reasons to quiet his antislavery views, the threat of violence chief among them. On August 27, 1797, he recorded "standing at my door" and hearing several guns unload, "one or two of which struck the house." Eight months later, on April 29, 1798, two men approached his home with a gun. On November 27, 1802, his house "was violent[ly] broke open and Several articles taken out." Two days after he died on October 9, 1806, an arsonist set his house on fire.[91] Banneker never speculated about why he was targeted, and several reasons unrelated to antislavery might explain it. But it would be impossible to discount his work in behalf of the antislavery movement as one probable cause. Antislavery activism was dangerous for anyone, but particularly for Black women and men. Banneker had the foresight, however, days before he died, to write a will. Never married and childless, he left what money he had saved to his sisters, Minta Black and Molly Morten. George Ellicott, meanwhile, received the only written record (which he kept close guard over), the only item that survived the fire: Banneker's astronomical journal. It seems to be what he wanted history to remember him for.

Franklin, Rush, and Banneker helped expand abolitionism's appeal beyond the movement's initial Quaker base. By offering a wide range of scientific arguments against slavery, and relying on scientific networks to circulate their work and strengthen their ties to the movement, they began to make antislavery appear not just moral, but rational—a movement backed by science. Whether it was Franklin's use of political arithmetic or embedding antislavery views in a scientific essay, Rush's numerous medical theories against slavery, or Banneker's use of astronomy to prove Black intellect, all three men showed how scientific knowledge could help broaden the movement. Yet their differences were also revealing. Neither Franklin nor Banneker was as committed to the organized movement, as is often presumed. Nor did the three of them share the same antislavery agenda. Rush and Franklin could not quite imagine free Black people as full and equal citizens, but Banneker, like many Black figures after him, would insist on it. Whatever uncomfortable questions their conflicting antislavery agendas made for the movement's leaders, their ability to critique slavery scientifically made them too valuable to ignore. In Britain, white abolitionists would also struggle to integrate the country's free Black population, and there too men of science would help make science and abolition seem like natural allies.

TWO

Full Steam Ahead

In 1787, the same year Franklin became president of the Pennsylvania Abolition Society, Britain's abolitionist leaders began searching for more scientifically minded allies. The British movement, initially centered on the Society for Effecting the Abolition of the Slave Trade (SEAST), grew in tandem with the American one, focusing most of its energy on ending the transatlantic slave trade, and for good reason. Before 1807, no country had transported more enslaved Africans across the Atlantic than Britain—3.3 million between 1662 and 1807, and at the slave trade's peak, on the eve of the American Revolution, an average of 42,000 enslaved people per year. In one of the SEAST's first meetings, on July 5, 1787, its members decided that, in order to raise awareness, they should design a logo that could be stamped on all its antislavery products: pamphlets, pins, bracelets, snuff-boxes. "Resolved," the society's minutes noted, "That a Seal be engraved for the use of this Society." The man they asked for help was Josiah Wedgwood, a wealthy potter—and also an accomplished chemist and a member of the Royal Society. Probably with Wedgwood's help, the SEAST came up with an image that featured "an African in Chains in a supplicating Posture, with the Motto 'Am I not a Man & a Brother' " on October 16. By the end of the year, Wedgwood was back at his factory in Stoke-on-Trent, about 170 miles northwest of London, working with his artists and hundreds of employees to create what would become the most recognizable piece of antislavery paraphernalia the movement ever produced.[1]

Fig. 4. Antislavery medallion, manufactured by Josiah Wedgwood, 1787.

Josiah Wedgwood, a chemist, abolitionist, and wealthy potter, manufactured thousands of these medallions for the antislavery movement. Both the image and the clay medallion drew on Wedgwood's extensive knowledge of chemistry, geology, natural history, and medicine. Courtesy of the Metropolitan Museum of Art, New York, Gift of Frederick Rathbone, 1908. Creative Commons.

Through the mass production of the "Am I Not a Man and a Brother?" medallions, Wedgwood helped popularize the antislavery cause. In the streets of London, Philadelphia, Birmingham, and New York, hundreds of antislavery supporters, many of them middle-class women, purchased Wedgwood items featuring the medallion. "Several wore them in bracelets, and others had them fitted up in an ornamental manner as pins for their hair," wrote Thomas Clarkson, a leader of the SEAST.[2]

Wedgwood's skills at marketing undoubtedly helped him seize the growing market for things fashionable and virtuous. But far less appreciated is the extent to which Wedgwood relied on scientific knowledge to marshal an audience for the antislavery medallions. Wedgwood was elected to the Royal Society, England's most prestigious scientific body, in 1783, and throughout his professional life he avidly followed the day's scientific research.[3] Chemists, geologists, physicians, and naturalists in fact played an important role in helping Wedgwood design and market the antislavery medallion. Chemists helped Wedgwood understand how to manipulate dyes and earthenware, which in turn produced more attractive medallions; geologists helped him search for raw materials, which he would use for the medallions. Physicians taught him about the physiology of emotions, influencing his decision to depict a supplicant enslaved man, begging for his freedom, which would better induce sympathy in the minds of his mainly white, often female consumers. The figure's scant clothing evoked the notion, put forth by natural historians, that indigenous Africans, before slavery, lived in a more peaceful, primitive state. Wedgwood absorbed many of these ideas by corresponding with a group of men of science, including Benjamin Franklin, Joseph Priestley, the discoverer of oxygen, and Erasmus Darwin, a physician, botanist, and grandfather to Charles Darwin, whose collective scientific work is often credited with laying the foundations for the Industrial Revolution.[4] Together these men were critical to crafting the narrative that slavery was antithetical to scientific progress.

Scientific ideas provided all these men with a language to critique established hierarchies. Wedgwood, Priestley, and Franklin had all been born to families of modest means and saw the rigid social hierarchies that defined eighteenth-century Britain and its colonies as obstacles to their own advancement. By the time of the American Revolution, they began to wed the image of science to their reformist politics. Wedgwood wrote that Priestley's and Franklin's experiments with electricity would make "the great ones of the Earth tremble," whereas Priestley, whose own experiments led to the discovery of oxygen, noted that even the sight of an "air pump, or an electrical machine" should give "English hierarchy . . . equal reason to tremble." There was nothing inherent in science that led these men to their antislavery views. But they folded the antislavery cause into their broader vision of social reform and depicted scientific progress as the central engine driving this social transformation. Darwin speculated that future discoveries in chemistry might one day make it possible to grow sugar in a laboratory rather than on a plantation. Wedgwood suggested that new machines sent to plantations might reduce the burden on enslaved laborers. Priestley described the abolitionist agenda in terms his readers had come to respect through his scientific writings. Scientific experimentation, he argued, required a respect for physical labor, gradual progress, empiricism, and modest arguments — precisely the values abolitionists shared themselves.[5]

The radical upheavals of the 1790s came as a surprise to these men. The French and Haitian Revolutions, coupled with women's growing self-assertion, threatened the model of gradual, orderly, male-led progress they believed was natural to the advancement of both science and antislavery. Wedgwood and Franklin died by the mid-1790s, but Priestley and Darwin lived another decade. Priestley fled to Pennsylvania in 1794 after a conservative mob attacked his Birmingham home and laboratory, while Darwin remained in Britain. But whether in England or America, both men would struggle to reconcile the decade's rapid and unruly changes with their vision of gradual, well-ordered progress, the vision on which their entire approach to science and abolitionism depended.

Little in Josiah Wedgwood's upbringing suggested an aptitude for science. He was born in 1730 to a long line of potters in the small town of Burslem, in midwestern England. The nearest school was seven miles away. Though his mother made sure he received some education, by the age of fourteen he began his first pottery apprenticeship. He eventually established his own pottery business and quickly realized the advantage that new machine technology, as well as chemical knowledge, could offer to his business. In 1763 he installed one of the first engine-powered lathes in a pottery factory; five years later he designed, together with his friend Erasmus Darwin, a horizontal mill that would pulverize stone for pigments—"an ingenious invention," as Wedgwood called it. All the while, he cultivated a taste for chemistry. According to one biographer, Wedgwood conducted nearly five thousand chemical experiments throughout his career; in fact, so confident was Wedgwood in his chemical skills that he routinely offered Priestley advice and supplied him with various experimental tools and materials: tubes, pipes, vanes, basalt, porcelain, and "*lava* from Vesuvius."[6]

Wedgwood, Priestley, and Darwin were part of a tightly knit group of scientific practitioners who lived in and around Birmingham. They called their group the Lunar Society, "because," as Priestley later recounted, "the time of our meeting was near the full moon." Between the 1760s and early 1790s, the members of the Lunar Society met regularly to discuss their research, much of it geared toward improving the efficiency of the manufacturers. For that reason, they are often viewed as helping pave the way for the Industrial Revolution.[7] Though membership was informal and fluctuated over the years, frequent guests included not only Priestley, Wedgwood, and Darwin, but also James Watt, the inventor of the steam engine, and Benjamin Franklin. For all of them, science's power lay not only in its practical utility but also in its ability to raise their social status. As their scientific reputations grew, they would use their newfound authority to challenge the established social hierarchy and secure their place atop the emerging industrial order that was quickly replacing it.

No one proved more adept at this than Benjamin Franklin. By the late 1740s, he had already made a fortune from his printing business, acquiring the luxuries, including enslaved people, that came with financial success. But he still craved scientific status, perhaps only the kind that British men of science, with their deeply rooted scientific institutions—the Royal Society was founded in 1660—could offer. Not coincidentally, he began to print his experiments on electricity at roughly the same time he began to pursue a political career in Britain. Peter Collinson, a close friend and natural historian in London, helped Franklin get his theories about electricity read before the

Royal Society in 1749, and he published them in London two years later. Almost overnight, Franklin became an intellectual celebrity all over Europe. In June 1752 London's *Gentleman's Magazine* noted that Franklin's experiments "have become famous" in France, and that King Louis XV "signified his pleasure of seeing the performance" of them. Two years later, the magazine published an ode to Franklin, likening his scientific work to something even greater, more noble than politics: "Let others muse on sublunary things / The rise of empires and the fall of kings"; this friend "to science, and to man" should receive whatever "honours that are virtue's meed." By 1757, his intellectual reputation firmly established, Franklin was appointed Pennsylvania's colonial agent in London. He spent most of the following two decades in England, reporting regularly to the king's deputies in Whitehall.[8]

Franklin's ability to transform himself into a virtuous, powerful gentleman on account of his scientific work proved immensely alluring to men like Wedgwood and Priestley. Priestley's own career as a chemist began when he decided to write a popular history of electricity based on Franklin's work. Titled *The History and Present State of Electricity* (1767), it presented Franklin as heir to "the glory of the great Isaac Newton," thereby securing Franklin's scientific fame. Franklin rewarded Priestley by helping him get elected to the Royal Society in 1766, calling Priestley a "very intelligent, ingenious and indefatigably diligent Experimenter." Seventeen years later, Priestley would do the same for Wedgwood. "I communicated your ingenious paper to Mr. Banks," Priestley wrote to Wedgwood in November 1780, regarding a new thermometer Wedgwood had designed: "Doubt not its being well received." By 1783 the Royal Society elected Wedgwood a fellow, lending him the kind of authority only science could confer.[9]

Scientific authority could be and often was leveraged for political ends. But in the late eighteenth century, men of science, particularly those affiliated with the Lunar Society, crafted an image of themselves that suggested just the opposite: that they were uncorrupted by politics. Relatedly, they suggested that they had unique access to the "secrets of nature," as Wedgwood put it, and were driven only by a desire to improve the well-being of all humanity. Wedgwood evinced all the hallmarks of this image in a 1766 letter commenting on Priestley's early experiments with electricity.

"Dr Priestleys very ingenious experiments" were not only "extensively useful," but almost superhuman: "What dareing mortals you are! to rob the Thunder of his Bolts, — & for what? — no doubt to blast the oppressors of the poor & needy or else to execute some public piece of justice."[10] If politicians accrued power for self-gain and by manipulating other men, then Wedgwood suggested that men of science wanted power only to do good and attained that power by manipulating a more awesome, mysterious, and divine source: nature itself.

Priestley's *History of Electricity* echoed and elaborated these views. Men of science were corrupted by neither politics nor money: let princes "fight for the countries when they are discovered," he wrote, "let merchants scramble for the advantages that may be made by them." Men of science alone enabled the "complete discovery of the face of the earth"; their "great inventions" allowed "mankind . . . to subsist with more ease." Priestley also suggested that studying science opened one's eyes to the suffering of mankind throughout human history, while also giving hope that scientific progress would help all societies overcome the "vices and miseries of mankind." History demonstrated that the "security, and happiness of mankind are daily improved" through the advancement of science.[11] The image these men crafted of themselves—politically uncorrupted, having access to nature's secrets, and imbued with an enlarged sense of humanity—reinforced one another. Unlike politicians, whose authority rested in their ability to command mere mortals, the authority of men of science derived from their ability to manipulate the entire natural world. By emphasizing that they served all humanity rather than any particular nation, they also reinforced the idea that they were above politics—an image that made men of science particularly useful to abolitionists, since the appearance of neutrality would strengthen the claims men of science would make in abolitionists' behalf.

Though men of science often portrayed their work as benefiting all mankind, there was nothing innate to science that inevitably drew them to the antislavery cause. As historians have long contended, arguments based in Enlightenment science could just as often condone slavery as they could undermine it. Moreover, Wedgwood and the network of manufacturers and men of science with whom he associated had long benefited from the Atlantic slave economy. The British West Indies had become the dominant market for British manufactured goods by the end of the eighteenth century: in 1700 the West Indies imported only 10 percent of all British exports; by 1797 it took in 57 percent. Though profits from slavery and the slave trade perhaps never made up more than 5 percent of Britain's national income in the eighteenth century, roughly 39 percent of slavery-derived profits were reinvested in activities that promoted industrialization, such as building new canals and purchasing steam engines.[12]

Wedgwood's private letters and receipts reveal his links to slavery firsthand. In 1766 he instructed his business partner, Thomas Bentley, to dump surplus "Green desert" dinnerware on clients in the "West India Islands." Ten years later, the slave-owning royal governor of Grenada, William Leybourne, placed an order for "the Yellow Ware." Even as Wedgwood's efforts in behalf of the antislavery campaign began in earnest, in the late 1780s, iron manufacturers near his Midlands factory continued to supply "muzzles or gags made at Birmingham for the slaves in our islands," as Darwin reported to Wedgwood in 1789. After the loss of the thirteen colonies forced Wedgwood to look for clients elsewhere, his company continued to sell to both West Indian planters and European slave traders in West Africa. In 1802, seven years after Wedgwood's death and his son now in charge, the Company of Merchants Trading to Africa, England's main slave-trading firm, placed an order for "Six Crates of Earthenware" and "Six Breakfast Setts," Wedgwood's financial records show; meanwhile, the Longlands of Jamaica ordered an "additional Cream mug" and a "Butter cup."[13]

Whatever the underlying realities, Lunar Society abolitionists continued to promote the idea that the spread of scientific knowledge would inevitably destroy slavery. Few were savvier than Priestley. Throughout the 1770s and 1780s, he wrote essay after essay arguing that advances in science would eradicate not only human suffering, but human evil itself. The diffusion of scientific knowledge would be "the means, under God, of extirpating *all* error and prejudice." Far from being the secular apostates that they are often made out to be, many Enlightenment men of science, and Priestley especially, saw no inherent tension between science and religion. In fact, science should only inspire divine reverence, Priestley argued. "The contemplation of the works of God," he wrote in the *History of Electricity*, "teach him to aspire to the moral perfections of the great author of all things." In *Experiments and Observations on Different Kinds of Air* (1774–77), he argued that all experimenters were but "instrument[s] in the hands of divine providence," working toward "some great purpose that we cannot yet fully comprehend." Certain scientific fields, such as astronomy, even elevated man's appreciation for his fellow beings—a godly act in and of itself. Anticipating the arguments that would appear in Banneker's almanac, Priestley argued in 1776 that astronomy, by inviting humans to see themselves as part of a larger cosmos, was "peculiarly calculated" to "give us a higher idea of the value of our being."[14]

Equally important, Priestley began to encode his scientific writings with the same values that underwrote his antislavery writings, bringing the two worlds into closer alignment.[15] One of the values Priestley stressed in his scientific writings was the importance of physical labor. By the mid-eighteenth century, scientific fields premised on active experimentation, as opposed to "passive" observation, had captured the public's imagination. The popularity of Franklin's experiments with electricity, and Priestley's in chemistry, epitomized this trend. Electrical and chemical experiments required instruments—air pumps, wires, flames, "an earthen retort filled with moistened clay"—and operating them required a lot of work. Priestley complained to Wedgwood that, though his experiments were difficult, "nothing of value is to be had without labour." In the *History of Electricity*, he argued that experimental science represented a supreme form of knowledge precisely because it combined "the hands and arms, as well as . . . the head" to arrive at deeper truths. Though the intellectual aspect was crucial, it was the physical dimension that taught experimentalists not only to admire nature but to harness its power "to the useful purposes of human life."[16]

By emphasizing the physical labor that went into scientific discovery, Priestley offered his readers a way to identify with enslaved people. The crime of slavery was that it devalued their physical labor, forcing the enslaved to toil without accruing any of its rewards. In a sermon denouncing slavery, published in 1788, Priestley told his audience to imagine themselves being "confined to hard labour all [their] lives," as enslaved people were. They would recoil at the thought, he said, because all people who labored shared a similar experience: "What they suffer . . . may in some measure be imagined by us." To be clear, Priestley did not want his readers to view scientific labor (or their own) as akin to slavery. On the contrary, what made the work of science morally worthy was that it required both body *and* mind; as such, it was ideally suited for "the middle ranks of life," he wrote. Priestley offered his middle-class readers an image of scientific work that mirrored not the labor of enslaved laborers, but the kinds of respectable labor they experienced in their own lives, as shopkeepers, homemakers, lawyers, and manufacturers. They worked for their keep and ought to feel good about it—unlike the landed aristocracy, and unlike the unfortunate enslaved laborers and servants, who, he wrote in *Miscellaneous Observations Relating to Education* (1778), ought to be "taught contentment in their station" while the middle ranks of life worked for their improvement.[17]

Another key value Priestley's scientific writings promoted was an appreciation for empiricism—of fact before theory, experiment before speculation. In doing so, he conditioned his readers to have modest expectations, not to let bold ideas run ahead of hard evidence.[18] In *Experiments and Observations on Different Kinds of Air,* he railed against his rival chemist, Antoine Lavoisier, who had begun to challenge traditional understandings of chemistry. Priestley remained committed to the older Aristotelian concept of matter, which was based on the idea that all matter derived from four essential elements: air, water, fire, and earth. But in the mid-1770s, Lavoisier, in part relying on Priestley's experiments, began to argue that matter was made up of even smaller elements, overthrowing the entire Aristotelian framework and providing the basis for the modern periodic table. Priestley defended his work by insisting that Lavoisier jumped to conclusions without adequate evidence: "But I chuse to wait for more facts, before I deduce any general theory," he wrote in his defense. "*Speculation* is a cheap commodity," he continued, "*New and important facts* are most wanted, and therefore of more value." It was precisely because scientific experimentation required "much *labour,* and *patience,*" he wrote elsewhere, that its truths were more reliable. Priestley's emphasis on facts, and the painstaking work it took to establish them, fit within his broader depiction of science's gradual progression. If scientific research took "much *time*," it followed that scientific progress, and the benefits mankind accrued from it, would come slowly.[19]

Priestley's valorization of the slow, gradual nature of scientific progress provided readers with a model for seeing how proper social change should be implemented in society as well. Often he made this link explicit. When, in the late 1760s, he advocated for the repeal of laws that barred Dissenters from holding public office and attending Anglican universities like Cambridge and Oxford, he argued that such a reform should be empirical, incremental, and conducted like a science. Civil society was "founded, as all arts are, upon science," he wrote in *An Essay on the First Principles of Government* (1768). As such, any experiment in social reform "must be performed by the help of *data* which with experience and observation furnish us." He also warned reformers against pushing for radical changes. Exercise "due caution," he told his fellow reformers, because, "like other arts and sciences," government-driven reform "improves slowly." Though gradual change was the "slowest method," it was also the "surest" and the most likely "to lead mankind to happiness."[20]

The early abolitionist agenda mirrored this model of social reform. The SEAST, like the PAS and white antislavery sympathizers in general, denounced immediate emancipation, pressing instead for incremental measures that would, at the very least, encourage planters to treat their captives more humanely. The Anglo-American movement's first great success, abolishing the Atlantic slave trade in 1807–8, was premised on this very idea: slave owners would treat their enslaved laborers with more care once they realized that the pool of imported African laborers was finite. Moreover, abolitionists, particularly in Britain, shied away from emphasizing the end goal of emancipation. To do so conjured the fear of having to live as political equals with emancipated Black people. Instead, they promoted alternative new colonies, especially in West Africa, where free African labor might prove more productive than enslaved labor. The early abolitionist elites also believed that enslaved people, long degraded by slavery and originating from a less well-developed civilization, were unprepared for immediate freedom; certainly, they were in no place to carry out emancipation themselves.

Instead, it was the responsibility of the upwardly mobile "middle ranks of life" to bring about their liberation. Priestley's 1788 sermon, the most extensive antislavery work he ever published, promoted this basic agenda. He argued for "a stoppage of the importation of slaves" and the creation of free-labor sugar colonies in Africa, and he cautioned against "immediate emancipation."[21]

But it was not just what measures he promoted, but *how* he promoted them, that made his sermon so revealing. By describing abolitionists' gradual approach in the same terms he used to describe the nature of scientific progress, his antislavery claims could be seen as resting on firm scientific ground. Priestley repurposed the core values his audiences came to value as the very essence of "good" science—an appeal to facts, careful experimentation, and gradual progress—and placed them in an antislavery context. For instance, he emphasized that cold, hard evidence would persuade skeptics of slavery's illegitimacy. "Nothing, my brethren, I am confident, will be requisite, besides stating the simple *facts*." He offered a deluge of information for his readers to circulate among their peers: the "half a million persons [who] are annually destroyed"; the kidnappings and wars fought in Africa just to attain enslaved laborers; the beatings, rapes, and "shocking indecencies to which the females are subjected"; the breaking up of "husbands and wives, parents and children." He gave them a history of the slave trade, beginning in 1551.[22]

Priestley also made a virtue of incrementalism. "Immediate emancipation" was "improper," he wrote; slavery was such a "complex and unnatural state" that undoing it would take ages, but "in time it will be done." The pace of change he imagined was Britain's centuries-long progress from feudalism to its present state, where all were now *"freemen."* Further counseling against immediate emancipation was slavery's degrading effects on the enslaved: "Those who have long been slaves would not know how to make a proper use of freedom." He suggested experimenting with Spain's relatively more liberal manumission laws, alluding to *coartación*, the Spanish policy in which enslaved workers signed a contract with their owner to buy their freedom at a fixed price. This would teach them the value of labor, and how to "make proper use of [their] freedom."[23] In sum, Priestley offered his readers a vision of how emancipation would unfold, one that comported with a broader Enlightenment vision of slow but steady scientific progress, and one that made a gradualist agenda appear not only as one path among many, but the "surest" path sanctioned by science.

Priestley did not only depict scientific progress as a model for antislavery reform. Along with his Lunar Society colleagues, he associated industrial technologies with science, inviting readers to view machine technologies as emblems of science, and therefore engines of emancipation. Lunar Society abolitionists routinely depicted the instruments and machines that came to define the Industrial Revolution—steam engines, cotton mills, electrical batteries—as markers of scientific accomplishment, ones that could be mobilized to attack slavery. In the process, they began to present slave societies as everything industrial societies were not: uncivilized, inhumane, a check on scientific progress. For Priestley, "philosophical instruments" were an "endless fund of knowledge," as well as the embodiment of science's ability to bring about social transformation: tyrants of all kinds had good "reason to tremble even at an air pump, or an electrical machine."[24] Wedgwood implied something similar when commenting on Priestley's and Franklin's electrical experiments, suggesting that electrical machines enabled men to "rob the Thunder of his Bolts," and in turn "execute some public piece of justice."[25]

Not to be outdone was Erasmus Darwin. Between 1789 and 1791, he published a widely read two-volume poem, *Botanic Garden*, that tightly linked science, industrial technologies, and abolitionism. Intended to teach women the rudiments of botany, *Botanic Garden* quickly veered off course, including a host of social commentary and laden with endless footnotes. For instance, Darwin depicted Wedgwood's factory, as well as the steam engine and the recently invented cotton mill, as the embodiments of scientific achievement, the pulling together of Franklin's, Priestley's, and Wedgwood's discoveries. Watt's "UNCONQUER'D STEAM!" might one day "Drag the slow barge, or drive the rapid car"; Richard Arkwright's cotton mill, which mechanically pulled apart cotton fibers, made "smooth the ravell'd fleece"; at Wedgwood's factory, "flint liquescent pours / Through finer sieves in whiter showers." All this "magnificent machinery" was rooted in scientific discoveries, he wrote, portending a more free and equal future. The electrical discoveries of "Immortal Franklin" powered not only Britain's industrialization, but also, allegorically, America's Revolution. Franklin unleashed a "patriot-flame," which "electrified man" and brought "the laurels of LIBERTY" to America's shores.[26]

In truth, there was nothing inevitable about industrial technologies being seen as harbingers of a more modern, egalitarian age. In other parts of Europe, tradition-bound scientific patrons depicted the steam engine as a machine that would restore order, simplicity, and "natural" hierarchies: an "Arcadian apparatus," as one historian has put it.[27] The idea that industrial technologies, and the science undergirding them, would flatten hierarchies and destroy slavery had to be invented. Darwin's *Botanic Garden* demonstrated how this was done. He wove antislavery passages into the poem's lengthy odes to Wedgwood's factory and Franklin's lightning rod, implying that science and technology would help bring about slavery's end. He not only held up Wedgwood's factory as the apotheosis of scientific accomplishment, but singled out Wedgwood's "Am I Not a Man and a Brother?" medallion as the noblest fruit of industrial technology: "The bold Cameo speaks," he wrote, "From the poor fetter'd SLAVE on bended knee / From Britain's sons imploring to be free." In the poem's footnotes, Darwin explained how Wedgwood's medallion would itself "excite the humane to attend to and assist in the abolition of the detestable traffic."[28]

These antislavery passages were joined by others that linked industrial technologies to the eradication of still other social evils. Watt's steam engine, Darwin wrote in a footnote, was now being used for a new "apparatus for Coining"; by preventing counterfeiting, Darwin explained, this new technology saved people from being unjustly executed for mistakenly using illegal money. Arkwright's cotton-spinning machine not only "abbreviated and simplified the labour" of cotton production: it also made it possible for cotton one day to be "the principal clothing of mankind."[29] Through Arkwright's invention, in other words, the world's poor would be clothed. Darwin's poem gave his readers a clear impression—that scientific knowledge created the day's technological marvels, and that these technologies would in turn eradicate social evils.

Darwin's poem was no outlier. Several Lunar Society men took it as a matter of fact that machine technologies would help spell the end of slavery. Wedgwood made this point explicitly in one of his most extensive private letters on slavery. "The introduction of machines & free labour," he wrote to the poet Anna Seward in 1788, would mean that plantation owners "could not be materially injured by prohibiting further importations." In the letter Wedgwood tried to convince the well-known Seward to join the movement and write a poem in its behalf. Seward declined, however, citing her anxieties about the "treacherous and bloody disposition of Negroes," and her concern that abolishing the slave trade might harm "our National interest." But Wedgwood persisted, arguing that the "introduction of machines" would prevent slave owners, and the British Empire more broadly, from suffering any financial loss. Wedgwood's letters to other Lunar Society friends suggested they felt similarly about technology's labor-saving potential. In 1788 he wrote to Watt: "I take it for granted that you and I are on the same side of the question respecting the slave trade," adding that he organized "a petition from the pottery for [the] *abolition* of it, as I do not like a *half measure* in this black business."[30]

In a 1788 book of lectures titled *Lectures on History and General Policy*, Priestley explained how slavery slowed the pace of scientific discovery. New technological inventions, he argued, came about only when human beings had a "considerable degree of *security* and *independence*," and slavery represented the exact opposite of security and independence. In addition, he argued that new inventions thrived on consumer demand, but because slaves could not spend money, slave societies were infertile grounds for technological innovation. Priestley then tied slavery's negative effects on technological innovation to the larger process of scientific discovery. "The connexion between arts and *science* hardly needs to be pointed out," he wrote ("arts" connoted technologies). Technological innovations spawned new scientific discoveries, and conversely, the "great improvement in the arts in modern times has certainly arisen from the late improvements of science."[31] By dampening the pace of technological innovation, then, slavery also slowed the emergence of new scientific knowledge. In the years to come, the association between industrial technologies and science would only grow tighter, and the moral value invested in science and technology only more secure. As that process developed, slave societies would come to

be seen as the opposite of industrial societies, and slaveholders as the enemies of progress themselves.

Wedgwood's antislavery medallion was not only the product of "magnificent
machinery." Its very design, of a supplicant enslaved man begging for his
freedom, rested on scientific ideas—specifically, scientific understandings
of emotions. By the eighteenth century, men of science began to see emotions as entwined with anatomy and physiology. Rejecting the mind-body
dichotomy of Descartes, Anglophone men of science viewed the mind and
brain as synonymous. Mind (or brain) and body were connected through
the nervous system, which was thought to consist of fine white fibers laced
throughout the body. The five senses received external stimuli, then sent
those sensations to the mind via the nervous system. For strict materialists
like Darwin and Priestley, thoughts, emotions, and ideas—the stuff of the
mind—were generated through the final vibration of nerves in the brain. In
this view, exposure to too many disturbing sounds, images, experiences, or
ideas could cause mental disease. Conversely, exposure to pleasing experiences could promote positive thoughts, healthy emotions, and even proper
moral behavior.[32]

Priestley helped circulate these ideas by editing the writings of Dr. David
Hartley, a renowned English physician and theologian whose work fused
the science of mind to moral behavior (and whose son, also named David,
would work with Franklin on the Treaty of Paris). In *Hartley's Theory of the
Human Mind* (1775), Priestley explained that Hartley offered a physiological basis of "the ideas of *moral right*, and *moral obligation*." Darwin, a practicing physician, also admired the "ingenious Dr. Hartley" and made a point
of citing him to explain how the images evoked in *Botanic Garden* worked
on his readers' emotions. In an interlude before one of his antislavery stanzas, Darwin wrote that artists should depict images that were neither too
strong nor too violent. Doing otherwise would be like a painter representing
a soldier's thigh "shot away by a cannon ball": the viewer would "turn from
it with disgust." By contrast, images that evoked pity would engender sympathy and, in turn, act as a natural stimulus to action. In this sense, Darwin's
suggestion that the "poor fetter'd SLAVE on bended knee" would "call the

pearly drops from Pity's eye" hinted at the scientific basis of the medallion's imagery.[33] The supplicant enslaved man was not a spontaneous artistic confection, but a scientifically informed image deliberately calculated to evoke the viewer's sympathy. Through physiology, then, Darwin was conditioning his readers to sympathize with helpless, sentimental images of enslaved people; anything else—a captive standing upright, defiant, breaking her chains—would naturally lead them to turn away in disgust.

Darwin's poem also suggested how the science of emotion enabled male abolitionist elites to manage women's place within the movement. As Clarkson wrote in his *History of the Abolition of the Slave Trade* (1808), "the ladies" played an important role in popularizing the cause, and their enduring involvement would continue to be crucial. Wedgwood manufactured hundreds of medallions for women—bracelets, rings, "pins for their hair"; he also imprinted the cameo on countless other products, from teacups to vases, purchased mainly by female consumers.[34] The challenge the movement's male elites faced was how to harness women's collective power without threatening men's leadership role, either in the movement or in society at large. In part, the answer lay in framing women's involvement in a way that did not challenge popular understandings of what women were inherently thought to be—innately sympathetic, loving, and emotionally fragile. Darwin's poetic odes to Wedgwood's medallion revealed how this process unfolded, showing how the science of emotion could be mobilized to solicit women's involvement while also keeping them in a subservient role.

In the eighteenth century, physicians believed that women had a more delicate nervous system than men; as a result, they were believed to have more active imaginations and to be more sensitive and emotionally volatile. If not properly regulated, women might suffer from nervous disorders considered more common among women than men, like hysteria. "In women," wrote the prominent Edinburgh physician Robert Whytt in 1765, "hysteric symptoms occur more frequently." Though Whytt insisted that hysteria was not exclusive to women, he argued that its greater frequency in women resulted from their having a "more delicate frame."[35] Darwin's *A Plan for the Conduct of Female Education* (1797) echoed these views. Though the text advocated for women's education at a time when they were not expected to receive any, it also reinforced the widespread belief in women's emotional volatility. He cautioned women against reading romantic novels, since their "high-wrought scenes of elegant distress . . . have been found to blunt the feelings of such readers toward real objects of misery." A proper education should be attuned to women's particular mental constitution, and it should cultivate their naturally sympathetic inclinations. As he had in the *Botanic Garden*, he argued that lurid scenes would "awaken only disgust in their minds," rather than "sentiments of pity and benevolence."[36]

Wedgwood's antislavery medallion, and Darwin's poetic odes to it, played to these allegedly innate characteristics. The enslaved figure's nonthreatening posture, bending on one knee, comported with the view that women were more emotionally vulnerable to violent, bold images. By depicting the figure in a way that evoked pity, rather than "disgust," it played to women's presumed sensitivity to despairing figures. Darwin's verses paid equal deference to women's allegedly innate mental and emotional temperaments. In one of the lengthier stanzas devoted to abolitionism, Darwin repurposed the biblical story of an infant Moses being cast into slavery, suggesting that women were more sympathetic to images of helpless children than were men. He depicted a weeping white woman saving the outcast infant, inviting female readers to see in her a model for their own behavior: she "Gives her white bosom to his eager lips / The salt-tears mingling with the milk he sips." Meanwhile, it was men in Parliament who would take bold, decisive action: "YE BANDS OF SENATORS!" he wrote, "Stretch your strong arm, for ye have the power to save!"[37]

It is tempting to view these men's embrace of women in the antislavery movement as part of a larger vision of women's political and social equality. But it is more accurate to see it as a careful attempt to leverage women's particular influence as moral authorities within the home, as mothers and wives, while keeping them confined to the domestic sphere. Darwin and Priestley may have advocated for women's education, putting them at odds with traditionalists, but they believed women's education served one primary purpose: to make them suitable wives and mothers. Darwin argued for women's education not on the grounds that it would help them enter the public sphere, but so they would become "a good daughter, a good wife, and a good mother." Education would help women cultivate the "mild and retiring" virtues, he explained, rather than make them "bold and dazzling" and thus unattractive to male suitors. Priestley felt similarly, writing that if women's "education has been virtuous and proper, and at all liberal, they will be valuable wives to men of liberal minds and better fortunes." None of these men expected women to step out of their domestic roles. "It is peculiarly necessary," Priestley wrote, that educated women "not flatter themselves with prospects, which there will be no probability of being realized."[38]

Nor would it be accurate to describe these men's views as radical. Some women's rights advocates had already begun to push for greater political rights. In 1792, five years before Darwin wrote *Female Education*, the British radical Mary Wollstonecraft argued in *A Vindication of the Rights of Woman* that women should no longer be "confined to domestic concerns." Though she conceded that educating women would also make them better wives, she expressed an equal commitment to women's "civil and political rights." By contrast, moderate reformers like Darwin, Priestley, and Wedgwood never made any claims about women needing political rights. What they did do was reject conservative attitudes that would prefer women to have no political opinions at all. Lunar Society men believed that women's domestic authority lent their political views weight and could make them useful anti-slavery foot soldiers. The highest virtue a woman should cultivate, Darwin wrote in *Female Education*, was "sympathy with the pains and pleasures of others" — exactly what the antislavery movement needed.[39] Since cultivating their innately sympathetic nature would also make them better wives and mothers, the movement's early female activists could be seen not as a threat to male authority, but as exemplars of middle-class womanhood.

Wedgwood's medallion drew on other scientific fields as well, one of the more important being natural history. The medallion's image of a supplicant, nearly naked enslaved man conjured a primitive simpleton, a "noble savage" violently stolen from an African Eden. It was an image straight out of a natural history book. In the mid-eighteenth century, one of Wedgwood's intellectual heroes, the French philosophe Jean-Jacques Rousseau, helped popularize the idea of the noble savage, in part deriving the idea from the French natural historian Buffon. Buffon relied on stadial theory to explain the progression of human societies from primitive to civilized, but Rousseau plucked the "savage" native out of the presumed natural order, then valorized him, using him as a tool to critique Western society. The noble savage image had become enough of a stock figure among eighteenth-century reformers that Wedgewood's selection of it was not particularly surprising. Yet Wedgwood may have been directly inspired by Rousseau, as well as his own passion for natural history. Wedgwood made a bust sculpture not only of Rousseau in 1779, the year after Rousseau died, but also of Rousseau's contemporary, the natural historian Carl Linnaeus. Three years earlier, in 1776, Wedgwood's business partner, Thomas Bentley, an important influence on Wedgwood's intellectual development, made a visit to the sixty-four-year-old Rousseau in Paris. "My heart expanded with joy when *Madame*"—Thérèse Levasseur, Rousseau's partner—"opened the door and desired us to walk in," Bentley wrote in his diary.[40]

Bentley's diary also points to how geology could be marshaled for antislavery purposes. During his visit to Paris, Bentley noted that Rousseau found "more pleasure in conversing with *Nature* than these artificial things" and was particularly eager to talk about geology—a field whose latest findings were increasingly being used for political ends. By the 1770s geologists had begun to challenge the biblical story of creation by pointing to recently unearthed fossil remains. John Whitehurst, a Lunar Society member with whom Wedgwood often went on "Fossiling part[ies]," and whose "arguments" Bentley mentioned in his meeting with Rousseau, played a particularly prominent role. Whitehurst's research showed that some land-dwelling fossils lay deeper in the Earth's surface than marine fossils, which implied that land animals might have been older than sea creatures, in contradiction to the biblical order of creation. Near Wedgwood's factory, Whitehurst found deposits of basalt, a rock formed from lava, which were taken as yet more evidence that the Earth's surfaces might have been formed from volcanic explosions, contravening the biblical story of a great flood.[41]

Whitehurst, averse to controversy, avoided spelling out the religious implications, but not Wedgwood. "Had the man a Bishop at his elbow?" Wedgwood wrote to Bentley on October 24, 1778, referring to Whitehurst: "He certainly was haunted with them somewhere." Rousseau did not mind the heretical implications either. "The world might have existed 5,000 or 5,000,000 of years for anything he knew," Bentley wrote (though Rousseau worried that "modern atheists" might use the findings to persecute people simply for "worshipping God"). Rousseau's and Wedgwood's remarks reveal the way geology was increasingly being used to undermine religious orthodoxy and, by extension, all established hierarchies, slavery included. Meanwhile, Darwin's *Botanic Garden* made these links—among geology, religious reform, and abolitionism—clear. The poem's lines extolling Wedgwood's antislavery medallion were immediately preceded by a lengthy passage explaining the volcano-centered view of Earth's creation, one that was by then widely seen as religiously subversive. The poem then described how the first volcanic eruptions produced all the known geological resources that Wedgwood needed to create his vaunted potted wares—limestone, flint, "ductile Clays," all fused together by "Nature's chemic toil."[42] The implication was that geology, antislavery, religious reform, even industrialization were all tied together—that the science of geology challenged religious and proslavery orthodoxies, while also providing the raw material that abolitionists could use to promote antislavery.

Wedgwood greatly approved of Darwin's poem. The two corresponded extensively about how Darwin was characterizing the antislavery medallion in his poems, and how he was portraying Wedgwood's contributions to the science behind pottery manufacturing. "After you have read the passage on the Slave Trade," Darwin wrote to Wedgwood on February 22, 1789, "I do not insist your reading any more." In June, Darwin asked whether "anything of *consequence*" had been done "in the medallion or cameo kind before you?" One month later, Darwin said he would include in his footnotes "some common mixture for pottery, and for the glazing and enamels, to have the appearance of conveying knowledge." Wedgwood was flattered with Darwin's final draft, praising Darwin's "learned notes" and likening him to a "powerful *magician*" who turned Wedgwood's humble "granite, & still harder flint" into "the softest poetic numbers." Their exchange suggested not only how much Darwin valued the scientific knowledge that went into Wedgwood's antislavery medallion, but also the weight these men placed on how their work was presented to the broader public. In fact, presentation often seemed to be more important than the reality: "What you have *really* done is no part of my question," Darwin wrote, regarding Wedgwood's contributions to pottery manufacturing: what mattered was whether "you *pretend* yourself to have improved the antique forms."[43]

Darwin's magic with words performed other wonders. The image of science he assisted in crafting—of science as a civilizing, humanitarian force— helped reframe the financial stakes these men had in abolitionism as in no way inhibiting. The loss of the American colonies forced British manufacturers to look for new consumers and raw materials. Wedgwood and Darwin showed particular interest in West Africa, even recommending an explorer to the Royal Society president, Joseph Banks, in 1789. Wedgwood thought that successful African expeditions might lead to the discovery of trading routes and new clays—and therefore, new profits—that could offset the losses stemming from American independence. The stock of scientific knowledge might expand as well. In June 1789 Darwin wrote to Wedgwood that George Gray, the explorer he found, would "improve science by new facts, or arts by new materials to work upon." The following year, Wedgwood sent Darwin updates on other African expeditions, which Darwin received enthusiastically, predicting they might soon "open immense sources of trade."[44]

These expeditions had little to do with abolitionism. In fact, the group of patrons Banks put together to fund these African expeditions, the African Association, included several slaveholders.[45] But the creation of an abolitionist colony, Sierra Leone, in 1787 help recast African exploration, and its ultimate end point—colonization—as a humanitarian mission. Abolitionists hoped Sierra Leone would function as a profitable and humane alternative to Caribbean slave plantations. If indigenous West Africans were turned into loyal British subjects and treated as paid agricultural laborers, the thinking went, then Britain could produce the same commodities as the Caribbean, but without the devastating human cost. In 1790 Thomas Clarkson redoubled abolitionist efforts to make Sierra Leone profitable, and on August 25 he offered Wedgwood "eight or nine shares" to become a board member of the Sierra Leone Company, which managed the colony. Three years later, he asked Wedgwood to help him find other investors, reminding Wedgwood to seek out investors who would "be better pleasd with Good resulting to Africa than from great Commercial Profits to himself."[46] By joining the Sierra Leone Company's board, Wedgwood was not so much hiding his financial interests in West Africa as recasting them in a nobler light.

Wedgwood's desire to find new raw materials cannot be understated. Since at least 1767, Wedgwood had relied on Native American clays, paying a South Carolina–based agent, Thomas Griffiths, to send him clays that Cherokees in particular used for their own pottery. "You & I shall do very well amongst the Cherokees," Wedgwood wrote to Bentley on May 20, 1767. By 1777 Wedgwood even considered profiteering from the War of Independence by spreading a rumor that one of his most valued artificial ceramics, jasperware, depended on "the Cherokee clay." There was nothing like "*scarcity*," he wrote to Bentley, "to make them worth any price you could ask for them."[47] But the Patriot victory cut off the supply of Cherokee clays, which led Wedgwood to search for alternatives in both Sierra Leone and Botany Bay. Founded in 1788, Botany Bay was created along similar lines as Sierra Leone, as a humane penal colony that could also function as a new profit center. Britain would send members of its growing incarcerated population to the Australian colony, turning them into productive laborers who would also help "civilize" the indigenous population. In addition, explorers would help scour the region for raw materials, enabling manufacturers in Britain to create new goods to then sell back to its Botany Bay subjects.[48]

Wedgwood's last scientific paper, published in 1790, captures the way these new colonies were being used to link science, humanitarianism, and profit, all of which were portrayed as being mutually reinforcing. After writing a chemical analysis of white clays from Botany Bay that Joseph Banks had sent him, Wedgwood concluded that these clays would make "an excellent material for pottery, and may be certainly made the basis of a valuable manufacture for our infant colony there." The paper tapped into a larger vision, partly rooted in stadial theory, which held that free trade would help civilize indigenous populations. Reform-minded men of science believed that indigenous peoples, particularly in warmer climates, were "lazy" since "their wants are few," as Priestley wrote regarding native Africans in his 1788 antislavery sermon. Yet they could be transformed into industrious, civilized subjects if Europeans introduced them to manufactured goods. Since the only way to attain these goods was with expendable income, the thinking went, indigenous peoples would gladly work as paid laborers. Natives would extract their own land's resources for their imperial rulers, then purchase the finished products, made in Europe, with the money they had made. Darwin's *Botanic Garden* spelled out this logic, connecting it to the Botany Bay and Sierra Leone projects. The same passage extolling Wedgwood's antislavery medallion included a similar ode to Wedgwood's so-called Hope medallion, which Wedgwood created to promote Botany Bay. "Made of clay from Botany Bay," Darwin wrote in a footnote, Wedgwood's Hope medallion, like the antislavery one, would "shew the inhabitants what their materials would do" and would "encourage their industry."[49] In Sierra Leone and Botany Bay, profit seeking, far from being something to hide, was being reconceived as virtuous precisely because it promoted science and humanitarianism.

By the 1790s the Lunar Society's vision of a peaceful, gradual end to slavery unfolding in tandem with scientific progress came into question. Immediate emancipation in Haiti, brought about by a slave rebellion and sanctioned by the French Jacobin government in 1794, posed a radical alternative to the Anglo-American movement's incremental approach. Meanwhile, British conservatives blamed abolitionists and like-minded reformers for stoking the Haitian Revolution and a host of events associated with it, from the American Revolution to the Reign of Terror. The *Anti-Jacobin, or Weekly Examiner*, a conservative British periodical, satirized Darwin's *Botanic Garden* in 1798, implying that reformers like him would soon be condoning the bloody overthrow of all British institutions: "Where nursed in seats of innocence and bliss, / REFORM greets TERROR with fraternal kiss." Priestley also became a target. In 1791, after some Lunar Society members fêted the falling of the Bastille, a mob attacked Priestley's home, "demolish[ing] my library, apparatus, and, as far as they could, every thing belonging to me," Priestley recalled.[50] Three years later, he fled to Pennsylvania.

Conservatives tried to delegitimize the Lunar Society's scientific work by associating it with a radical political agenda. Lunar Society members made it easy, demonstrating the political links to their scientific work: "I feel myself becoming all French in both chemistry and politics," Darwin wrote to Watt in 1790, upon the outbreak of the French Revolution.[51] But conservatives helped create an image of these men as something they were not: antislavery *radicals*, unflinchingly committed to slavery's demise, regardless of the pace or on whose terms it was carried out. In truth, their support for abolition shifted depending on the circumstances, and according to the degree to which it conformed to an idealized model of gradual progress, one that they had come to associate with their own beliefs about how science progressed.

Throughout the 1790s, Darwin continued to argue that science would help reduce the need for slave labor. His last scientific work, *Phytologia* (1800), a study of how science could improve agriculture, offered a host of speculative "future experiments" that might "strengthen the country" and enable it to live in accordance with the principles of "justice and humanity." Some of Darwin's ideas were intended to foster the growth of Britain's population, which was seen as a marker of civilizational progress. Machine technology, such as a drill plow he invented, would be "less liable to be out of order" and, by increasing the food supply, would help Britain's population grow. Other ideas would reduce Britain's reliance on imports: growing trees in Scotland, he suggested, would reduce Britain's imports of foreign timber. Still other ideas would end Britain's reliance on slave labor. Britain could reduce its reliance on slave-grown sugar, Darwin wrote, by experimenting with the "chemical production" of sugar in a laboratory: "If sugar could be made from its elements without the assistance of vegetation, such abundant food might be supplied as might tenfold increase the number of mankind!" In the meantime, he hoped sugar "may soon be cultivated by the hands of freedom." The same suggestion appeared in his final poem, *Temple of Nature* (1803), published a year after his death: "If our improved chemistry should ever discover the art of making sugar from fossile [*sic*] or aerial matter without the assistance of vegetation," he wrote, humans "might live upon the earth without preying on each other."[52]

Darwin's *Temple of Nature* not only suggested a chemical solution to slavery: it also offered a version of stadial theory that preserved the idea that science would foster slavery's gradual demise, while implicitly rejecting the kind of immediate emancipation enacted in Haiti. Like *Botanic Garden*, *Temple of Nature* was a didactic poem, with copious footnotes explaining the scientific theories underlying each verse. The poem offered a unified theory of evolution, one that wed the evolution of individual species to the evolution of human society as a whole, an idea that would later influence his grandson Charles. In *Temple of Nature*, Darwin contended that all matter emerged from the same primordial origins, each successive species steadily improving on the one before it. Humans stood at the apex of this organic chain and would shepherd their own species toward social and moral improvement. Physiology guided the process. According to Darwin, the human nervous system was primed to lead human beings to seek out pleasurable sensations and avoid painful ones. In pursuit of pleasure, human beings would devise ideas—often scientific ones—that would alleviate human suffering. In this way, science would promote moral progress, a sentiment captured in the lines "The plans of Science with the works of art; / Give proud Reason her comparing power, / Warm every clime, and brighten every hour."[53]

The tumultuous events of the 1790s posed a challenge to this idea of science-driven human progress. Guillotines, a slave-led emancipation in Haiti, and slavery's steady growth elsewhere—the stuff of reality—were not supposed to exist in a world of steam engines, American independence, electricity, and chemistry. Darwin's first unpublished draft of *Temple of Nature*, titled *The Progress of Society* and composed in 1798 and 1799, shows him struggling to reconcile these anomalies. In the unpublished draft, Darwin jotted down five distinct stages in society's evolution, a rubric based on stadial theory. All human societies, he suggested, started out in an "Age of Hunting," then progressed to an "Age of Pasturage"; next came an "Age of Agriculture," followed by an "Age of Commerce," which he suggested European society was living in. Last came the idealized, final state, the "Age of Philosophy." He then listed the key attributes of the latter three stages, and "Slavery" and "Manufacturing" appeared together only in the Age of Commerce, but not in the final Age of Philosophy. Meanwhile, "Science" appeared only in the Age of Philosophy, alongside "Liberty," "Peace," and "Moral World."[54] By removing "Science" from the Age of Commerce and putting it in a yet-to-be-realized Age of Philosophy, Darwin was trying to disassociate science from slavery. Yet in decoupling "Science" from "Manufacturing," he was also revising his earlier belief that science and manufacturing, a synecdoche for industrialization, were mutually reinforcing, and that both would help alleviate human injustices, slavery included. In short, Darwin was coming to the realization that industrialization was no panacea for slavery, but he still could not conceive of a world where science and slavery coexisted.

Darwin's *Temple of Nature* also depicted slavery not as a contingent development within human societies, but as a natural and inevitable part of societal progress. Human-made and environmental disasters served as natural checks on population growth, he suggested: "But war, and pestilence, disease, and dearth, / Sweep the superfluous myriads from the earth." Though Darwin did not include slavery in the line describing "war" and "dearth," he discussed slavery in the same canto, inviting readers to make the link themselves. Yet he offered a hopeful ending, suggesting that all matter "from the grave returns," and therefore helped subsequent species evolve toward a more peaceful, improved future.[55] Darwin was suggesting that no death was senseless, that in the larger view of things, even what were momentarily understood as senseless acts of violence might ultimately serve some positive evolutionary function.

Darwin's attempt to put a positive gloss on human-made disasters was hardly unique. During the 1790s, many Enlightenment thinkers began to incorporate human suffering into their theories of societal progress. With so much suffering around them, alongside so much alleged civilizational progress, many began to think that suffering must have some useful evolutionary role. Representative of this trend was Thomas Malthus, whose *Essay on the Principle of Population* (1798) argued that if the population continued to grow unchecked, it would outstrip the food supply. As he wrote, "Misery is the check that represses the superior power of population."[56] Malthus greatly influenced Darwin, but there was an important difference between the two. After proslavery theorists used Malthus's theory to justify slavery on the grounds that slavery served a necessary ecological function, Malthus, an antislavery sympathizer, explicitly rejected the idea in the revised edition of *Theory of Population* (1803).[57] By contrast, Darwin let it stand, allowing readers to see him as endorsing the view that slavery might serve a natural purpose. Moreover, his unpublished notes for the poem suggest that he saw slavery as a tragic but inevitable stage in society's progress, one that would be eradicated by the time humankind reached the Age of Philosophy.

Darwin's poem was equally significant for what it did not mention. Most important, it said nothing about Haitian emancipation, or its official sanction by the radical Jacobin government in France. Of course, to celebrate Haitian emancipation would play into the hands of his conservative critics, who frequently invoked Haiti to delegitimize the abolitionist movement. At the time, British abolitionist leaders kept their distance from the Black republic, disassociating their own goal of gradual slave-trade abolition with immediate emancipation.[58] But other antislavery supporters were less demure. Throughout the Atlantic world, enslaved and free Black women and men openly celebrated the Haitian Revolution, and within the United States, there were even pockets of cautious white support. In 1797 President John Adams opened formal diplomatic ties with the Haitian republic, and some white abolitionists even argued for immediate emancipation. American and British newspapers not infrequently depicted Haiti's revolutionary leader, Toussaint-Louverture, in a favorable light: an "extraordinary man," New York's *Balance* called him on July 16, 1801. One year later, the *Mercantile Advertiser* of New York republished a London newspaper article that read, "Were Toussaint L'Ouverture to be put to death, it would be thought the *blackest* murder of the whole Revolution."[59] For Darwin, the inability to see in Haiti a model for emancipation was not a failure of imagination: it was the deliberate avoidance of a path that did not comport with how elite white abolitionists thought it should unfold, a path that was too rash, too unplanned, too unscientific.

Priestley displayed a similar unease with Haitian emancipation. Not long after he arrived in the United States, on June 4, 1794, Democratic-Republicans embraced him as a natural ally. Benjamin Rush, who began to align himself with Democratic-Republicans in the 1790s, unsuccessfully courted him for a professorship in chemistry at the University of Pennsylvania. Meanwhile, in 1794 a New York Democratic-Republican society welcomed Priestley in an open letter, noting that, like him, they believed "a republican representative government was not only best adapted to promote human happiness" but was the "only rational system." Federalists were as hostile toward Priestley as Democratic-Republicans were fawning. The Federalist propagandist William Cobbett portrayed Priestley as a delusional radical, associating his political and scientific ideas with anarchy in France and a hasty, blood-soaked emancipation in Haiti. Priestley supported the "mad plan" of immediate emancipation in Haiti, rather than the "much more sincere desire of seeing all mankind free," Cobbett wrote, as epitomized in the U.S. Constitution; to Cobbett, the Constitution embodied the process by which freedom should unfold—that is, through the creation of a careful legal framework rather than a bloody revolution. At the same time, Cobbett lampooned Priestley for contending that political institutions operated by scientific principles: a government "presents nothing like a *system*; nothing like a thing composed, and written in a book."[60]

Cobbett's caricature of Priestley was not entirely baseless. He was correct in implying that Priestley tried to legitimate certain political reforms by associating them with scientific laws. But in suggesting that Priestley approved of immediate emancipation, or was even a firmly committed abolitionist, he could not have been more wrong. During the entirety of Priestley's ten-year stay in America, until his death in 1804, Priestley seldom spoke out against slavery. When he did, he offered at most tepid support for abolitionism. For instance, in response to the Democratic-Republican society's open letter, which in part solicited his support for abolition, Priestley offered faint approval, suggesting only that slavery would soon fade into history: a vestige of "former times," he called it, one that "may be expected soon to die away." In 1799 he responded to another Federalist attack, this one smearing him as a radical on account of the honorary French citizenship he received from the revolution's initial leaders. Priestley countered by reminding readers that the meaning of the French Revolution, as well as of abolitionism, had been vastly different in those early, halcyon days: "Attend to the circumstance of the *time* in which I was made a citizen of France," Priestley wrote. The revolution's initial leaders, he wrote, planned only to reduce their government to "a limited monarchy resembling that of England"; they also gave William Wilberforce honorary citizenship because he advocated not for emancipation per se but rather for "the abolition of the slave trade." Even the universally vaunted George Washington, he noted, was given honorary citizenship—"and surely you do not for this suspect *him* of being your enemy."[61]

In private, Priestley also showed a certain tolerance for slavery. In 1795 he hired an enslaved servant to take care of his ailing wife. "We only hire a black slave by the week," he wrote to his brother-in-law, John Wilkinson, from his rural Pennsylvania home in December 1795. This was fifteen years after Pennsylvania enacted one of the nation's first gradual emancipation laws, and when enslaved people made up less than 1 percent of the entire population. Priestley was hardly alone: several prominent British émigrés who opposed slavery while in England found it impossible to get by without it once in America. Priestley's fellow chemist, friend, and émigré, Thomas Cooper, once an ardent abolitionist in Britain, tried to purchase enslaved workers shortly after his arrival; by the 1820s he had become one of the South's most prominent proslavery defenders. One of Priestley's sons became a successful sugar planter in Louisiana; his daughter married a sugar planter as well. To be sure, Priestley cannot be faulted for the company he kept, and he did privately confess unease with slavery. In 1796 he grumbled to the wife of a British diplomat in Philadelphia, "The Servants alone are sufficient to render a native of Britain miserable in this Country."[62] But in light of his meek defense of abolition and his own use of an enslaved servant, these comments represent at best evidence of a man conflicted, or at worst the disingenuous remarks of a man trying to burnish his abolitionist credentials.

Priestley may have avoided public references to slavery once in the United States, but he nonetheless continued to present science and freedom as natural allies. In 1799 he published an essay titled "Maxims of Political Arithmetic," which described his vision of America's future, essentially endorsing Jefferson's yeoman republic ideal. Like Jefferson's version, Priestley's yeoman republic implicitly had no enslaved people, and it both protected and rested on a healthy scientific culture. Priestley argued that Federalists wasted tax dollars on military defense rather than investing in education, using science as a symbol of education: imagine "what solid advantage, might be derived from half the expence in sending out men of science for the purpose of purchasing works of literature and philosophical instruments," he wrote.[63] The implication was that science and a yeoman republic, science and freedom, were mutually reinforcing.

Priestley had a strong ideological kinship with Jefferson. He enthusiastically endorsed Jefferson's presidency, dedicating his final work to "Thomas Jefferson President of the United States" in 1803. Three years earlier, Jefferson had solicited Priestley's advice for the "selection of the sciences" curriculum at the university Jefferson was devising, and the two maintained a regular correspondence until Priestley's death. Both men were attracted to each other for obvious reasons. They were Francophiles and science enthusiasts who felt that the diffusion of scientific knowledge would help foster freer and more equal societies. Both men's views on slavery also followed a similar arc. Though Jefferson fought, in the 1780s, to limit slavery's expansion in the northwest territories, by the turn of the century he lacked the political will to end it.[64] Similarly, in the 1770s and 1780s, Priestley proudly endorsed abolitionism, but when faced with public ridicule and a radical alternative—immediate emancipation, of the kind achieved in Haiti—he backed away, fearful of defending a now unpopular cause, and refusing even to acknowledge Haiti's existence. The belief that the diffusion of scientific knowledge would gradually bring human beings to their senses, that men of science might one day devise "future experiments" to alleviate the need for slave labor, proved too comforting. There was no point defending abolitionism if it meant being associated with radicalism, and certainly not if scientific ideas could be relied on to suggest that slavery would eventually fade away.

On February 6, 1804, Benjamin Rush received notice of Priestley's death. The chemist Thomas Cooper, Priestley's fellow émigré, brought the news to Rush: "Dr. Priestley died this morning about 11 o'clock, without the slightest degree of apparent pain." He added, "I am sure you will sincerely regret the decease of a man so highly eminent and useful in the literary and philosophical world, and so much a personal friend."[65] Rush would indeed regret the loss of such a close friend, but if he had hoped Priestley might help him find a solution to slavery, he would have been mistaken. Already, by the beginning of the nineteenth century, more promising help seemed to be coming by way of Sierra Leone. There, botanists, physicians, naturalists, and explorers were amassing scientific evidence that seemed to prove, beyond question, that free-labor colonies in West Africa, rather than slave plantations in the Americas, offered the most practical way to gradually end slavery. Yet for all the work men of science would do to legitimate Sierra Leone's existence, their story remains barely known.

THREE

A Natural History of Sierra Leone

In August 1783 Benjamin Rush's fellow physician and abolitionist ally John Coakley Lettsom introduced Henry Smeathman, a British naturalist and explorer, to Benjamin Franklin. Smeathman was in Paris that summer looking for high-profile patrons to fund a scientific expedition to Sierra Leone. For more than a decade, antislavery advocates in Britain and America had been floating the idea of establishing a free-labor colony in West Africa, worked by paid indigenous laborers and overseen by white officials, in the hopes that it would prove free-labor colonies in Africa could be more humane and more profitable than slave plantations in the Americas. Franklin, who was in Paris negotiating a peace treaty with Britain, agreed to meet Smeathman for breakfast. Over their meal, Smeathman played up his scientific credentials, not only highlighting his previous experience exploring Sierra Leone, but also showing Franklin his aeronautical designs for a hot-air balloon he hoped to patent. Later that month, Smeathman used his meeting with Franklin to attract other deep-pocketed funders: "Dr. Franklin . . . has paid me some great complements [sic]," he told one potential supporter. The scientific evidence that Smeathman and other men of science published about the region was vital to the founding of Sierra Leone, the abolitionist movement's first free Black colony, established in 1787. Europeans had only a few firsthand accounts of the region before the 1780s, and therefore natural knowledge about the region—the healthfulness of its climate, the fertility of its soil, the disposition of its indigenous inhabitants—would be crucial to gaining public support.[1]

Little attention has been paid to the role that botanists, physicians, natural-ists, and explorers played in creating an image of Sierra Leone as ideally suited for a free Black settlement. But between 1770 and 1807, the work these men of science published in behalf of the colony helped legitimate the project—and, by extension, the broader antislavery movement. Sierra Leone was a key component of the early abolitionist agenda. In addition to proving that free Black labor could outperform enslaved labor, abolitionists argued, Sierra Leone's commercial success would undermine the African slave trade by encouraging indigenous Africans to trade in "legitimate" goods rather than enslaved people. White resistance to free Black integration in America and Britain provided another rationale for the colony. Long before the American Colonization Society was founded in 1816—the all-white organization that promoted the removal of free Black Americans to West Africa and created Li-beria in 1821 for that purpose—white antislavery advocates on both sides of the Atlantic pioneered the idea of colonization—the resettling of freed Blacks outside white societies—seeing it as a viable alternative to integration.[2]

To be sure, the first abolitionist societies emphasized educating free Blacks and protecting their fragile freedoms; they did not advocate Black removal, and they hoped that, eventually, white societies would come to accept Blacks as equals. But on an individual level, many white abolitionists remained skep-tical of Black people's prospects. Often with the support of free Black elites, white abolitionists hoped that establishing successful free Black colonies would offer emancipated Blacks an escape from the racism endemic to white societies, while also allowing abolitionists to focus their attention on ending slavery, rather than worrying about the fight that came next: Black citizen-ship. In the 1780s Samuel Hopkins, a prominent New England abolitionist, expressed the view of many white abolitionists in a series of letters to Gran-ville Sharp, the abolitionist founder of Sierra Leone. "Their circumstances are, in many respects, unhappy, while they lived here among the whites," he told Sharp: white people look "down upon them . . . treat them as under-lings." Hopkins therefore believed that voluntarily resettling freed Blacks in Africa was their best option, and he gave Sierra Leone his unbridled support. As he wrote in a 1793 pamphlet: "This appears to be the best and only plan to put the Blacks among us in the most agreeable situation for themselves."[3]

Integrating free Blacks proved equally difficult in England. The fact that nearly all Britain's enslaved population lived in the Caribbean did not stop metropolitan Britons from fretting over Black integration. The Somerset decision in 1772 led to an uptick in England's free Black population, as enslaved Africans in England increasingly escaped their owners and assumed a free identity. In addition, after the War of Independence, hundreds of the enslaved Americans who won their freedom fighting for the British, known as Black Loyalists, found their way to England. By the mid-1780s, an estimated 15,000 Blacks, the majority of them free, were living in London alone. But equality proved as elusive for these women and men as it did for their American counterparts. Confined to the worst-paying jobs, disparaged and mocked, many sank into poverty. It was from among these destitute Black Londoners that Sierra Leone's first settlers would come.[4]

Much like white abolitionists in America, Britain's white abolitionist leaders catered to domestic racial anxieties. During the Somerset case, for instance, Somerset's attorney argued that, in addition to abolishing slavery in England, Parliament should "prevent the abominable number of Negroes being brought here." That would deter enslaved Blacks from immigrating to England and declaring themselves free. Without such a restriction, he warned, "I don't know what our progeny may be I mean of what colour." Granville Sharp catered to fears of racial mixing as well. In a 1768 case similar to Somerset's, Sharp allayed concerns that English emancipation would encourage Black immigration by arguing that outlawing slavery in England would diminish, rather than increase, the number of Blacks. "There are now in town upward 20,000 Negroes," he wrote, and "there would soon be 20,000 more" if slavery remained legal in the country. Abolishing slavery in the metropole would discourage planters from bringing their enslaved laborers to England, he reasoned, ultimately reducing the number of Blacks. As one scholar has argued, when it came to racial equality and the concomitant fear of interracial marriage, British abolitionists and their proslavery adversaries "were not as far apart as we might suppose."[5]

For all these reasons, portraying Sierra Leone's natural environment as perfectly suited to free Black settlement was a high priority for early abolitionists. Several of the men of science who traveled to the region in abolitionists' behalf had close ties to slaveholders. But rather than being a liability, their ties to slaveholders helped bolster claims of neutrality—after all, if their research was supported by both slaveholders and abolitionists, accusations of bias would be difficult to prove. The men of science sent to Sierra Leone also relied heavily on the colony's Black settlers and the region's indigenous inhabitants to conduct their research. Yet that did not stop the colony's men of science from blaming their Black aides when their sanguine portrayals of the region's natural environment proved wanting. However unreasonable the expectations men of science placed on the colony's Black settlers, the value their research added to the movement was too great for antislavery leaders to ignore.

The idea for an antislavery colony in Sierra Leone emerged in the 1770s from the mind of John Fothergill, a botanist, physician, and abolitionist. Trained in medicine at the University of Edinburgh and a member of the Royal Society, Fothergill became involved in antislavery politics in the late 1760s. In 1769 he wrote to Granville Sharp to thank him for the *"humane"* antislavery pamphlet Sharp had sent him, and three years later he offered to pay the legal expenses for James Somerset's freedom suit. Though Fothergill, a Quaker, died in 1780, and Sharp would ultimately oversee the colony's initial settlement in 1787, abolitionists routinely credited Fothergill with laying the intellectual groundwork. Yet Fothergill seems to have been inspired by indigenous African sources. According to Lettsom, Fothergill pursued the idea of a free Black colony in Sierra Leone only after hearing the story of a West African leader, King Agaja of Dahomey, who, decades earlier, had tried to halt the enslavement of his own people but lacked a lucrative trading alternative. Fothergill believed that trading in locally grown goods, harvested by free Black laborers in a British-run colony, and not enslaved laborers, would offer a solution.[6]

Fothergill's interest in Sierra Leone converged with his scientific interests. By the 1770s, Fothergill had amassed an enormous collection of exotic plants that rivaled that of the Royal Gardens at Kew, many of them from North America. Before the American Revolution, American men of science, many of them fellow Quakers, provided him with seeds, plants, and fossils. In 1769 John Bartram, the renowned American botanist and Quaker, sent Fothergill a "box of plants" and "American seeds"; Peter Collinson, a New York botanist and Quaker, taught Fothergill "to love plants," as he told Carl Linnaeus, the era's most revered naturalist. The scientific exchange went both ways: in 1745 Benjamin Franklin, still unproven as a man of science, relied on Fothergill to defend his theory of electricity at the Royal Society in London.[7] Ultimately, the rising political tensions between Britain and its thirteen colonies forced Fothergill to search for new sources for his plants, and he would soon turn to Africa.

Americans not only helped Fothergill collect botanical specimens; they also influenced his antislavery views. In the mid-1750s Fothergill's younger brother Samuel traveled throughout the thirteen colonies preaching against slavery with Israel Pemberton, the father of the later PAS president James Pemberton. With Samuel's help, Fothergill invited the American Quaker John Woolman to attend the Quaker Yearly Meeting in London in 1772, where, for the first time, England's Quakers resolved to protest slavery beyond their community. "The Americans help us much," Fothergill wrote to Samuel after the meeting, giving special credit to Woolman for his "solid and weighty" remarks. Philadelphia's antislavery Quakers may have also further fueled his interest in Sierra Leone. Anthony Benezet, the pioneering Philadelphia abolitionist, solicited Fothergill at roughly the same time he sought out Rush and Franklin. In 1771 Benezet compiled an influential compendium of older natural histories about West Africa titled *Some Historical Account of Guinea* — "instrumental, beyond any other book ever before published, in disseminating a proper knowledge," Clarkson would call it — hoping that it would prove the feasibility of establishing free-labor colonies in Africa.[8]

Yet one of the problems with Benezet's *Historical Account* was that the natural histories it cited were decades old. One of the few authoritative accounts about West Africa had been written by the French naturalist Michel Adanson, based on his expedition to Senegal between 1749 and 1753. Benezet's *Historical Account* borrowed heavily from Adanson and underscored his scientific intentions: Adanson was invested "wholly in making *natural* and *philosophical* observations," Benezet wrote, and he traveled under the sponsorship of the "Royal Academy of Sciences at Paris."[9] Fothergill undoubtedly knew of

Adanson's work, but he may have also realized that it was high time for another. After all, Benezet had no scientific standing, and he stacked his book with out-of-date observations. Fothergill could offer something original: credible, and new, scientific evidence about Sierra Leone's natural environment. With that goal in mind, he quickly organized a scientific expedition to the region in 1771. It would be led by an eager, young explorer named Henry Smeathman, and its mission would be to collect authoritative scientific knowledge about the region's climate, environment, and people.

Fothergill was not the only man of science interested in funding the expedition. Several of Britain's most influential men of science began to look to the region in the 1770s, seeing it as an untapped source of scientific and commercial materials: botanicals, medicines, fossils, gold. Seizing on the uptick in interest, Fothergill convinced Marmaduke Tunstall, a naturalist and Royal Society member; Dru Drury, founder of the Society of Entomologists of London; and Joseph Banks, an explorer, naturalist, and soon-to-be Royal Society president, to help fund Smeathman's expedition. Some of these patrons, like Tunstall, appeared sympathetic to antislavery, but others were more ambivalent. Banks, for instance, adhered to a studied neutrality on slavery, leery of antagonizing the slaveholders he had long depended on for his scientific collecting. While Banks conceded that slavery was "impolitic," he also thought that "a slave well provided for & humanly treated is as certainly happier than a free man who has a choice only of bad master or none."[10]

Rather than a detriment to abolitionists, the dubious antislavery convictions of Fothergill's fellow patrons proved to be an asset. They provided Fothergill and Smeathman with a buffer against accusations of bias, since patrons like Banks had no clear vested interest in abolitionism. Smeathman in fact highlighted Banks's apolitical, scientific interests whenever he sensed the antislavery agenda might undermine the credibility of his findings. Banks funded his 1771 trip, Smeathman wrote a decade later, solely "impelled, by the ardour of science." And though Smeathman acknowledged Fothergill's antislavery politics, he stressed that Fothergill was primarily motivated by "the encouragement of arts and sciences [and] the advancement of medicine." Indeed, what all Smeathman's patrons had in common was that they were "lovers of science." Fothergill, for his part, also made sure that his antislavery views would not color Smeathman's findings. In 1774, three years into Smeathman's expedition and anxious that he might not return—already, one of Smeathman's co-explorers, Anders Berlin, had died from a fever—Fothergill asked Linnaeus if he could recommend another explorer. Fothergill made no mention of the explorer's needing to be an antislavery supporter; the only prerequisite was that he be "thoroughly imbued with the love of plants and of natural history."[11]

If there was any doubt about the scientific credibility of Smeathman's initial expedition, Fothergill could point to other instances in which he himself worked with planters on scientific projects. Most obvious was his lobbying of Parliament, on behalf of Banks and Caribbean planters, to transplant breadfruit trees to the West Indies. Banks discovered the fast-growing, nutritious breadfruit on a 1769 expedition to Tahiti with Captain James Cook and realized that it could feed enslaved Caribbean laborers cheaply and efficiently. Caribbean planters welcomed Banks's breadfruit project not because they saw it as a step toward emancipation, but just the opposite: it might placate abolitionists and ultimately prolong slavery. It was part of a larger strategy that Caribbean planters adopted in the 1780s, called *amelioration*, in which they implemented certain measures to better care for their enslaved workers as a hedge against the potential closing of the slave trade. But the extent to which abolitionists worked with planters on amelioration projects is seldom appreciated.[12] Fothergill left no record indicating why he supported the breadfruit initiative, but he probably saw it in abolitionist terms—that is, as a way to increase enslaved people's longevity, and in his view, their relative comfort, while at the same time diminishing the need for the slave trade.

Fothergill's ties to planters ultimately bolstered his abolitionist work. At times he used these relationships to advocate for antislavery measures in terms slave owners could understand. For instance, in 1773, while working on the breadfruit project with John Ellis, a Royal Society fellow and the agent representing Dominica's planters in Parliament, Fothergill tried to convince him to experiment with free-labor coffee cultivation. The low start-up costs of coffee production, he argued, would enable poorer whites to start their own free-labor farms. That would diminish the number of slave plantations, increase the number of white people, and in turn help "quell the insurrections of [planters'] negroes." In any event, Fothergill added, "Negroes, who, being lazy, ignorant, and generally ill-disposed, either cannot or will not" grow coffee as productively as whites.[13]

Fothergill was not the only man of science involved with Sierra Leone to have close ties to slaveholders. Lettsom, who worked closely with Smeathman on Sierra Leone, and who provided "surgeons" and medical "equipment" for the first settlers in 1787, was born to a wealthy slaveholding family in Tortola, in the British West Indies. Lettsom readily acknowledged his links to slavery, but he learned to recast them in a way that burnished his antislavery credentials. He proudly noted that after inheriting his father's plantation in 1767 he "gave [his slaves] freedom, and began the world without fortune." But Lettsom also avoided blanket condemnations of planters, careful not to appear too "irrational." In his eulogy of Fothergill, Lettsom praised Major John Pickering, a slave owner in Tortola and close friend to him and Fothergill. Lettsom noted that he named his son after Pickering, and he described the alleged loyalty of Pickering's enslaved workforce as evidence that not all masters were monsters: "I frequently accompanied [Pickering] to his plantations," Lettsom wrote, and every time Pickering passed "his numerous negroes" they "saluted him in a loud chorus of song."[14] By publicizing his ties to slaveholders, even occasionally speaking about them in favorable terms, Lettsom, like Fothergill, protected his scientific work from accusations of bias. The upshot was that Sierra Leone could be seen less as an imminent threat to planters and more like a rational experiment designed with planters in mind.

Henry Smeathman was no less entangled with slavery. Smeathman was part of a generation of ambitious young men enthralled with overseas exploration, many of whom had been inspired by Captain James Cook's 1768 South Sea voyage. In the years before his own African voyage, Smeathman held a series of odd jobs — cabinetmaker, upholsterer, insurance agent — but he developed a passion for science in childhood and, in the late 1760s, joined the Aurelian Society, a natural history society where he met Dru Drury. Drury in turn introduced him to the men of science who, in 1771,

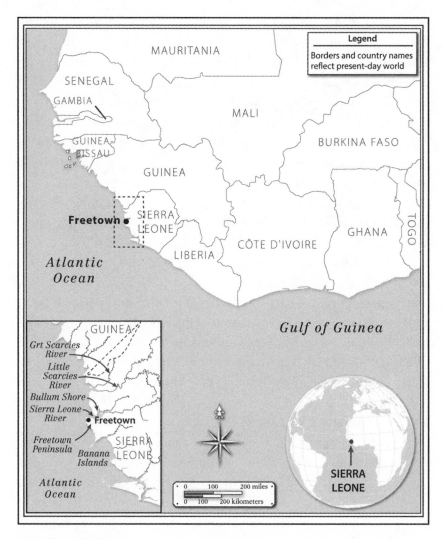

Map of Sierra Leone.
Designed by Gerry Krieg.

chose him to lead the Sierra Leone expedition. Not long after arriving in Sierra Leone, in December 1771, Smeathman realized that the only way to gain access to the region's interior was to ingratiate himself to local leaders, the most dominant of whom were slave traders. To that end, and following local custom, he married three African wives, one of them the daughter of the region's most powerful slave trader, James Cleveland, a man of Afro-European descent. Smeathman even briefly worked as an agent for the Liverpool slave trader William James in 1774. Smeathman's ties to slavery have led some historians to question his antislavery convictions, and they might well have been opportunistic. But that was to abolitionists' advantage: it allowed Smeathman's observations to appear less political, more scientific. Indeed, Smeathman never tired of describing his interest in Sierra Leone as being, above all, about "the promotion of science," a way to "furnish me with many interesting facts relative to Natural History."[15]

Fothergill left no letters explicitly linking Smeathman's voyage to abolitionism. But Smeathman's private letters clearly show him playing to the abolitionist sympathies of some of his patrons. When Smeathman first arrived on the Banana Islands, just off the coast of Sierra Leone, in late 1771, one of the first things he did was ask for permission to investigate a slave ship, named *Africa*, bobbing off the coast. Smeathman climbed on board and explored every nook and cranny, its stern, decks, lower bowels. He estimated that the *Africa* held about four hundred captives, and he described with patent disgust the noxious smell, the cramped slave quarters, the grotesque sounds echoing from belowdecks—"Our ears were struck at some distance, with a confuse noise of human voices and the clanking of chains which on board affects a sensible being with inexpressible horror." Then there were the quieter moments of devastation, like the two enslaved women, "infants yet at the breast," who suffered "a silent grief." Five years later, in 1776, while exploring the West Indies for the same patrons, he described in scathing terms the brutality of Caribbean planters: "The excess to which tyranny is sometimes carried over the poor slaves is disquieting to a mind in the least tinctured with humanity," he wrote to Tunstall. Smeathman even told Tunstall that he was ready to publish a pamphlet about the slave ship's inner workings—"The Oeconomy of a Slave Ship"—though the project never came to fruition.[16]

Yet like Lettsom, Smeathman cautiously modulated his antislavery views, careful not to muddy his scientific integrity. In his journal, intended for his patrons' perusal, he described the enslaved children he saw on board the *Africa* as "more happy than the constrained disciples of a country grammar schoolmaster." Though a few enslaved mothers were clearly anguished, "the greater part of the women . . . seemed rather cheerful." Elsewhere he depicted the ship's captain, Captain Tettle, as judiciously working to maintain the enslaved people's health. Enslaved children, especially little girls, were allowed to play "upon the deck at pleasure being all without fetters"; scaffolds were put up on the ship's sides and mast "for the convenience of washing and cleansing the ship."[17] Smeathman's journal thus had enough to appease all sides in the slave-trade debates.

Ultimately, what mattered most to Smeathman's patrons was his assessment of Sierra Leone's natural environment. They needed evidence of the region's suitability for agriculture, the healthfulness of its climate for settlement, and the capacity of its indigenous inhabitants to become "civilized." On all these fronts, Smeathman offered promising news. His journal described Sierra Leone's coast as overflowing with beans, berries, pineapple, palm trees—a place that "teemed with natural wonders." In addition, "the whole country abounds with Iron as is evident from the rocks on the shore" and might be "used in paving the streets of London." He was happy to report that he, a white man, survived an initial bout of sickness and recovered because he "put my faith in Dr. Fothergill's prescriptions and my own drugs." Smeathman also noted that Africans could possibly help treat common illnesses. "They use with great skill" their own medicines, he wrote, though he added that they "will not trust" their own healers "if they can procure a white Dr."[18] The emphasis on the region's relative healthfulness was critical. At a time when Europeans increasingly worried about whether white bodies could survive in tropical climates, knowing that medicine could overcome tropical diseases helped counter one common argument against new colonies in West Africa.

Smeathman's views on native medical knowledge, a mix of fascination and denigration, echoed his broader descriptions of the region's inhabitants. During the 1770s and early 1780s, explorers who traveled to Africa, regardless of their views on slavery, tended to discuss racial differences between Africans and Europeans in terms of culture, religion, and politics—in a word, character, not color. Though the early debates over slavery heightened scientific interest in physical and mental differences, these differences were still seen as malleable and somewhat superficial. For explorers, the core questions in regard to race were whether West Africans could become civilized and what explained the alleged primitiveness of certain societies. If proslavery explorers portrayed African societies as hopelessly savage, places where human sacrifice, cannibalism, and indigenous forms of slavery were rampant, then antislavery explorers offered a more sanguine view. They did not deny African primitiveness, or the presence of warfare and slavery. But they placed the blame on European demand for enslaved laborers, not on features innate to African cultures or on the natural environment, from which cultures were thought to arise.[19]

Antislavery explorers also amended, if not quite rejected, the idea of "tropical exuberance," the notion that the natural abundance of tropical climates made indigenous populations lazy. Proslavery writers argued that slavery, and the violence that came with it, was justified in part because African people, as a "tropical" race, were not used to hard work: "Slavery, and the authority resulting from it," wrote the Scottish physician and astronomer Alexander Wilson in 1780, was "in a certain degree necessary to counteract the natural causes of inactivity in hot latitudes."[20] Antislavery explorers countered that while West Africa was indeed teeming with natural commodities, and Africans were indeed unaccustomed to hard work, a change in the social environment—that is, the introduction of commerce, Christianity, good government, and education—could easily entice them to work. If anything, they contended, it was the slave trade that was keeping them backward, not the climate.

Smeathman was among the first African explorers to comment on indigenous African societies at precisely the moment when the slave trade debates began to transform ideas about racial difference. Like other antislavery explorers after him, he focused on differences in character and culture, and he emphasized that both were shaped chiefly by the social environment, not the natural one. In 1783 he explicitly rejected Wilson's claim that "the extremes of heat" accounted for Africans' perceived slothfulness. Instead, he underscored that "all men in regard to their dispositions" are "formed more by artificial than natural causes"—and that "a change in government and education would make a wonderful difference in their disposition."[21]

Smeathman spent little time discussing physical features or refuting ideas about deep-seated racial differences, which reflects how the debates over slavery had not yet completely dominated debates over race. Far more important was for West African societies to show signs of civilizational development, or at least the potential for it. Smeathman's journal contained precisely this kind of evidence. On the one hand, he described the Mandigo people as a "kind of bastard Mohatematens," depicting their conversion practices as petty financial scams. A Muslim preacher would sell a charm, or *greegree*, to the destitute solely for profit, he wrote, and by the "same insidious and diabolical methods as the founder of their religion." Yet he also noted that Islamic converts, far more than Africans who practiced non-Abrahamic religions, zealously embraced their freedom: African Muslims "have a strong notion that they have a right to freedom . . . and are ever plotting to cut off [slave ship] vessels in which they are shipped and often succeed too well."[22] Though Smeathman's comments reflected the widespread assumption that West African cultures were less civilized, they also underscored the potential of Mandigos to become more like Europeans. In his eyes, their embrace of Islam was one step closer to conversion to Christianity, and their love of freedom evidence that they had the capacity for civilization.

During Smeathman's four-year stay in Sierra Leone, he conducted an impressive amount of research. He planted a botanical garden, experimented with rice irrigation, and closely observed the region's geography, plants, and people. He hoped to publish a comprehensive natural history on West Africa, but the outbreak of the American Revolution "frustrated" his plans, he later recalled. While sailing back to England in early 1775— aboard a slave ship, which made a detour to the Caribbean to sell its human cargo—an American privateer captured his ship. In the mayhem, Smeathman lost many of his notes. Ultimately, Smeathman's ship was allowed to continue to Tobago, in the Caribbean, where Smeathman remained for the next four years. The saving grace was that he was now able to study a comparable tropical climate, and in turn strengthen his claims about Sierra Leone's suitability as a free-labor alternative to slave plantations. In fact, Smeathman lamented that he had not gone to the West Indies *before* West Africa, telling Tunstall in 1776 that "it would have been a vast advantage to me." Many of the commodities that grew in the West Indies—"Plantains, bananas, Cam-manyok or sweet Cassavas"—either did, or could, grow in Sierra Leone, he wrote. All that was needed for a successful colony in Africa was "the knowledge of how to cultivate or prepare them"—disregarding, or perhaps ignorant of, the fact that enslaved Africans introduced many of those commodities to the Caribbean.[23]

His stay in the West Indies—and the comparisons it enabled him to draw to Sierra Leone—also strengthened his claims of scientific legitimacy. During his African expedition, Smeathman spent months studying termites, and when he arrived in the West Indies, he found planters struggling with their own entomological problem: ants, specifically sugarcane-eating ants, which had been destroying their leading cash crop. The colonial assembly of Grenada promised payment for anyone who proposed a "means of removing the calamity," Smeathman told Tunstall, adding that he would soon "declaim my discoveries and claim the rewards." Smeathman's solution—drowning ants in whale-fat oil—did not work, but the competition helped refine his entomological expertise. In 1781, two years after returning to England, he published his first essay in the Royal Society's *Philosophical Transactions*, titled "Some Account of the Termites, Which Are Found in Africa and Other Hot Climates." The reference to "other hot climates" was significant, for it indicated how Smeathman's scientific work in the Caribbean corroborated his work in Africa. He even framed the essay as a gift to planters, calling termites their "greatest of plagues."[24]

For Smeathman, thinking about termites may have also influenced his thinking about African colonization. Smeathman's main scientific contribution was that, contrary to common opinion, termites served a positive ecological function. Termites, he realized, ate through wooden homes not to antagonize human beings but because they mistook homes for decaying trees, which, when eaten through, hastened the wood's decomposition. As one scholar has argued, Smeathman's views of Africans reflected a similar underlying logic: like termites, Africans should be respected and understood in their own environment, on their own terms, not treated as beasts of burden or disposable pests. Smeathman actually spoke about termites in human terms, suggesting that he saw resemblances between the two: termite societies organized themselves into "*labourers*," "*soldiers*," and "the *nobility or gentry*," he wrote. And the lowest order, the "*labourers*," actually did some of the most important work, securing the safety and survival of the entire colony. Smeathman even noted how termite villages, with their "rooted turrets" colored "Black," resembled "*Negro Heads*."[25] Smeathman's ideas about termites ultimately echoed and anticipated his views of Africans: if understood properly, if managed properly, they might become a British society in miniature.

After publishing his essay on termites, Smeathman began to solicit funders for a colony in Sierra Leone, as well as a second expedition to the region. For years he had complained that his travails during the War of Independence left his scientific work incomplete. But occasionally he blamed his native African aides for the scientific work he had lost. "The stupid carelessness or curiosity of the ignorant natives," he wrote, explained why Fothergill received so many ruined plants while he was still in Africa. For his part, Fothergill attributed the ruined items to the slave trade. Writing to Linnaeus in 1774, he noted that Smeathman had sent him plenty of insects, but little else survived: "No plants, no seeds—not a single one." The reason, Fothergill told Linnaeus, was that Smeathman's ship had to sail first to the West Indies to sell "the wickedest of cargoes"—enslaved people; "because of this long detour, everything dies." The difference between Fothergill's and Smeathman's explanations—Fothergill blaming the slave trade, Smeathman indigenous Africans—obscures a closer ideological kinship, however. Fothergill was equally capable of dismissing Africans as "lazy" and "ignorant," as his letter to John Ellis illustrates.[26] For both men, casual disparagement of Africans was not uncommon, particularly when it provided a convenient alibi for the shortcomings of their research. But the excuses could also prove strangely useful, keeping alive the notion that more science, better science, would redeem Sierra Leone.

To be sure, few of Sierra Leone's scientific founders doubted the ability of science to help solve society's problems. They were living in a moment in which new scientific inventions—steam engines, diving bells, hot-air balloons—were helping humans overcome every seeming limitation. Moreover, the two worlds, of scientific invention and antislavery, often overlapped. Granville Sharp, who occasionally attended Royal Society meetings, proposed a "diving bell drawing" to the society in 1775. In 1784 Smeathman patented his own hot-air balloon design, a "dirigible aerostatic machine," as he called it, and hoped the money from the patent could help fund an antislavery colony. When he met Franklin in Paris, Smeathman solicited his support for both the colony *and* his balloon design. In addition to "tell[ing] Dr. Franklin of his plan of civilizing Africa," Smeathman detailed the mechanical principles of his hot-air balloon. Franklin said he had "no doubt" his colonization plan would be supported "at Boston," then indulged his balloon ideas—"launch[ing] half a sheet of paper obliquely in the air," Smeathman told Lettsom, "observing, that that was an evident proof of the propriety of my doctrines."[27]

Eighteenth-century men of science did not simply share an overlapping interest in invention and antislavery; often, ideas from one domain inspired ideas in the other. Smeathman told Lettsom, for instance, that the wing design for his hot-air balloon came from observing "many insects," which indicates how his research in Africa influenced his later invention. The language Smeathman used to describe the sea creatures he discovered during his African expedition—marine worms that "twinkle like stars," the sparks they emitted appearing like "flashes from an electrical machine"— also evoked skies and machines, conjuring images of mechanical inventions putting humans in flight. Romantic poets of the period, who, like Erasmus Darwin, were sometimes men of science themselves, linked scientific inventions to humanitarian projects. In 1810 the poet Percy Bysshe Shelley tied emancipation to hot-air balloons specifically: the "shadow of the balloon" cast over Africa, Shelley exclaimed, would soon "emancipate every slave." But perhaps of greatest benefit to abolitionists was Smeathman's insistence that patronizing Sierra Leone and expeditions to the region was not only about abolitionism: these projects had scientific value in themselves. As Smeathman told Lettsom in 1785, a new expedition to Sierra Leone "cannot fail to enrich science in general."[28]

Smeathman's hopes for a second expedition to Sierra Leone came closer to reality in 1786. In January, a group of London philanthropists formed the Committee for the Relief of the Black Poor, a charity whose initial aim was to provide food and shelter for the city's impoverished Black population. Realizing an opportunity, Smeathman in February 1786 approached the committee and the city's free Black leaders, asking the latter if they would consider encouraging other Black Londoners to resettle in Sierra Leone, with the support of British philanthropists.[29] London's free Black leaders were central to the plan's execution. It was Black leaders, for instance, who brought in Granville Sharp, whom they trusted, and asked to author the colony's legal system.[30] Black leaders also pushed Smeathman to guarantee, in writing, that Black settlers would never be sold into slavery—that, as Smeathman wrote, "they and their posterity may enjoy perfect freedom."[31]

But the scientific legitimacy that Smeathman conferred on the project also helped get the colony off the ground. The natural knowledge of the region he provided helped convince British officials and philanthropists—many of whom never thought to consider the settlers' input—that Sierra Leone might make a plausible place of settlement. To be sure, both the Committee for the Relief of the Black Poor and several early Black settlers doubted Smeathman's intentions. But they nonetheless recognized his value—as a man of science with firsthand experience in the region, he could testify to the region's environmental suitability with an authority that few others had. In May 1786 the Committee for the Relief of the Black Poor asked Smeathman to publish a formal document to garner public support for the antislavery colony. The document was titled *Plan of a Settlement to Be Made Near Sierra Leone*, and its title page highlighted Smeathman's firsthand experience—the author had "resided in that Country near Four Years," it read—demonstrating the value of Smeathman's earlier scientific voyage.[32] Smeathman's *Plan of a Settlement* not only helped generate public support for Sierra Leone, but also became the blueprint for the colony and its antislavery objectives.

Central to Smeathman's *Plan of a Settlement* were his lengthy remarks about Sierra Leone's environment. As he had in his private letters to patrons, he presented the region as an African Eden: "Rice, and a species of Indigo superior to any other" grew everywhere, he wrote, as did "Cotton and Tobacco equal to those produced in the Brasils." Only the slightest effort would make these crops grow: "It is not necessary to turn up the earth more, than from the depth of two or three inches, with a slight hoe, in order to cultivate any kind of grain." The native inhabitants, moreover, would readily welcome new settlers: "The peaceable temper of the natives, promise the safest and most permanent establishment of commerce." Nor did settlers, Black or white, need to fear the climate. For Blacks, he wrote "the climate is very healthy," assuming, as most whites did at the time, that Black people acclimated more easily to tropical climates. He assured the white overseers and poor white women who would also join the settlers that the climate was far less forbidding than they might fear. The reason "many white people" died in Africa, he wrote, had nothing to do with the climate and everything to do with the bad habits of previous white travelers, who lived "intemperate lives." In any event, he promised all settlers the best medical care money could buy, "a Physician, who has had four years practice on the coast of Africa" as well as medical aides trained in "Surgery, Midwifery, Chemistry, and other medical arts."[33]

Smeathman's *Plan of a Settlement* may have helped convince officials to support the colony, but it also set unreasonably high expectations. Smeathman died suddenly of a fever on July 1, 1786, but Black Londoners were committed to seeing the plan through. By April 8, 1787, with the help of Olaudah Equiano and Sharp, Black leaders managed to attract 456 settlers—290 Black men, 41 Black women, 11 Black children, 70 poor white women, 6 white children, and another 38 officials and artisans—to set sail for Sierra Leone. The settlers arrived a month later, and almost immediately the picture Smeathman had painted—a healthful climate, fertile soil, friendly indigenous inhabitants—failed to match the reality. "This country does not agree with us at all," the settlers wrote to Sharp on July 20, 1787. "Not a thing, which is put into the ground, will grow more than a foot tall." But the greatest problem by far was the number of deaths. More than fifty died on the voyage to the colony alone, and by September 1787 almost one-third of the settlers—Black and white combined—had died, most from malaria, yellow fever, or dysentery. Though white settlers died at a higher rate than Black settlers, Black settlers were still extremely vulnerable. Thirty Blacks and twenty whites died within the first two months, for instance, and nearly 150 other settlers, of both races, remained seriously ill.[34]

The colony's scientific and abolitionist backers vigorously denied that the deaths had anything to do with the climate or related ideas about differences in racial immunity. Modern scientists now know that yellow fever killed Black settlers less frequently than white settlers not because of *innate* racial immunity, but because of *acquired* immunity—that is, if one was stricken once and survived, the chances of dying from a second bout were greatly diminished. Many Black settlers may have been stricken with yellow fever while enslaved in the Caribbean, or in West Africa, if they were born there, and they thus would have been less vulnerable to a second attack. As for malaria, many people of West African descent have acquired a genetic adaptation that staves off the disease—as have many *other* populations in regions throughout the world where malaria has long been present, such as the Mediterranean Basin and India. This genetic mutation, then, is not unique to any one "race," but common to many populations.[35]

Of course, eighteenth-century writers did not know the modern science. The colony's defenders therefore had to either explain the deaths in a way that avoided implicating the region's climate, or fall back on what were then increasingly common beliefs about racialized immunity, which held that tropical climates were simply too dangerous for white people. The problem with that theory was that, for Sierra Leone's financial backers, it would have made the colony impossible to support, since it was assumed that white officials needed to live among and oversee the settlers. Placed in this situation, the colony's defenders chose to absolve the climate of responsibility and, instead, echoing Smeathman, blame the deaths on the recklessness of the settlers, Black and white. Sharp, writing to Franklin in January 1788, used the exact language of Smeathman, blaming the deaths on the "wickedness and *gross intemperance* of the Settlers themselves." The deaths were "not to be attributed to the climate," he repeated to Lettsom, but ascribed to "intemperate" behavior.[36]

The colony's scientific supporters legitimated this defense when using it themselves. Carl Wadström, a prominent naturalist who traveled to the region in 1787, and who became an ardent defender of Sierra Leone, also downplayed the climate, stressing personal behavior: Black settlers died because they were "daily intoxicated," whereas "Europeans have their own conduct, more than the climate, to blame for their unhealthiness in Africa," he wrote. Lettsom did not even try to explain the deaths, instead sounding a note of optimism: things might turn around, he told a sympathetic physician on October 20, 1787, if only someone of Smeathman's "enterprising spirit" could be found to oversee the colony.[37]

If Smeathman's overly optimistic depiction of the climate helped set the settlers up for failure, so did his portrayal of the "peaceable temper of the natives." When the first settlers arrived, the colony's overseers naively believed that they had purchased the land outright from the region's main ethnic group, the Temne. But King Tom, the local leader who signed the treaty, did not have the authority to relinquish the land; nor did the colony's overseers abide by long-standing diplomatic custom, whereby the Temne allowed settlers to stay only on the condition of the ongoing exchange of gifts and military protection. Even though the British slave traders operating in the region had long understood this, the colony's white overseers conducted themselves as they pleased, perhaps believing that their benevolent intentions— ending the slave trade—relieved them of any obligation to local custom. Some of the colony's supporters in London were ignorant at best, willfully blind at worst, of the tensions between the overseers and Temne leaders. But rather than investigate the issue, they tended to blame the Temne for being greedy; otherwise, they faulted Black settlers for foolishly trading away what few goods they had. In October 1787 Lettsom explained that the Temne came to the colony "merely to get clothes and provisions without labour," while the settlers gave their clothes away "very foolishly," asking for nothing in return. Wadström sounded a similar note, blaming the colony's lackluster agricultural output on the "generally profligate" habits of the Black settlers, if not their "indolent and depraved dispositions."[38]

Promises that the soil would grow spontaneously also failed to materialize. A large part of the reason was that nearly half of the settlers had died, including the botanists sent to assist them. But the colony's patrons continued to insist that the soil was exceptionally fertile, hoping to keep interest in the colony alive. On October 13, 1787, Sharp told Lettsom that, whatever the other problems, the "natural products are equal to the most sanguine hopes," highlighting the "fine cotton, the best indigo in the world, [and] sugar-canes" that grew "wild upon the mountains." Lettsom failed to mention any possibility that the soil was less ideal than advertised, explaining the lack of commercial crops on, ironically enough, "voracious ants." Wadström followed suit, touting the region's agricultural promise, and in turn providing cover for the colony's struggles. The "sugar-cane . . . grows spontaneously in many places," he wrote, and the cotton would "not disgrace their best workmen," referring to British textile factory workers.[39]

By 1791 abolitionists felt things had reached a nadir. That year, abolitionist leaders, led by Thomas Clarkson, formed the Sierra Leone Company, which took control of the colony and whose chief aim was to make the colony profitable. Among the company's stockholders were Wedgwood and Darwin. The four years before the Sierra Leone Company took over had certainly been trying, the causes many: patrons who ignored indigenous diplomatic customs and refused to take settlers' input seriously; slave traders surrounding the colony who undermined efforts at "legitimate trade"; settlers who struggled to keep up morale. But Sierra Leone's scientific supporters also contributed to the difficulties. Their scientific claims about the soil, climate, and inhabitants set unreasonably high expectations, and when the reality failed to match the descriptions, they faulted the settlers, deflecting attention away from their own culpability. To be sure, most settlers overcame these early trials, creating "a fragile but functioning government and civil society." They instituted a militia to guard against marauding slave traders, elected local officials, served on juries, farmed, traded with natives, held church services, married. They told Sharp that their difficulties had nothing to do with "want of industry" or "gratitude," and at least something to do with the "many injuries done to us from aspersions."[40] The problem was not that they were wrong, but that they had to compete with a much too powerful narrative: that science, not settlers, would be the key to Sierra Leone's survival.

In 1792 the Sierra Leone Company recruited 1,200 Black Nova Scotians to immigrate to the colony. Most were Black Loyalists, and without these settlers Sierra Leone may well have vanished. But the colony's survival also depended on the continued portrayal of Sierra Leone's natural environment as fertile and healthful, and of the colony as a project that, with more scientific research, could be made more profitable than slave plantations. Between

1792 and 1807, the men of science who either traveled to or lived in the colony would provide abolitionists with exactly this kind of evidence.[41]

Among the most important scientific travelers was Carl Wadström. A Swedish engineer, Wadström sailed to the region in October 1787 on a scientific expedition sponsored by the Swedish king Gustaf III, a "great lover of natural history." He was joined by Anders Sparrman, a student of Carl Linnaeus, and Carl Axel Arrhenius, a military lieutenant trained in chemistry. The trio spent three months exploring the region, looking for "mineralogy, antiquities, and in general what regards the state of man in that country." A fervent abolitionist, Wadström moved to London in 1789 and immediately introduced himself to Thomas Clarkson, the organizing force behind the Sierra Leone Company. Clarkson knew he had stumbled on a gold mine: "I perceived the great treasure I had found," Clarkson wrote in his history of the abolition movement. Since Smeathman's voyage, few men of science had traveled to West Africa, especially on expeditions ostensibly for scientific purposes. Wadström and Sparrman "had been lately sent to Africa . . . to make discoveries in botany, mineralogy, and other departments of science," Clarkson wrote, implying, and hoping, that the scientific nature of their work would give it the valence of objectivity.[42]

Wadström wasted no time involving himself in abolitionist politics. Upon arriving in London, he published an essay about his three-month expedition, *Observations on the Slave Trade and a Description of Some Part of the Coast of Guinea* (1789), which emphasized the region's suitability for free-labor colonies and how the slave trade was impeding progress toward that end. Though Wadström made clear his political agenda—that European nations would "unite in establishing colonies on the coast of Guinea" and help alleviate "the cause of suffering humanity"—he also underscored his work's scientific integrity. "All the observations I have been able to make," he wrote, came from when "I went to the coast of Africa, not with any commercial views, but for the sole purpose of inquiry and observation." He added that rapid advances in science would make colonization easier, eradicating diseases and making the growth of new commodities more feasible: "The moment is now arrived, when mankind will begin to make a real use of their great scientific acquirements"— that is, for the purpose of creating slave-free African colonies.[43]

Wadström also defended the natural history claims of his predecessors, re-fusing to let Sierra Leone's troubles stand in the way of its promise. He flatly dismissed the notion that Africans were inherently lazy. Like all "raw na-tions," he claimed, they were simply in a "state of infancy," "their faculties" not yet "cultivated." He provided firsthand evidence that Africans had a taste for finer things, that if they were taught that wage labor would help them ac-quire material goods, they would work harder. A local African leader he met was enamored of the bejeweled buttons Wadström showed him: "This fond-ness for European baubles," he wrote, "proves that an advantageous com-merce might be established" among the natives. To the extent that political instability mitigated against their social progress, he placed the blame squarely on the slave trade: "The *Wars* among the negroes," he emphasized, "originated in the slave trade." As for the climate, he argued that the region had several different climates, some more healthful than others, and stressed instead that most white deaths on the coast arose from slave traders' own ruthless pursuit of profit. In order to cut costs, "monopolizing Companies"—by which he meant slave-trade companies—failed to provide their European traders with adequate medicine or food. The African soil, for its part, begged for cultivation: all the slave-grown commodities—sugarcane, tobacco, cot-ton, indigo—could be "cultivated on the coast with very little trouble, and in a profusion perfectly astonishing to an European."[44]

Observations convinced Clarkson that Wadström would be the perfect witness to testify before Parliament during its hearings on the slave trade. Conducted between 1788 and 1791, the hearings focused on whether free-labor African colonies could undermine the slave trade in Africa, whether free-labor African colonies could be as profitable as Caribbean plantations, and whether they might one day function as replacements. For two days in April 1790, members of Parliament grilled Wadström about all things related to natural history: the character of indigenous Africans, the quality of the soil, the healthfulness of the climate. "What have you formed of the capacity of the Negroes?" members asked him during his first day of questioning, on April 28, 1790. Wadström refused to acknowledge gestating ideas about in-nate inferiority; nor did he default on the climate to explain Africans' char-acter. Instead, he focused on the reigning theory of social development, stadial theory, which presumed an essential equality among all races and focused on the potential for all groups to progress from a primitive to a civilized state. Speaking of African natives, he stated that they are "as capable of being brought to the highest perfection, as those of any white civilized nation." When the committee pressed further and asked whether "their indolence would be such as to prevent their supplying the market" with goods grown on their own continent, he reiterated what was by that point abolitionists' stock reply: the slave trade worked as a disincentive. Because the "minds of the natives" were "excited by merchants to engage in [the slave-trade] business," they had "no encouragement to improve their country, and cultivate the productions."[45]

The remainder of Wadström's testimony underscored West Africa's natural abundance. African indigo was "in all respects equal to the best Carolina indigo"; the coast not only had an abundance of cotton, but the natives also manufactured it "with uncommon neatness." Wadström also "brought with [him] a collection of minerals" to show the committee members, items essential for Britain's growing number of steam-powered cotton mills and pottery factories. Then there was all that remained to be discovered. He mentioned the thousands of specimens he collected for the "cabinet of natural history, of the royal academy of Stockholm," as well as the "*Materia Medica*" — or pharmaceutical drugs; all of it was meant not just to document what he had found, but also to encourage further research.[46] Emphasizing the scientific nature of his expedition had the added benefit of obscuring his antislavery motives and sustaining an image of scientific impartiality.

Wadström's testimony was quickly published on both sides of the Atlantic. Abolitionists in the United States and Britain printed abstracts of the hearings, which highlighted abolitionists' best arguments and often included Wadström's testimony. Benjamin Rush cited the hearings in a 1793 essay in which he argued that cultivating maple trees in New York and Pennsylvania would help wean Americans off slave-grown Caribbean sugar. Published in the American Philosophical Society's *Transactions*, Rush's essay relied on scientific testimony to prove that in other parts of the world, "free people" were already cultivating commodities that most Westerners knew only as slave-grown. Though Rush did not cite Wadström's testimony, he cited other scientific testimony from the hearings, as well as evidence from an earlier African expedition that showed that indigenous Africans already produced sugar themselves.[47] Wadström's evidence, then, was not an outlier. It was part of a larger effort in which abolitionists relied on scientific authorities to defend African colonization projects, and with them the broader antislavery movement.

The ongoing French and Haitian Revolutions had, by 1794, put abolitionists on the defensive. Bills to end the slave trade in the United States and Britain would not get a serious hearing for another decade, and once the prospects for slave trade abolition had dimmed, defending Sierra Leone's reputation took added importance. As a result, Wadström's support for the colony became ever more consequential. In 1794 and 1795 he published the two-volume *Essay on Colonization, Particularly Applied to the Western Coast of Africa*, which repeated many of the arguments he had made in *Observations* and the parliamentary hearings. Despite the title, *Essay on Colonization* read more like a work of natural history than a political pamphlet, replete with observations about West Africa's climate, culture, and soil, which suggests that his scientific claims were still his greatest asset. Wadström may have even included in the book parts of a natural history that he intended to write but never finished: in *Observations*, he promised readers that he would soon publish a separate work of pure natural history, tentatively titled *Two Views of the Coast of Guinea*, and he may have folded parts of it into this new essay.[48]

Wadström knew that the antislavery politics infused in the *Essay on Colonization* might undermine its scientific components. He therefore told his readers, right at the start, not to conflate his politics with his science. "The reader has no doubt, by this time, discovered that the person who now addresses him" — Wadström — "is a zealous friend to the Africans." But, he reassured his readers, his abolitionist "zeal is not inconsistent with sober truth." The sections devoted to West Africa's natural produce, the fertility of its soil, and its climate repeated the arguments he had made earlier. "The vegetation is luxuriant to a degree unknown in the most fertile parts of Europe," he wrote, which only "confirms the observations of M. Adanson, my fellow travellers and myself"; settlers had their "own conduct, more than the climate, to blame for their unhealthiness."[49]

But worth particular note is his explanation of Africans' alleged backwardness. Here he gave a fuller explanation of Africans' alleged primitiveness, placing greater emphasis on the slave trade than before. Wadström argued that continuing the slave trade "will prove *the* grand obstacle to their improvement and civilization." Like Smeathman, Wadström focused on the social environment, rather than the natural one, because it allowed him to explain Africans' alleged simplicity on something humans could control—the slave trade—not the whims of nature. By emphasizing the slave trade's role in preventing Sierra Leone's success, Wadström also challenged an increasingly prevalent proslavery argument, one that insisted on the deep-rootedness of African savagery. By the 1790s slavery's defenders increasingly argued that Africans' barbarity, their "violent spirit of revenge," doomed any attempt to impose "civilized" government on them. Wadström countered that the political instability and violence sometimes witnessed in Africa arose from the slave trade itself. Indeed, the entire "state of anarchy and blood" that Wadström claimed characterized African societies stemmed from the pressure certain African groups felt to procure enslaved laborers for European slave traders.[50] Though many nonscientific abolitionists also made this argument, Wadström lent it scientific legitimacy both as a firsthand witness and by embedding the claim in a work intended to read like a work of natural history.

Wadström's essay was quickly translated into French and German, and it soon became an international hit. In 1795 he moved to France, joined the French abolition society and, with Napoleon's support, helped convene the first conference on African colonization in 1799. He died that same year, and though his antislavery colonization ideas were at times seen as naive, "perhaps sometimes delusive," wrote Helen Maria Williams, a British antislavery sympathizer, it was also true, she added, that no one did more to give "rise to the foundation of Sierra Leone."[51]

While Wadström waged a public campaign to defend Sierra Leone, another Swedish man of science, Adam Afzelius, quietly tried to revive Sierra Leone's fortunes in the colony itself. In 1792 Joseph Banks recommended the forty-two-year-old student of Carl Linnaeus to the directors of the Sierra Leone Company, who had been struggling to grow commercial-grade crops. Afzelius had in 1789 left his post as professor of botany at Uppsala University to travel to England, where several other Swedish men of science, including Wadström, had settled. Banks welcomed these men into his scientific circle, in part seeing their interest in Sierra Leone as a way to gather scientific information about the region. "Afzelius is here still & Looks vastly well & Fat," Banks wrote to Olof Swartz in 1790, who was also a student of Linnaeus. "He is in truth an excellent Botanist. . . . I wish the kind Fates would always allow me a Swede or two to Study in my Library."[52]

From 1792 to 1796, Afzelius was Sierra Leone's official botanist. Stationed in the colony, he went on several small expeditions into the interior, experimented growing indigo, cotton, coffee, and rice, and sent back scores of rare African goods to European men of science like Banks—medicines, dyes, yams, guava, a plant "not unlike mint." Afzelius's search for medicinal, commercial, and scientific materials was a form of "bioprospecting"—the search for knowledge and control over natural objects in non-European lands. And like other European bioprospectors at the time, Afzelius relied extensively on indigenous Africans and Sierra Leone's Black settlers, a fact made plain in Afzelius's unpublished correspondence and private journal.[53] Yet Afzelius seldom acknowledged the intellectual work of these aides in his published writings, an elision that both reflected and reinforced the perception that Black settlers and indigenous Africans had not yet become civilized. And yet these omissions may have had an unintended benefit: by erasing Black contributions, abolitionist elites could uphold the image of men of science as heroic figures, single-handedly working to solve the colony's ills.

In his journal Afzelius frequently notes the settlers and indigenous Africans who helped him. Women appear as often as men: "A Native woman who had some powdered bark fastened in the middle of her forehead" told Afzelius that the bark was used for headaches. Mrs. Logan, a Black settler, probably from Nova Scotia and possibly a former enslaved Virginian, demonstrated how to turn the ashes of leaves into soap. The sister of one of his closest assistants, an Angolan settler named Peter, "brought to me a Banisteria [plant], which she called *Marimba*"; natives used it for medical and religious purposes, and Afzelius sent some seeds back to England. Male settlers, meanwhile, worked as Afzelius's personal aides. Tarleton Fleming, a Black Nova Scotian who was probably once enslaved in either South Carolina or Virginia, was in charge of upkeep for Afzelius's experimental garden. Thomas Cooper, a Black settler since 1787, often obtained botanical specimens from indigenous Africans and shared them with Afzelius, including the "beer of the root called *Ningee*," which, Afzelius wrote, "resembled table beer."[54]

These men and women did not simply take orders. Several of them offered important information. In May 1795 Fleming warned Afzelius that "there was not to be found any timber" if the settlers kept cutting the trees down, advice that proved prescient: by 1829, timber accounted for 69.2 percent of the colony's exports. Many of the native plants that Afzelius collected were also, of course, already used by local indigenous groups. The fact that Afzelius took copious notes on how indigenous peoples used local crops indicates that they, not he, were the main knowledge producers. That some of these indigenous plants made their way back to Banks also demonstrates the value of their work to men of science—and by extension, the value of indigenous African knowledge to the broader antislavery movement. Upon receiving a package of plants from Afzelius in 1794, Banks wrote to Swartz, "May heaven assist him!"; from what Afzelius had sent, "we have been assured, of his good luck."[55]

Other times, settlers functioned both as knowledge producers and as go-betweens—that is, as intermediaries brokering exchanges between Europeans and Africans. For instance, Thomas Cooper, a Black settler, volunteered to go on an expedition with Afzelius in March 1796, and at least once he brokered a meeting between Afzelius and a nearby African leader, Mong Kerrapha. But Cooper was also a knowledge producer. He wrote a letter that was eventually forwarded to Banks, corroborating information about a trading route that Banks had learned about from an earlier European explorer, Simon Lucas. In the letter Cooper wrote that he had spoken with an Arab merchant who had been on Lucas's route, and that the merchant, not Lucas, "ought to be the best judge as he has walked it more than once." But if men like Banks, and Fothergill and Smeathman too, often collaborated with slave traders, whether for the sake of a scientific mission or for a larger antislavery purpose, Cooper would not. Afzelius recorded how Cooper, during one leg of their journey, refused to buy slaves from a group of slave traders they encountered, making "the slave-traders . . . as inveterated [sic] against him as ever."[56]

For abolitionists, the interest Banks took in Cooper and Afzelius's expedition helped legitimate the colony, offering a rationale for its existence—the promotion of science—that need not be humanitarian. Indeed, at the same time Banks was working with Afzelius and Cooper, he was funding expeditions into the African interior with a different group of patrons, called the African Association. The African Association comprised abolitionists and slaveholders alike; abolitionists hoped that the knowledge gained from its expeditions would help Sierra Leone prosper, and proslavery members saw the expeditions as a hedge against the possibility of the antislavery movement's success. Should slavery end, proslavery patrons reasoned, knowledge from these expeditions could be used to create new free-labor African colonies. Occasionally, the association's slaveholding patrons used its expeditions to sabotage the antislavery agenda. In 1798, for instance, Bryan Edwards rewrote parts of Mungo Park's expedition journal—the association's greatest expedition—to highlight the ubiquity of the slave trade in Africa: given the extent of indigenous slavery, abolishing the slave trade "would neither be so extensive or beneficial, as many wise and worthy persons fondly expect," Edwards wrote in Park's manuscript. In light of these manipulations, the Sierra Leone Company's own expeditions became even more crucial, since they could provide evidence of the region's environment that was not likely to be undermined by duplicitous proslavery editors.[57]

But if men of science often relied on Cooper's research, they rarely gave him, or Black and indigenous aides like him, recognition. To be sure, most travel narratives from the period mention the help of local aides, but they treat them as ancillary, rendered subordinate even when they made it into print. For his part, Afzelius often mentioned the Black settlers and natives who helped him, but only in his private letters. Writing to the Sierra Leone Company directors in 1794, he noted that the natives "at Anamaboo" had shown him the cotton seeds he included in the accompanying package; Mr. Kooda, "a Black man and from Gambia," gave him medicinal herbs that natives used "with great success against cold, cough and all sort of rheumatic complaints." Afzelius also wrote that natives supplied him with cola, yams, wild figs, and a new species of coffee, which, he told the directors, was "worth while" to cultivate.[58]

But men of science in Europe mostly ignored these contributions. Nowhere is this clearer than when looking at how the species *Afzelia africana*, named after Afzelius, received its formal scientific name. *Afzelia africana* remains the tree's official name to this day, but local African ethnicities have many different names for it: in Yoruba, it is called *bilinga*; in Edo, *apa*; in Temne, *kontah*—itself a derivative of the common English name "counterwood." Native ethnic groups have used the tree for everything from making djembe drums to creating a drug that promotes pregnancy. It is particularly good at resisting termites. Afzelius spent much time searching for various species of the *Afzelia* genus, and he almost always relied on indigenous inhabitants and settlers: a man named Amarah gave him a bark that the Susu tribe called *Serig Baillee*; Peter, his Angolan aide, found three different species in January 1796; on his expedition with Cooper, Afzelius noticed that indigenous Africans near Tooka Kerren used the tree's bark, but for a ceremonial custom he could not understand.[59]

Afzelius probably hoped that one of the species might help cure the diseases that continued to fell settlers. He and Thomas Winterbottom, the colony's chief physician in the 1790s, spent years experimenting with barks that they hoped might function as an alternative to Peruvian bark, the most common treatment for malaria at the time, but which the Spanish Empire zealously hoarded. In 1803 Winterbottom cautiously suggested that one of the barks they experimented with seemed like a promising alternative. The "African bark," he wrote, is "an object of importance for us."[60] Ultimately, the African bark proved ineffective, but no doubt Afzelius hoped one of the species of *Afzelia* he collected might be used for some kind of medicine.

The naming of plants, whatever their potential uses, was significant: it conveyed a subtle but clear message about who and how scientific knowledge was created. Banks bolstered Smeathman's scientific reputation by naming a species of plant he found near Sierra Leone after him (*Smeathmania*). Now it was Afzelius's turn. Afzelius and John Gray, a former Sierra Leone official, exchanged several anxious letters about whether the bark samples Afzelius sent to England were all the same species. As the deadline approached for the formal naming, a panicked Gray wrote in 1798, "I am afraid the *doubt* about the Bark has not yet been cleared up"; if they were not all the same species, "we shall be more embarrassed in our researches." Afzelius also sent samples to James Edward Smith, president of the Linnaean Society in London and the one in charge of naming plants. On January 8, 1798, Smith promised Afzelius, then in London, that the forthcoming issue of the Linnaean Society's *Transactions* would include, *"very soon,"* the first official description. Smith kept his word; the 1798 issue of *Transactions* included the first appearance of the name *Afzelia africana*. All in Latin, the article names prominently the man who allegedly discovered it, the "celeberrimi" (celebrated) Afzelius.[61] Yet there is no mention of Afzelius's aides: not Amarah, Cooper, Peter, or anyone else. The omission was not simply about credit and intellectual honesty: it reinforced the notion that Africans were sitting on a mountain of untapped resources, resources that only European men of science could help them discover.

Afzelius's scientific reputation was built partly on the work he did in Africa. But he failed to achieve the goal for which the colony's directors had sent him. None of the crops or cures they hoped to profit from ever materialized: not the "specimens of African cotton" Thomas Clarkson's brother, John, had requested; not the "blue Dye" that the wealthy East Indian merchant John Prinsep hoped to grow; certainly not the "African bark" and its purported healing powers. Part of the problem seems to have been the directors' stinginess. Afzelius complained to them that not even the most talented botanist could grow plants with the second-rate seeds they had been sending him: the directors, Afzelius wrote, should "pay less attention to the cheapness of the seeds than to their goodness." Another factor was the mismatch between the cultural preferences of the Black settlers and the colony's London-based directors. The directors wanted Afzelius to experiment growing English foodstuffs in the hope that the colony might provide Britain with the food staples Britons preferred: cabbage, turnips, carrots, "English potatoes." But Afzelius wrote that these crops "grew very well, but not so copious, large and dry as in Europe." Black settlers grew their own foods—yams, cassava, red rice, black-eyed peas—all of which, Afzelius wrote, "seem to be in a thriving state."[62] The tropical climate obviously favored the foods that the settlers preferred, but in insisting that their English foods take priority, the directors wasted valuable resources.

But perhaps nothing impeded Afzelius's scientific work more than the colony's political instability. In September 1794 French sailors raided the colony, not long after Britain went to war with France. Seven French boats, in addition to an American slave ship, ransacked the colony, destroying Afzelius's experimental garden in the process. "As soon as they got possession of the Governor's and my house," Afzelius wrote to the colony's governor, Zachary Macaulay, "they went into the gardens cutting down and pulling up all the plants they either knew were useful or which had a showy appearance." After the raid, Afzelius sailed to England, where his patrons promised to help him return. John Sims, a botanist and abolitionist, denounced the "mischievous" French and "American slave traders" who attacked him, telling Afzelius that he would soon send him "a case of Instruments" to replace the ones he lost.[63]

Conflicts between the colony's Black settlers and white authorities—mainly the governor, Zachary Macaulay, and the Sierra Leone Company directors—also made Afzelius's work difficult. The Black Nova Scotian settlers, who made up the majority of the colony's inhabitants after 1792, felt cheated by their lack of political power and by the directors' refusal to honor a guarantee of free land: "We want nothing but what you Promised," wrote the settlers Cato Perkins and Isaac Anderson in a letter to the directors on October 30, 1793. When the authorities continued to charge rent, many were forced to work for the company's public works, in effect turning over most of their wages to the white directors. To many, it felt as if they were slipping back into slavery. Tensions boiled over in June 1794, not long after Macaulay fired Scipio Channel and Robert Keeling, two Black settlers who worked for the governor, for threatening a slave ship captain who came into the colony. The colonists staged a protest to defend Channel and Keeling, which led Macaulay to station a cannon at his front gate, which in turn set off a day of rioting. The settlers continued to solicit the directors for help:

"Our present Governor [Macaulay] allows the Slave Traders to come here and abuse us," wrote Luke Jordan and Isaac Anderson on June 28, 1794, to the directors. "We are sorry to think we left America to come here to be used in that manner."[64]

Afzelius left the colony in 1796, but the problems continued. In September 1800, several settlers, fed up with the lack of a political voice, wrote a new legal code, which the directors promptly dismissed. In response, fifty settlers—including George Washington's former slave Harry Washington—staged an armed revolt. The new governor, Thomas Ludlam, recruited 550 Jamaican maroons, who had led their own revolt in the island five years earlier and had recently arrived in Sierra Leone, to put down the uprising. But between 1801 and 1808, the year the British government took over control of the colony from the Sierra Leone Company, political instability continued. Settlers struggled to gain power from the colony's white authorities, African leaders attacked the colony when its directors reneged on agreements or encroached on their land, and slave traders stymied attempts at a "legitimate trade." Whatever promise Afzelius's research had held out, political instability ensured limited gains.[65]

Yet none of the realities prevented the colony's directors from promoting Afzelius's research. In fact, the Sierra Leone Company's annual reports, published in England and the United States throughout the 1790s, made great use of his work. "It is hoped that some future benefit, either to the Company or colony, as well as some useful accession of botanical knowledge, may result from the labours of this gentleman," they wrote in a 1794 report, in reference to Afzelius's botanical experiment. They also included, in an appendix, the list of crops for which Afzelius had been bioprospecting: from the rice that grows "as luxuriantly as in Carolina," to the sugarcane that, though not growing abundantly, "will thrive exceedingly, as soon as the land in which [the seeds] have been planted shall have been some time in cultivation."[66]

Of course, the public was fully aware that things were hardly as good as the directors' reports suggested. But much as the colony's early founders had done, the new directors diverted blame away from themselves, focusing it instead on the settlers. The same reports that praised Afzelius's work, for instance, ridiculed the Nova Scotian settlers for the colony's problems. Rather than showing gratitude, the directors wrote in their 1794 report, the new settlers clung to "the false and absurd notions which the more forward among them have imbibed concerning their rights as freemen." In 1801 the directors again blamed the settlers for low crop production, calling them "idle, turbulent, and unreasonable." Abolitionist-friendly journals, like the *Critical Review*, embraced this view, attributing the colony's struggles to the settlers' unreasonable demands, such as the "right to nominate judges from among themselves."[67] Regardless of whether Afzelius agreed, his failure to publicly acknowledge the aid he received from settlers denied them an opportunity to prove the public slanders wrong. Afzelius would ultimately leave Sierra Leone with his scientific reputation intact, even improved. The settlers, meanwhile, would stay behind, struggling under the weight of dismissive directors and the high expectations that had been set by the colony's scientific promoters.

Not all men of science blamed settlers or the indigenous population for Sierra Leone's problems. The colony's official physician, Dr. Thomas Winterbottom, stood out for the lengths he was willing to go to defend the settlers' achievements and to promote the indigenous population's potential to become "civilized." Educated at the University of Edinburgh, Winterbottom arrived in Sierra Leone the same year as Afzelius, in 1792, and stayed until 1799. Only twenty-six when he arrived, he came with few of the attachments to slavery that many of his scientific predecessors had had. He had grown up in northeast England, in South Shields, a city that the Industrial Revolution was quickly turning into a hub for coal miners and glassmakers; his father ran an apothecary shop.[68] A few years after he returned to South Shields, in 1799, he produced one of the most pro-abolitionist natural histories to that date. Published in 1803, *An Account of the Native Africans in the Neighborhood of Sierra Leone* made many of the same arguments that its predecessors had made: the favorable climate, the fertile soil, the peaceable natives.[69] But Winterbottom's text devoted far more attention to denouncing ideas about innate racial difference, ideas that were quickly gaining attention.

Winterbottom did not hide his motives. He opened his two-volume *Account* by highlighting his "hopes that [his work] may at least tend to remove some prejudices respecting its inhabitants, whose customs have, in various instances, been misrepresented." He gave numerous examples of how the slave trade itself was impeding the spread of civilization—but also instances in which civilization had begun to progress, especially after Africans themselves prohibited slavery. He presented "remarkable proof" of a Muslim sect of Mandingos near Sierra Leone that had, seventy years earlier, ended slavery as a punishment for crimes. By the time he encountered them, "a great comparative degree of civilization, union, and security was introduced." He then refuted several natural histories that cited instances of cannibalism, which slavery's defenders used to show that Africans were hopelessly depraved. Winterbottom conceded that the authors of some of these texts were "highly respectable," but he impugned their assertions of cannibalism on the grounds that they were based on hearsay. His evidence, by contrast, was based on "ocular demonstration alone." Neither in nor around Sierra Leone, he observed, did "this horrid practice . . . exist."[70]

Winterbottom's most significant contribution was his defense of Africans' innate equality. The emphasis on racial equality in part reflected Winterbottom's training as a physician, but it also reflected the degree to which the debates over slavery had, by century's end, generated increased scientific interest in studying biological difference. European professors of comparative anatomy, a new field at the time, were among the most prominent scientific thinkers to shape the era's rapidly changing racial discourse. Though most anatomists insisted that Africans were part of the human species, and often denounced those who suggested otherwise, their work also suggested that physiological features—hair, lips, nose, and most significantly, the shape of the skull, seen as a proxy for intelligence—reflected a hierarchy of the human species. To be sure, the environmentalist framework still held sway, meaning that physical and mental differences were still seen as resulting from differences in climate and the natural environment. But the intensified interest in physiological differences meant that physical markers of difference could no longer be dismissed as superficial. As a result, men of science who wrote about indigenous Africans would increasingly have to dwell on physical differences.[71]

In *An Account of the Native Africans*, Winterbottom gave ample space to denouncing anatomists who he suspected were using their research to condone slavery. He took to task, for instance, Pieter Camper, a revered professor at the University of Groningen in the Netherlands, who argued that differences in facial angles confirmed the differences between the races, and whose research implied that Europeans sat atop the racial hierarchy. Though Camper publicly opposed slavery and insisted that the four main "races"—African, European, Asian, and American—were all part of the same human species, he described Africans as the least physiologically developed, closer to apes than to Europeans. The 1794 edition of his work, the first one published in English, suggested that, judging from his "assemblage of craniums," there was a "striking resemblance between the race of Monkies and Blacks." Winterbottom rejected Camper's facial-angle theory, arguing that it "will probably not be found to stand the test of experience." In his defense, he cited Johann Blumenbach, another leading anatomist, whose research showed that Africans had as many varieties in their facial features as Europeans did, and who argued that "the negroes, in regard to their mental faculties and capacity, are not inferior to the rest of the human race."[72]

But Winterbottom could provide something most European anatomists could not: firsthand observations of indigenous Africans. From his own measurements, he showed that as much variety existed between the facial features of individual Africans as "is to be met with the nations of Europe." He also denounced Camper's belief that Grecian sculpture provided the objective baseline of beauty, which Camper used to suggest Europeans' aesthetic superiority. The notion of an "ideal beauty," Winterbottom countered, "never existed." Winterbottom then attacked the racial research of William Falconer, a member of the Royal Society and once a close friend of Fothergill. In 1781 Falconer argued that the hot climate caused African minds to develop unevenly; like plants that grew too rapidly in the hot sun, Falconer contended, the brains of Africans matured too quickly, resulting in their inability to control their baser passions. To counter this notion, Winterbottom pointed to the educational achievements of young African students in Sierra Leone's schools, concluding that "African children are no more inferior to English children than men in Africa below men of the same age in England."[73]

Yet even as Winterbottom underscored the achievements of indigenous Africans, he occasionally presented them as far behind European standards.

As a physician, Winterbottom focused considerable attention on African medicinal practices, devoting an entire volume of *An Account* to indigenous cures. If there was any doubt of the value of non-Western medical knowledge to Europeans, he cited Benjamin Rush's own research on Native American medicine as proof of its worth. "Dr. Rush, who is so deservedly eminent as a physician and philosopher," Winterbottom wrote, understood that "we are indebted to the experience of nations, more rude than those of Africa . . . for some of our most valuable remedies."[74]

Despite praising certain African medicines, Winterbottom argued that their underlying theories and basic practices were still primitive. In Africa, he wrote, the functions of physician and priest, surgeon and sorcerer, were rolled into one, as was the case in medieval European practice. He viewed *obeah*, a West African healing practice that fused medicine and religion, as a form of witchcraft, one that "may be properly considered a mental disease." Of course, Winterbottom was alert to the potential abuse of these statements, and he carefully noted that African medicinal practices, however primitive, were not that different from what many Europeans still believed in. For that matter, even Paracelsus—a revered sixteenth-century naturalist—"appears to have believed implicitly in the power of witchcraft."[75] Yet Winterbottom's insistence that African healing practices represented a lack of civilizational development reinforced the belief in European superiority. In doing so, his work both justified a continued investment in the Sierra Leone experiment and obscured some of the root causes of the colony's problems—most notably, European indifference, and even hostility, to indigenous customs.

Sierra Leone was hardly thriving in 1808, the year it became an official royal colony. Of its roughly two thousand settlers, almost no one grew commercial-grade crops, and most of its basic foodstuffs had to be imported. But for every story of despair, there were dozens more of settler enterprise. Take Mary Perth. A former enslaved woman in Virginia, Perth had earned her freedom by running to British lines with her daughter, Patience, during the War of Independence. After the war, she resettled in Nova Scotia and in 1792 struck out for Sierra Leone. She sailed to the colony with her husband, Caesar, and a daughter, Susan, but shortly after their arrival Caesar died. Yet Perth soldiered on, opening a boardinghouse and a store and preaching in a local church. In 1794 she began working as a housekeeper and teacher in Governor Macaulay's home, giving her students "simple but solid lessons of wisdom," according to Macaulay's private journal.[76] For all the colony's struggles, women like Perth ensured Sierra Leone's survival.

But the colony's scientific supporters played another important role, burnishing Sierra Leone's image with their depictions of its natural abundance. A year after Winterbottom published *An Account*, the transatlantic abolitionist campaign was revived. By 1804 Napoleon's conquests in Europe and his attempts to reestablish slavery in the French Caribbean had breathed new life into the British movement, turning support for abolitionism into something like a statement of patriotism. The resurgence of British abolitionism also gave a second wind to the American campaign, and, after years in abeyance, the SEAST and PAS renewed correspondence.[77]

Both organizations focused their energy on abolishing the transatlantic slave trade, not rallying support for Sierra Leone. But many individual American abolitionists began to explore similar colonization projects, hoping to encourage recently freed Black Americans to voluntarily resettle outside the United States. In 1805 Thomas Branagan, a white antislavery advocate in Philadelphia who had been inspired by Sierra Leone, argued for the creation of a free Black colony in the western frontier, seeing it as the only viable path to emancipation. The alternative—emancipating and "admitting Africans . . . to the rights and privileges of citizens," he wrote—was "not only obnoxious to the judgment and principles of all advocates of slavery, but also a very large majority of the advocates for the emancipation of slaves."[78] Branagan's plan was part of a wave of similar colonization schemes that proliferated in the early American republic. And as in Britain, these projects would count several of the new nation's leading men of science as avid supporters.

FOUR

Trials in Freedom

The creation of Sierra Leone gave Benjamin Silliman hope. Appointed Yale's first chemistry professor in 1802, Silliman grew up admiring the first generation of antislavery men of science, people like Rush and Priestley, both of whom he met while training to become a chemist in Philadelphia. But as he set about modernizing Yale's curriculum, placing the sciences at its core, he quickly saw how dependent he had become on slaveholders. His former student John C. Calhoun, one of South Carolina's wealthiest slaveholders and an eventual states' rights champion, helped him launch the *American Journal of Science* in 1818, which soon became the nation's premier scientific journal. To create Yale's natural history museum, Silliman raised money from slaveholding alumni and collected minerals from deep within slave country. Silliman's friend and fellow abolitionist William Maclure found himself in a similar situation. A wealthy British geologist who moved to Philadelphia in 1796, Maclure relied on slaveholders to house, clothe, and feed him on his many geological expeditions.[1]

As two of the early American republic's most prominent men of science, Silliman and Maclure faced the uncomfortable position of being abolitionists who were deeply dependent on slaveholders for scientific patronage. The specific antislavery agenda they endorsed—colonization coupled with emancipation—had many antislavery supporters in the early nineteenth century, but as the one antislavery agenda Southerners might tolerate, colonization also offered these men a convenient political position. For Silliman especially, supporting colonization allowed him to cast his views as occupying a moderate middle ground, particularly when radical abolitionists began to reject colonization in the 1830s. Dependent on slaveholders to build the nation's scientific institutions, both men learned to frame their antislavery views in ways their patrons would find least objectionable.

In the first three decades of the nineteenth century, colonization drew considerable support from antislavery advocates. Scholars have come to see colonization, once depicted as a proslavery ruse, as drawing support from a diverse set of figures—free Black leaders, northern white abolitionists, southern slaveholders—albeit for very different reasons. For some Black leaders, colonization provided an opportunity to prove Black people's equality. By creating self-sufficient colonies, whether in Sierra Leone, Haiti, Central America, or along the western frontier, they might counter the image of Black servility and help build the case for citizenship. Other free Black Americans saw colonization, or *emigration*, as historians sometimes call Black-led colonization efforts, as a way to escape the nation's suffocating racism. But Black supporters did not view all colonization plans equally. In 1817 roughly three thousand free Black Philadelphians famously rejected the help of the American Colonization Society (ACS), the nation's leading and exclusively white colonization society, which created Liberia in 1821. Given the ACS's refusal to explicitly tie colonization to emancipation, coupled with its heavy slaveholder membership, many Black Americans believed that the ACS would only deport troublesome slaves while keeping slavery intact.[2]

Northern white abolitionists supported colonization for different reasons. In the first decades of the nineteenth century, northern states enacted a host of laws restricting Black freedoms. In 1807 New Jersey become among the first northern states to bar free Blacks from voting. In 1818 Connecticut expanded voting rights for whites while simultaneously denying them to Blacks. In 1821 New York eliminated property requirements for white male voters while adding them for Black men. In the Northwest, new states such as Ohio (1803), Indiana (1816), and Illinois (1818) not only denied free Blacks the right to vote, they also restricted Black immigration through new laws called Black Codes. Legal discrimination only reinforced de facto discrimination. Free Blacks throughout the North were forced into the most menial jobs and denied access to white-run schools and churches. Those who sank into poverty and homelessness were treated like criminals. By 1810 in Philadelphia, free Blacks made up 45 percent of the city's main prison population, despite numbering 10 percent of the general population.[3]

The first abolitionist societies tried to address these problems in part by patronizing free Black institutions. They hoped that by assimilating Black people through Black schools and churches, white people would gradually come to see them as equals. Nevertheless, several abolitionist societies— and many more individual members of them—simultaneously supported colonization as an alternative to these integrationist strategies. Some white antislavery supporters of colonization simply believed that citizenship should be reserved for white people, while others believed that anti-Black prejudice was so entrenched that Black Americans could achieve full freedom only outside the United States. Support for colonization could also be justified on purely pragmatic grounds. Many white abolitionists felt that fighting for Black citizenship alongside emancipation would doom emancipation altogether. Slave owners would never accept emancipation, they believed, unless they could be reassured that Black people would be removed from their midst once free. Whatever the rationale, white abolitionist support for colonization before the 1830s was undeniable. Though the Pennsylvania Abolition Society gradually distanced itself from the ACS, many individual PAS members—thirty-one in 1830—simultaneously joined or donated to the colonization society's Pennsylvania branch. In 1829 the American Convention, the umbrella group for the nation's white-led abolitionist societies, endorsed colonization on the condition that any plan be explicitly tied to emancipation.[4]

For Silliman, who joined the Connecticut branch of the ACS in 1828, the organization's Liberia colony seemed like the best option. But other scientific thinkers backed less well-remembered colonization plans. In the 1820s Maclure supported a short-lived experimental plantation in western Tennessee, called Nashoba, which sent its enslaved laborers to Haiti. The plantation was created by Frances Wright, a fellow British émigré, science enthusiast, and early women's rights advocate. Wright portrayed Nashoba as a model for emancipation; she hoped to show that if plantations were managed scientifically, and enslaved children given a science-based education, enslaved people would be more productive, which would enable them to purchase their freedom as well as the cost of their own removal. Though Nashoba was plagued with difficulties, Wright refused the help of the more conservative ACS. What united Silliman, Wright, and Maclure was not the specific colonization projects they supported, but their mutual support of colonization itself—as well as the belief that colonization could best be achieved with the support of science.[5]

In 1796, for his graduation speech at Yale, Benjamin Silliman composed a poem. Titled "The Negroe," it depicted slavery in the most unsavory terms: "the uncur'd gangrene of the unreasoning mind," he called it. Nine years later, in 1805, Silliman visited a slave ship anchored in Liverpool, an experience that left a searing impression. "Liverpool is *deep*, very *deep* in the guilt of this abominable trade," he wrote in his memoir of the trip. Yet Silliman was in deep as well. To fund his college education, Silliman's mother, Mary Fish Noyes, had sold two of the family's twelve enslaved laborers. Shortly after he graduated, Silliman gained legal custody over at least one of them, Cloe. In 1802 Silliman considered separating Cloe from her husband, Iago, relenting only after Cloe and Iago protested. "Iago is determined to prevent it," Silliman's brother-in-law wrote to him in 1802, asking for his advice: "She says that she was yours."[6] Though Silliman was an antislavery Northerner, his ties to slavery were not uncommon. Like Franklin and Rush before him, many elite Northerners who became abolitionists either owned or hired enslaved laborers. But Silliman's growing abolitionist sympathies would eventually pose a challenge to his scientific ambitions. As an idealistic college student, he probably cared little about what slaveholders thought of his antislavery views. But in the coming years, that luxury would vanish.

In 1799 Yale's president, Timothy Dwight, asked Silliman if he would consider becoming the university's first chemistry professor. At the time, Yale's curriculum was geared toward the classics and theology, but Dwight wanted to modernize the curriculum and felt a new scientific professorship would help. Silliman could not refuse: "It excited almost as much surprise as if I had been named President of the United States," Silliman recounted in his journal. To prepare, Silliman enrolled in the University of Pennsylvania, the only American university with a chemistry curriculum, briefly taking a course taught by Benjamin Rush. "His voice was musical," Silliman recalled, "his diction clear and emphatic" (though he ultimately dropped the class, citing a heavy course load). While in Philadelphia he also met the aging Joseph Priestley. The two spoke at length at a dinner party hosted by Dr. Caspar Wistar, a university medical professor and future PAS president, and Silliman remembered the conversation as "very gratifying."[7]

Silliman finished his preparation in Europe, attending chemistry and geology lectures at the University of Edinburgh. While in Europe, he also began building Yale's natural history museum, a prerequisite for any serious scientific institution. Silliman was deeply embarrassed that, before his appointment, there had been only a "few minerals . . . in the drawers of the old Museum at Yale College." So while studying in Edinburgh he began acquiring natural history objects, continuing the project upon his return to America. Building the museum, the origins of today's Yale Peabody Museum of Natural History, became his first lesson in the messy realities of doing science in a slaveholding republic. His most significant American donor was Colonel George Gibbs, a wealthy Rhode Island slaveholder, mineralogist, and scientific patron. By 1805 Gibbs had amassed thousands of minerals from Europe, the East Indies, and China, among other places, creating a collection that became known as the Gibbs Cabinet. Silliman's brother Gold Selleck Silliman Jr., a business colleague of Gibbs, arranged for his brother to inspect the collection when he returned from Europe in 1806. What Silliman saw astonished him: "What I now saw I had never seen before, excited in my mind a strong interest to see and examine the whole."[8]

What he chose not to see was where Gibbs's money came from. No northern state had closer ties to the Atlantic slave trade than Rhode Island, and no city fought harder to resist the 1807 slave-trade ban than Newport, Gibbs's hometown. In the twenty years before the slave-trade ban, Rhode Island slave traders shipped roughly 40,000 enslaved Africans to the Americas, more than any other state, north or south. The Gibbs family invested in many of these slave-trading ventures and profited handsomely from their

Newport rum distillery, which was dependent on slave-grown Caribbean sugar. Gibbs himself owned several enslaved laborers, most of them at his estate in Long Island, New York, and some in Newport. Rhode Island passed a gradual emancipation law in 1784, which promised to free all men born that year and onward at the age of twenty-one, and women at eighteen. But as of 1800 there were still 380 enslaved people in the state, 103 of them in Newport. One of them, Scipio, was owned by Gibbs and waited on Silliman while he inspected Gibbs's collection. "An intelligent coloured servant, Scipio, was always ready to admit me," Silliman wrote in his journal. Whatever displeasure this may have given Silliman, he kept quiet. His mission was clear: convince Gibbs that Yale would make a better home for the collection than Harvard, which Gibbs had offered the collection to first.[9]

Convincing Gibbs required delicate diplomacy. Gibbs demanded that Yale build a separate hall for the collection to lend it grandeur. When a Yale trustee balked at the request, Silliman had to convince the trustee of the enormous boost it would give the college's reputation. Was there "not a danger that . . . these physical attractions will overtop the Latin and the Greek?" the trustee worried. Silliman replied: "Sir, let the literary gentlemen push and sustain their departments. It is my duty to give full effect to the science committed to my care." Silliman won, the trustees relented, and Silliman was satisfied with the result: "This cabinet doubtless exerted its influence upon the public mind in attracting students to the College," Silliman later recalled. What's more, it attracted many prominent statesmen to campus (potential donors) as well as "trains of ladies."[10]

The minerals on display, over twelve thousand in total, were more than teaching tools: they were entertainment, a marvel for the eye as much as the mind. There were quartzes of every imaginable shape, color, and size: violet blue amethyst, "apple-green" crystals, iron-infused minerals colored yellow, red, and "light rose." But the beauty masked an unseemly reality. As Silliman and Gibbs added more and more minerals to the collection, they increasingly took specimens from places where slavery dominated, especially Brazil and the West Indies. Some also came from American slave country: "Red River, Louisiana," Maryland, "good size pieces from Georgia, Tennessee and Virginia." It is likely that enslaved laborers played a role in collecting them, too.

Silliman occasionally went on collecting expeditions with William Maclure around New Haven, and he noted that Maclure "travelled in a private carriage with a servant," probably a euphemism for enslaved laborer. When Silliman visited gold mines in Virginia in 1836, he described how he was "met with slavery everywhere": they "were employed to crush the quartz for us in heavy iron mortars," "they also broke the quartz from the veins." He even recorded the sounds of an enslaved worker being whipped in the plantation he stayed on: "We heard through our open windows, the sharp reverberations of the lash rapidly repeated and accompanied by loud cries of distress."[11]

By 1825 Gibbs no longer wanted to loan the collection to Yale: he wanted the college to buy it. He set the price at $20,000, roughly two-thirds of Yale's annual revenue. Unable to pay for it themselves, the trustees initiated a fund-raising campaign, targeting their most affluent alumni—many of whom were slaveholders. Alumni from South Carolina, many of them plantation owners, made up the third-largest donor group, raising $700, trailing behind only New York and Connecticut alumni. John C. Calhoun, one of Silliman's first chemistry students and a major slaveholder, was among the larger donors, pledging $100. "The loyalty of our *alumni* and the liberality of the friends of science of the College" made the purchase possible, Silliman remembered proudly. The Gibbs Cabinet also changed the way the sciences were taught at Yale. Silliman originally planned to use the collection for chemistry lectures. But as he began to label the minerals, students became interested in learning more about the study of mineralogy. By 1813 Silliman had began to teach "separate lectures on mineralogy and geology from the chemical course." The Gibbs Cabinet thus not only raised the stature of Yale; it also changed the nature of scientific education—and none of it would have been possible without slaveholder support.[12]

The Gibbs Cabinet was not the only project tying Silliman's scientific work to slavery. Equally important was his relationship with Eli Whitney, a fellow Yale alumnus, class of 1792. Though best remembered for inventing the cotton gin, Whitney earned most of his money from his gun manufacturing business. In 1798 Silliman lobbied Yale officials to sell university property to Whitney for a firearms factory he wanted to build; that same year, Silliman also wrote contracts for Whitney to sell firearms to the federal government. But the real benefit to Silliman lay elsewhere. The gun factory gave Silliman access to first-grade chemical materials, which helped his research. In 1808, for instance, Silliman walked to the factory to get a "tube of block tin, for the purpose of drawing, through an innocent metal, the soda water." Even more beneficial was the access the factory gave Silliman to government patrons. As Silliman recalled, Whitney knew all the "leading members of both houses of Congress," and as they came to New Haven to do business with Whitney, Silliman steered them to campus. The access Whitney's factory gave Silliman to government patrons mattered because, at the time, the federal government generally funded only scientific projects with obvious practical value, such as surveys and continental expeditions. But the utility of natural history museums, of air pumps, pyrometers, and blowpipes—the stuff of academic science—was less straightforward. Therefore, Silliman's support for the Whitney family (he took his son, Eli Jr., on the Virginia gold mine expedition), and his gun factory, especially, gave him access to the kinds of patrons that a mineral collection itself could not.[13]

Silliman's support for Whitney also meant he was sometimes drawn into Whitney's other business: the cotton gin. Patented by Whitney in 1793, the cotton gin solved a major bottleneck in cotton production. Picking seeds out of cotton fiber was time-consuming, a problem Whitney noticed firsthand while working as a tutor on a Georgia plantation after graduation from Yale. Whitney's idea was to invent a machine that separated the seed from the fiber in a fraction of the time. Though much else contributed to cotton's astronomical rise in the South—the seizure of Native American land, easy credit from northern and European bankers, the sheer violence involved in the expansion of the domestic slave trade—Whitney's cotton gin was of undeniable importance. In 1791, two years before its invention, American cotton production stood at a paltry 2 million pounds per year; ten years later, in 1801, it was 40 million pounds per year. To meet the demand, slave traders dramatically increased the sale of human beings: in Georgia, where cotton plantations sprang up seemingly overnight, the enslaved population doubled between 1790 and 1800, to 60,000; in South Carolina, the enslaved population went from 21,000 in 1790 to 70,000 twenty years later.[14]

Silliman was loosely connected to Whitney's cotton business. In 1799, after helping with the firearm contracts, Whitney offered to take Silliman to Europe "for the purpose of obtaining patents for the Cotton Gin." (Silliman declined, citing his "youth and inexperience.") Decades later, Silliman became a vocal advocate for the Whitney family, demanding that they see more of the profits from nationwide cotton sales. "Both justice and honor required that compensation should be made," Silliman wrote in an 1832 obituary, published in the *American Journal of Science*. Whitney fought a long, losing battle to stop southern planters from counterfeiting his design, and though Silliman knew those "frustrated" profits came from slave labor, he ignored that fact and instead cast Whitney as a scientific genius robbed by a greedy, narrow-minded few. Only "the highest intellectual vigor—the brightest genius" could create Whitney's cotton gin, Silliman eulogized, a fact that underscored the injustice of a man "not more than compensated . . . for the enormous expenses which he had incurred."[15]

Whitney was not the only one whom Silliman relied on for government contacts. Silliman's former chemistry students were helpful as well. Most significant was John C. Calhoun, who formed a close relationship with Silliman throughout the 1810s and 1820s. The son of one of South Carolina's wealthiest slaveholders, Calhoun returned to his family's plantation after graduating from Yale and soon embarked on a lifelong career in politics. By the late 1820s, Calhoun had become the intellectual architect of "nullification," the legal justification for a state's right to disregard federal laws it deemed unconstitutional. Though Calhoun's states' rights doctrine would later define his career, for much of the 1810s and 1820s he supported a strong central government. As a South Carolina congressman in 1815, he drafted a bill to create the Second Bank of the United States and pushed for internal improvements. When he became secretary of war under President James Monroe one year later, he turned that same nationalist impulse to a two-part enterprise: building up the nation's scientific institutions alongside its military.[16]

Silliman became an obvious source of help. In May 1822 Calhoun wrote to Silliman, inviting him to inspect the science curriculum at West Point. Calhoun wanted to make sure that West Point graduates received an education in "all of the branches of the sciences," and he said he would be "highly gratified to receive a report of your observations upon the actual state and progress of the institution." Silliman happily obliged, issuing a report that same year. Calhoun had been an avid supporter of science for much of his career. As secretary of war, a position he held until 1824, he was a chief proponent for federally funded geological surveys of the western frontier. He also helped fund the *American Journal of Science*, which Silliman created in 1818 and edited from his Yale office. "The utility of such a work, particularly in this country, must be apparent," Calhoun told Silliman upon receiving a prospectus in 1818; he had "every reason to feel the strongest gratitude to Yale College, and shall always rejoice in her prosperity." Calhoun was also, of course, a Gibbs Cabinet donor.[17]

Silliman formed strong ties to other powerful slaveholders as well, including Andrew Jackson. By the early 1820s, Jackson was already a national hero for his military campaigns in New Orleans and Florida. But he was also a parent, and he occasionally relied on Silliman to get his foster child, Anthony Butler, a Yale student, out of trouble. In 1821, writing from his desk as military governor of Florida, Jackson pleaded with Silliman to readmit Butler after he had been expelled for bad behavior. That Butler acted with "impudence" Jackson did not deny, but the expulsion had taught him a lesson, Jackson wrote, and, if not overturned, it would harm "his future life and studies." (While he was at it, he asked Silliman to "erase the sentence" from Butler's record.) The close relationships Silliman formed with the nation's slaveholding elites would influence his antislavery views. To varying degrees, men like Calhoun and Jackson provided the financial and social capital necessary to build up the nation's scientific institutions. Rather than risk their patronage, he would advocate the only solution that elite planters found tolerable—colonization. That Silliman joined the ACS in particular was not surprising either. Founded in 1816, it counted Andrew Jackson as an early supporter, though he later distanced himself from the organization, as well as Benjamin Rush's son, who was a charter member.[18]

Andrew Jackson, in fact, facilitated one of the most important federal commissions Silliman ever received, one that highlighted the challenges abolitionists faced when relying on slaveholders for scientific support. In 1831, during Jack-

son's first presidential term, his treasurer "desired me to take charge" of a federal investigation into the nation's sugar-making industry, Silliman wrote in his
journal. The nation produced little sugar at the turn of the century, but the
growth of a northern urban working class, increasingly reliant on inexpensive
sugar-laced foods, coupled with slavery's uncertain future in the West Indies,
opened an opportunity. Between 1827 and 1831, sugar plantations in Louisiana,
the nation's main sugar-growing region, more than doubled, from 308 plantations to 691. The problem was that the technologies used by southern planters
and northern refineries were woefully outdated. In 1830 Congress passed a bill
ordering Jackson's treasurer to study the issue and suggest "modern improvements." Jackson's treasurer put Silliman in charge, who in turn recruited several Yale colleagues, including faculty from "the chemical department of Yale
College." Silliman himself traveled with Yale's natural history professor, Charles
Upham Shepard, to plantations throughout the South. Yale's president, Jeremiah Day, promised Silliman "any moneys which may be due" for his research.
The government, meanwhile, paid Silliman the hefty sum of $1,200.[19]

Silliman must have thought twice: he was being asked to help an industry
that was synonymous with slavery. Yet the prestige and the connections it
would bring Yale, its scientific faculty, and science writ large ultimately proved
irresistible. When he delivered his finished report to the Treasury Department
in 1833, he was invited to the White House to dine with Andrew Jackson. Silliman in turn invited Jackson to see "the [Yale] Colleges, which he said he
should be very happy to see." But Silliman also fretted, during the two-year
study itself, about the ways slavery hung over the whole project. Silliman's unpublished letters to Shepard, written while he visited various plantations in
1832 and 1833, reveal Silliman struggling with the moral implications. In addition to the scientific questions he asked Shepard to investigate ("do they use
any alkali?"; "do they use thermometers?"), he asked several moral ones as well:
"The slaves on the sugar estates—do they appear hard worked, dispirited, and
oppressed?" He then added, in a parenthetical whisper: "(Open your eyes and
ears to every fact connected with the actual condition of slavery everywhere—
but do not talk about it—hear and see everything but say little)."[20]

Shepard followed orders. Yet the planters with whom Shepard corresponded were hardly the backward, science-averse Luddites that abolitionists

increasingly made them out to be. Quite the opposite: they were hungry consumers of scientific knowledge. "I could talk to you forever on the improve[ment] of Sugar," one New Orleans planter wrote to Shepard, now back at Yale, on March 3, 1833. He asked Shepard to send an "able chemist . . . to investigate the properties of our cane juice," adding that he "subscribed to no nostrum or discovery without knowing its basis." Another planter thanked him for "leaving me your chemical Book" and hoped "god willing we shall meet early next fall if your business or affairs will permit you to leave the north." Shepard also received at least one detailed response about the treatment of enslaved laborers, per Silliman's request. Stephen Henderson, one of Louisiana's largest sugar planters, provided detailed reports on everything from the number of enslaved workers he owned to how childbearing decreased an enslaved mother's productivity: "A woman does no hard work for 6 weeks before and 6 weeks after her confinement," he wrote. He added, "At 65 negroes are of no further value." Silliman's subsequent letters to Shepard are silent on the issue, making it impossible to know what he made of this information. But the early interest he took in enslaved people's treatment shows that Silliman was keenly aware of the ways the sugar commission tied him to slavery.[21]

Silliman would ultimately use the report to insert a role for science in any solution to slavery. Titled *Manual on the Cultivation of the Sugar Cane* and published in 1833, the 126-page document carefully avoided mentioning the word *slavery*, but it presented the adoption of new technologies and alternative agricultural commodities as ways to reduce the reliance on slave labor. Central to Silliman's advice was for planters to invest in a recently patented sugar processing machine—a steam-powered vacuum pan—in order to cut back on human labor. Invented in 1813, the vacuum pan was the brainchild of Edward Charles Howard, a British chemist and Royal Society member; it was intended to reduce the impurities and amount of sugar lost in the traditional cane juice–evaporation process. But American planters had yet to adopt it. Silliman insisted that Howard's machine would both quickly pay for itself and, critically, save "considerable labor." Of equal importance was for planters to adopt steam engines. "The greatest improvements of modern times," he wrote, "is the introduction of steam and vacuum process into the plantations." Silliman chided planters for their meager investment in steam engines: only 240 of Louisiana's 691 sugar estates used steam-powered mills, he calculated. He then scolded planters for their hidebound ways: "The account given nearly forty years ago, by Bryan Edwards, the celebrated historian of the West Indies is, substantially, a statement of what happens to this hour."[22]

Near the end of the report, Silliman suggested that beets and maple trees, if grown in the North, could provide an alternative to southern-grown sugarcane. "It would be a great national loss if our maple forests—now rapidly disappearing—should be finally extirpated," he wrote. Advocating for maple and beet sugar carried an implicit antislavery message. In 1820 Silliman published an article in the *American Journal of Science* citing Benjamin Rush's article from years earlier, which made the antislavery link explicit—maple trees grown in the free-labor North, Rush wrote, would reduce "the quantity of human suffering." In Silliman's memoirs, he seemed to delight in the implicit rebuke of slavery: "I called attention of the government to the cultivation of the beet and the maple trees for their sugar." Yet the report itself was hardly the unabashed antislavery critique he remembered it to be. Even as he wove in implicit antislavery arguments, his refusal to use the word *slavery* explicitly spoke volumes. He alluded to the enslaved only when it was absolutely necessary, and only in planter euphemisms, calling them "the force" or "hands." His reticence to use the word stood in marked contrast to his explicit discussion of wage laborers in northern refineries. Silliman praised paid sugar factory workers in the North, applauding their industry, sobriety, and "how patient and skillful" they were. Though he no doubt hoped that readers would intuit the message—wage labor was superior to slave labor—his fear of offending slaveholders made him circumspect. Soft scientific innuendo ultimately allowed him to placate his slaveholding patrons without compromising his principles.[23]

The most important speech Silliman made on colonization came in the middle of his investigation of the nation's sugar industry, and it seems to have captured his belief that colonization offered the only antislavery position that would not jeopardize slaveholder support. Silliman delivered the speech in New Haven on July 4, 1832, speaking as vice president of the Connecticut

Colonization Society, which he had joined four years earlier. While he called slavery "a national evil," he also tried to curry favor with moderate slaveholders. He had their best interests in mind: "No person can be more sensible than myself of the great amount of personal excellence which is found in the states where slavery exists." If Southerners did not free and remove their captives soon, he argued, they risked uncontrollable insurrections and the specter of a Black majority. He presented colonization as "the golden mean" between "two extremes": diehard proslavery ideologues were on one side, and upstart radical abolitionists on the other. Only one year earlier, William Lloyd Garrison, a colonization supporter throughout the 1820s, sparked a seismic rift within white antislavery circles by insisting that emancipation be immediate, uncompensated, and unconnected to colonization, a position he adopted from Black activists. Garrison's radical new approach to abolitionism, coupled with the proslavery backlash, allowed Silliman to depict colonization as eminently reasonable. It was irresponsible to immediately free enslaved people without first educating and Christianizing them, Silliman chided Garrisonians. "The prejudices of the country" would "forever, exclude them from social equality with the white." In any case, slaveholders would never accept emancipation otherwise. "Cut off the hope of colonization," he said, "and all the moral machinery for emancipation . . . will be clogged and crippled."[24]

In making his antislavery intentions explicit, Silliman was keeping with the Connecticut Colonization Society's larger mission. Unlike the national ACS organization, but like other northern ACS branches, the Connecticut chapter made gradual emancipation central to its colonization agenda. The Connecticut society's inaugural address, published in 1828, said that colonization would *"make abolition of slavery through the world a thing inevitable."* At the same time, it insisted that slaveholders' rights be respected and forcefully rejected federal intervention. "The National Government has no control over the subject," it wrote, "for the right of the slave-holder to his property is guaranteed by the very compact on which the National Government rests." Like many white antislavery colonizationists, members of the Connecticut society hoped slaveholders could be persuaded to voluntarily manumit their enslaved laborers, and they kept faith in the good intentions of *"benevolent masters."*[25]

Silliman hoped his speech would staunch more defections to Garrison's radical antislavery wing. But his speech was also directed to related events in New Haven. In 1831 Black and white abolitionists tried to create the first college for free Black Americans near Yale's campus. James Forten, then an aging leader of Philadelphia's free Black community, published a petition arguing that the college would "cultivate habits of industry . . . while pursuing a classical education," and planned for the school to be "located at New-Haven, Conn." But a committee of city elders, including many Yale law professors and alumni, rejected the proposal. A Black college was "incompatible" with the town's already existing educational institutions, the committee wrote. In addition, it was too closely associated with immediate emancipation—"an unwarranted and dangerous interference with the internal concerns of other States." Silliman's colonization speech was his attempt to manage these tensions. By insisting that one could be both antislavery and anti-integration, he offered those who opposed slavery a way to maintain their moral integrity without having to live with emancipation's social consequences.[26]

Of course, Silliman's rejection of Garrison's radicalism and his insistence on colonization was a long time coming. For decades, he had worked with slaveholders to build Yale's—indeed, the nation's—scientific institutions. Whether the goal was to found Yale's natural history museum, create the *American Journal of Science*, or win the prestigious sugar commission, slaveholders were an unavoidable source of patronage. By supporting colonization, Silliman was able to maintain his antislavery beliefs without offending his slaveholding donors. Even as Garrison began to radicalize more and more white abolitionists, Silliman remained a steadfast colonizationist, and he would remain one until the Civil War. When faced with the choice of appeasing his slave-owning patrons or embracing a newly radical antislavery agenda, Silliman chose the safer middle ground of colonization. For him and many other white antislavery Northerners, it was a most rational choice.

Silliman was not the only man of science to embrace colonization. His close friend William Maclure was another. Born in Scotland to an affluent family in 1763, Maclure spent his early adult years selling slave-grown Virginian tobacco for one of London's largest merchant firms, Miller, Hart & Company.

The work made him extraordinarily wealthy, and in 1796 he left Britain for Philadelphia, where he soon retired and shifted his attention to science. Like Silliman, Maclure played an outsize role in developing the nation's scientific institutions. In 1817 he donated hundreds of books and minerals to the fledgling Academy of Natural Sciences in Philadelphia, an upstart rival to the more elite American Philosophical Society, and not long after was named president. In 1819 he was elected president of another scientific body, the American Geological Society, and he would spend the next decade contributing rare stones to Silliman's Gibbs Cabinet, writing to Silliman in 1820, for instance, that a "cabinet of specimens" was on its way from Europe.[27]

Perhaps even more important was his scientific work along the western frontier. In 1825 Maclure invested in a utopian settlement called New Harmony, in Indiana, turning it into a hub of scientific research. The settlement gave him a population on which he could test his science-based educational theories, as well as a center from which to publish scientific and political works and outfit geological expeditions.[28] It was also at New Harmony that Maclure met Frances Wright, who strongly influenced his antislavery views. Wright had recently set up her own experimental plantation, called Nashoba, about four hundred miles south of New Harmony, in western Tennessee. Like Maclure, Wright saw Nashoba as a social laboratory, populated with a captive labor force on which she could test her scientifically informed theories about education and labor management; she was guided by the hope that it would provide planters with a model for freeing and removing their enslaved laborers on their own, without government intervention and without losing a profit. Wright's ruthless scientific management of her enslaved workforce convinced Maclure that gradual emancipation—coupled with colonization—would secure abolition's success. Though Maclure and Wright kept their distance from the ACS, both saw colonization as the only means by which emancipation could be achieved. What's more, both relied heavily on scientific knowledge to promote their antislavery projects.

Before Maclure met Wright, he was not very active in public debates over slavery. He had never joined an antislavery society, and he did not make his first public antislavery comments until 1817. Initially, Maclure may have felt muzzled by his business—buying and selling slave-grown tobacco. But he also may have been inhibited by his precarious position within the scientific community. When Maclure moved to Philadelphia in 1796, he was seen as a businessman, not a man of science. Rather quickly, he set about remaking his image, hosting salons in his Philadelphia home and inviting leading men of science to attend: the Comte de Volney, a French naturalist; Giambattista Scandella, an Italian naturalist. At one dinner, on March 1, 1798, the Polish poet Julian Niemcewicz recorded an "instructive and interesting" discussion about husbandry and agriculture, a conversation that "made the time fly by quickly." But if Maclure wanted to gain real scientific credibility, he needed to publish, and the most important person in that regard was Thomas Jefferson. Elected president of the American Philosophical Society in 1797, Jefferson would hold the position for the next eighteen years, making him a nominal gatekeeper for the society and its journal, *Transactions*.[29]

Maclure and Jefferson formed a close bond in those years. Jefferson frequented Maclure's salons, and when Maclure left for Europe in 1800, where he lived for the next eight years, he continued to foster the friendship. Jefferson appointed Maclure to a commission to settle the claims of American sailors against French pirates; Maclure used their correspondence to exchange scientific information. In a letter to Jefferson on November 20, 1801, Maclure gave a detailed report of the climate similarities between Europe and the eastern states, sending along cabbage, turnip, and beet seeds: they were "for you to make the experiment" and see whether they grew well in Virginian soil, Maclure wrote. He wrapped the seeds in "an abstract of the Kantian Philosophie," which, though much in fashion, neither he nor anyone else seemed to understand: "I neither comprehend it nor have met with any one that appeared capable of explaining it." Jefferson would have found Maclure's casual disdain for high theory endearing. In the early republic, American men of science cast scientific theorization as the embodiment of European elitism. American science, by contrast, would privilege facts over theory—the embodiment of the nation's democratic ideals.[30]

Maclure may have flattered Jefferson's hope for a more democratic science, but he conspicuously avoided any mention of slavery. In the same letter that mocked the pretensions of Kantians, Maclure described new tobacco cultivation methods in Germany, and if there was ever a moment to explicitly attack slavery, this was it. But Maclure avoided the subject and stuck only to the facts, discussing how to improve the soil so that tobacco would not be such an "impoverishing crop." Having not yet achieved scientific respect, perhaps he wanted to avoid offending the man who would help him attain it. Maclure's courtship worked. In 1809 the American Philosophical Society invited Maclure to deliver a paper on the geological survey of the United States he had conducted the previous year. Published later that year, the report was not only Maclure's first scientific publication, but also the nation's first comprehensive geological survey, a milestone in the nation's scientific history. After its publication abroad in 1811, Maclure secured a measure of scientific legitimacy. But unlike the expanded 1817 edition, which he published on his own, the 1809 *Transactions* edition excluded all political commentary. Instead, it simply detailed, in dry scientific prose, the main types of soil and mineral content covering the nation's surface, and left it at that.[31]

Maclure was equally reluctant to mention slavery when he published another important geological survey of the West Indies in 1817. For the two years preceding its publication, Maclure explored the islands with the French geologist and illustrator Charles-Alexandre Lesueur. Maclure published their survey, *Observations on the Geology of the West Indies*, in the journal he founded for the Academy of Natural Sciences. (One year later, Silliman published it again in the *American Journal of Science*.) Like the 1809 U.S. geological report, the West Indies survey was meant to improve agricultural production by detailing the soil's mineral content. Yet the West Indies survey made no mention of slavery—a glaring omission in light of the fact that Maclure had spoken about slavery with a planter he met on the expedition. In his unpublished journal from the expedition, Maclure recorded a story of "30 negroes on Mr. Grants plantation [who] were burned in their houses or scalded to death by the liquid mud and boiling water." Everything surrounding the quote was published, except for the slavery anecdote. Perhaps Maclure felt including it would have been a distraction—or perhaps, like Silliman, he felt that mentioning slavery might offend his slaveholding patrons.[32]

Whatever the reason, Maclure did not keep slavery out of his scientific work for long. In fact, the expanded version of the 1809 U.S. geological article, which he published independently in 1817, represented Maclure's first attempt to make a scientific case against slavery, albeit a subtle one. Much like the 1809 piece, the new version detailed the main classes of rock (and by extension the quality of the soil) that were found in the major geographical zones of the North American continent. Yet the new version, nearly six times as long, included explicit political commentary. Maclure tried to show how each region's soil and topography supported, or could support, an independent yeoman republic. And the key axis he focused on was not horizontal, across the Mason-Dixon line, but vertical, up the spine of the Allegheny Mountains. The eastern half of the nation, north and south, was dominated by a primitive class of rock that made for poor soil and, by implication, provided a poor basis for a long-lasting yeoman republic. In contrast, the land west of the Allegheny Mountains was dominated by a nutrient-rich soil, "equal in magnitude and importance, if not superior, to any yet known." As to the southeast, still the heart of the nation's agriculture, Maclure conceded that certain parts of it had superior soil; but, he argued, too often it was exhausted by "so ruinous a system of culture" — by which he seemed to imply slavery.[33]

Maclure did not argue that the northeast was superior, however. On the contrary, he emphasized that, despite the modest advantages the original southern states had in terms of soil quality, the real trouble for both regions lay with the Atlantic Ocean. Exposure to the ocean meant that all seaboard states were exposed to the constant threat of invasion. That geographical fact made it likely that all states east of the Allegheny Mountains would be vulnerable to tyrannical politicians, since officials would forever be given to raising taxes for the purpose of funding an army. As Maclure put it, "The Atlantic states" would always be "liable to be governed by the few or the minority." By contrast, lands west of the Allegheny Mountains were more likely to be "governed by the majority," since they both were removed from the threat of invasion and had the Allegheny Mountains as a protective barrier.[34] Maclure ultimately depicted the western frontier as the natural geological home of a future, freer American republic, a point made clear by the accompanying map. The map colored the entire landmass east of the Appa-lachian Mountains a wavy blend of pink, orange, and yellow, each color sig-nifying a different class of rock; the west was entirely blue. The eye immediately sees the point: America's future lay out west, all open blue, whereas to the east, the sun was setting.

Maclure also tacitly rebuked slavery in a short section about the Erie Ca-nal. Construction began on the canal in 1817; it was intended to give free-labor farmers in the Midwest access to eastern and foreign markets. But Maclure opposed it on the grounds that, for it to be profitable, "slave states" would have to provide foodstuffs for northwestern farmers. Alternatively, the canal would require plantations, where "the labourers are slaves," to take root in northwestern states themselves. Though Maclure's condemnation of slavery was indirect, his point was clear: slavery was not worth the cost of the canal's construction. Maclure's antislavery comments were notable not just because he couched them in scientific terms. They were notable because he made them at a time when the national debate over slavery was still muted. Not until the start of the Missouri Crisis in 1819 would slavery pro-voke a nationwide debate. But Maclure seemed to be addressing an issue that flew just beneath the radar: slavery's expansion into the Northwest Ter-ritories. Though Indiana's state constitution, ratified in 1816, outlawed slav-ery, it would, like Illinois's constitution two years later, allow enslaved people to remain in the state if their owners brought them in before state-hood was achieved. In addition, lawmakers in both states found clever ways to permit slavery, most commonly by legalizing indentured servitude and extending the term of service for decades. By arguing, however modestly, that the very geography of the western frontier was naturally endowed to support free labor, that the future of American democracy depended on it, Maclure showed a willingness to critique slavery at a moment when few of his scientific colleagues would. In doing so, he glimpsed a nascent antislav-ery activism that would blossom in the decades to come.[35]

In the 1820s Frances Wright became a critical influence on Maclure, pushing him to take bolder antislavery stances. The two met in 1826 at New Harmony, Indiana, the utopian settlement that Maclure had become in-volved with one year earlier. In 1825 New Harmony, a settlement of about

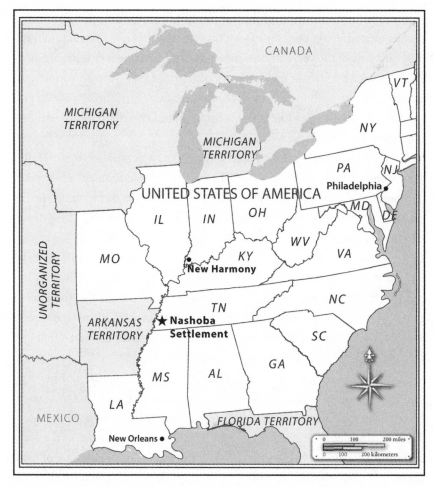

Map showing the location of the Nashoba Settlement, ca. 1826.
Designed by Gerry Krieg.

eight hundred mostly German immigrants, was taken over by Robert Owen, a British cotton manufacturer and social reformer already famous for his quasi-socialist ideas. Owen hoped to turn New Harmony into a model factory town, believing that the promotion of scientific research, free education, and communal ownership would make the settlement profitable and its workers model citizens. Owen enlisted Maclure to run the educational and scientific aspects of the settlement, and Maclure in turn asked Wright to teach at New Harmony's school.[36] Wright declined because she was already invested in her own social experiment—Nashoba—a plantation modeled on many of the very same scientific and utopian ideals that had inspired New Harmony. Wright's meeting with Maclure at New Harmony marked a critical phase in both their lives. Over the next five years, the two would promote the idea that scientific reforms implemented on plantations could secure colonization's success.

Born in Scotland in 1795, Wright was part a new generation of well-born, reform-minded European thinkers who saw the United States as a beacon of democratic reform. In her first published essays, written in the 1810s, she depicted the United States as a modern Athens, the epitome of enlightened democracy. But the persistence of slavery troubled her. After an initial visit to the country in 1819–20, she wrote to her friend Julia Garnett that she wished the South would "follow [the North's] example" and free its enslaved population. Four years later, in 1824, when she moved to the United States, she committed herself to the antislavery cause. "This plague spot so soils the beauty of the robe of American liberty," she told Garnett, "that I often turn in disgust from the freest country in the world."[37]

Wright's commitment to abolitionism was never in doubt, but her support for colonization as the best means to achieve it is seldom appreciated. In endorsing colonization, Wright was influenced by three powerful men. The first was the Marquis de Lafayette, who had mentored her in the 1810s and traveled with her to the United States in 1824. An early advocate of colonization, Lafayette spent decades trying to create his own free Black settlement. After the War of Independence, he asked George Washington if he would be interested in "Purchasing a small Estate Where We May try the Experiment to free the Negroes." Washington declined, but Lafayette pushed forward. In 1785 he bought a plantation in Cayenne, on the northern edge of South America (modern French Guiana), hoping that his enslaved laborers would quickly be able to purchase their freedom and sustain a free Black settlement in its wake. The settlement struggled, however, and in 1802 Lafayette sold it. But he did not give up on the idea. In 1824 he considered turning land Congress had granted him in Florida into another free Black settlement, but again he changed his mind. Though the reason is unclear, he might have realized that it would have competed with the ACS's Liberia colony, established three years earlier, and he probably did not want to trouble an organization that in 1824 named him to the "perpetual vice-presidency."[38]

The second important influence on Wright was Thomas Jefferson. Wright saw Jefferson's support for colonization as offering essential validation—not least because he was a respected slave owner and Founding Father. After visiting him at Monticello in November 1824, she suggested to Garnett that his advocacy of colonization, "sketched in the Notes"—*Notes on the State of Virginia* (1787)—reflected widespread planter attitudes, and that he was very "anxious" that "some steps . . . preparatory to the abolition of slavery" be taken soon. In the summer of 1825, as Nashoba took shape, she asked for Jefferson's blessing: "I feel very anxious for your opinion," she wrote. "Since my interesting visit to Monticello" she had been "engrossed by this one object."[39]

Yet Wright kept her distance from Liberia and the ACS, which Jefferson supported, and instead believed that freed Black people should be sent wherever they wanted to go. In this regard she seems to have been swayed by a third figure, Jonathas Granville, a Haitian diplomat. In the 1820s Haiti's president, Jean-Pierre Boyer, put Granville in charge of an immigration campaign aimed at attracting free Black Americans to Haiti. In the 1820s alone, between 6,000 and 13,000 Black Americans took up the offer, far surpassing the 3,160 Black Americans who emigrated to Liberia in roughly the same period. Wright met with Granville in Philadelphia in 1824, and he seems to have convinced her that Haiti was far better than Liberia, not least because Black Americans actually supported it. "The near vicinity of Hayti," Wright wrote to Garnett after meeting Granville, afforded "a safe & convenient haven for the Black population of the U S."[40]

Wright came up with a specific plan for how to run Nashoba after visiting New Harmony in the summer of 1825. New Harmony had gained international attention only months before, when Owen spoke to Congress about his utopian settlement the previous winter. New Harmony, Owen told Congress, would be run according to what he called the "new science of circumstances," in which scientific principles promoting "virtue, intelligence, affluence and happiness" would guide every aspect of the community's organization. Though Owen kept his distance from abolitionism—"their condition, in a great many respects, is much preferred to . . . a large majority of the working class," he told a Cincinnati audience in 1829, referring to enslaved people—Wright left New Harmony convinced that his ideas about cooperative labor, scientific management, and a secular, scientific education could be adapted to Nashoba. "When I first visited [New Harmony]," she wrote to Garnett on June 8, 1825, "a vague idea crossed me that there was something in the system of united labor . . . which might be rendered subservient to the emancipation of the South." By the end of the year, she had used money she inherited to purchase a few hundred acres fifteen miles west of Memphis, Tennessee, for her new plantation. Working with her sister, Camilla, she then bought eight enslaved people ranging in price from $500 to $1,500: Willis, Jacob, Grandison, Henry, Redick, Nelly, Peggy, and Kitty. At least three more were donated by a South Carolinian well-wisher.[41]

Wright's announcement of Nashoba's objectives displayed her earlier influences. Published on October 1, 1825, in the *New Harmony Gazette*, Wright's "Plan: For the Gradual Abolition of Slavery in the United States" channeled Jefferson when it argued that planters would accept emancipation only if it was tied to colonization: it was "indispensable that emancipation be connected with colonization," she wrote. The "Plan" also reflected her belief, akin to Owen's, that for colonization to work, enslaved laborers needed to be reeducated in a "school of industry," and have a "co-operative model" of ownership over plantation profits. That would increase productivity and, in turn, shorten the time to emancipation. But unlike Jefferson, and acting in line with Granville's campaign and the preferences of many Black Americans, she rejected Liberia and argued that all destinations should be considered for colonization: "Independent of Hayti," she wrote, there was also "the Mexican territory of Texas . . . and a fine region beyond the rocky mountains, within the jurisdiction of the United States."[42]

Another important part of the "Plan" was its allusion to scientific management theories. By the turn of the century, northern factory managers, as well as southern plantation managers, began to keep detailed records of every aspect of a worker's life—how long it took to complete a task; how much it cost to house, clothe, and feed laborers; how productivity changed with age—all in an effort to increase efficiency. Owen's "system of united labor," as he called it, applied scientific management principles to the entire community's organization. At New Harmony, Owen required all children, boys and girls, to be educated in free schools. He offered payment in credits, not common currency, which were redeemable only at a communal store. Laborers worked in strict shifts, and in jobs that matched their age and skills. Most property would be owned collectively. Wright adopted New Harmony's "co-operative labor" system, and she emphasized her plantation's scientific management by including a detailed chart that calculated exactly how long emancipation would take if every planter adopted her system. By her estimate, it would take eighty-five years—which meant that slavery would have died out by 1910. But the chart's deeper significance was that it helped demystify the emancipation process. By offering a detailed projection of how long emancipation would take if planters adopted her plan, she took abolition out of the realm of vague abstraction and into the realm of quantifiable reality.[43]

The initial results of Wright's Nashoba experiment proved wanting. By the summer of 1826, several enslaved workers had caught malaria, as had Wright and her sister, Camilla. Once Frances recovered, she tried to salvage the project and turned to Maclure for help. The two met at New Harmony in the spring of 1826, almost by happenstance. Owen had lobbied Maclure to come to New Harmony, promising him free rein over New Harmony's schools and the ability to use the settlement's facilities for scientific research. Owen planned to build not only "a complete school, academy, and university" but also a hall with "lecture rooms [and] laboratories," all of it designed by "men of great science and practical experience." Maclure agreed to give it a try, leaving Philadelphia for Indiana on November 28, 1825, and arriving two months later. He brought with him some of the Academy of Natural Science's luminaries—Charles-Alexandre Lesueur, a geologist; Thomas Say, an entomologist; Gerard Troost, a chemist and mineralogist—all of whom traveled together on a steamboat that Owen dubbed the "Boatload of Knowledge."[44]

Maclure had little interest in the communal aspect of New Harmony, but the educational and scientific opportunities appealed to him. Like Priestley, Maclure embraced the Enlightenment belief in education, science, and the free flow of knowledge as the antidote to all social evils. If slavery was kept in place through ignorance, he wrote in an 1829 essay titled "Fear, a Concomitant of Slavery," then education was its natural cure: "Knowledge of the reality dissipates the fear, and converts the craven passion into moral courage." Or, as he put it in another essay, from 1826, channeling Francis Bacon, "KNOWLEDGE IS POWER." Since the early 1800s Maclure had been funding experimental schools based on the ideas of Johann Heinrich Pestalozzi, a Swiss educational reformer, who also believed in education's redeeming power. Pestalozzi argued that children learned primarily through the senses, and that they would learn how to think abstractly if exposed to tactile experiences first—touching wooden blocks, seeing real trees, birds, fish. Language, thought, and complex ideas would naturally follow. He also prohibited physical punishment, arguing that the psychological scars of abuse would last a lifetime. Maclure saw a Pestalozzian school firsthand in Paris, and he had taken one of its teachers, Joseph Neef, back to Philadelphia to start a version of the school in America. When Maclure went to New Harmony, he took another Pestalozzian educator, Marie Duclos Fretageot, with the aim of putting her in charge of his new Pestalozzian school at New Harmony, the job Wright had turned down.[45]

Maclure's planned school at New Harmony made a few important changes to the Pestalozzian model. First, students would be trained in a practical trade alongside more traditional subjects. Second, adults would have access to education as well and would be able to attend free scientific lectures given by the men of science who came on the "Boatload of Knowledge." Third, and most important, scientific subjects would be at the center of the curriculum. "Natural history, ought perhaps to be the foundation on which instruction is bottomed," he wrote in the *New Harmony Gazette* in 1828, "consisting of simple ideas obtained by the direct exercise of the senses." Silliman helped Maclure promote the school, publishing "Maclure's Outline, or Course of Study, for the New Harmony Schools" in the *American Journal of Science*. The outline emphasized the scientific curriculum, from "the children [who] are to learn mechanism" to the use of "a machine called the arithmemoeter" to learn arithmetic. The appeal of scientific subjects, in Maclure's mind, was that they matched Pestalozzi's theories. Like Pestalozzi, Maclure believed that true knowledge was arrived at inductively: fact before theory, sensory experience before abstract truths. In addition, Pestalozzi's emphasis on early childhood education—and his concern that childhood abuse was particularly damaging—easily transferred to Maclure's antislavery views. Expose "infant minds" to "the vices and propensities [of] slaves" and they might never recover, Maclure wrote in 1828, in an essay published in the *New Harmony Gazette*, which Wright helped edit.[46]

It would be up to Wright, however, to test these ideas on enslaved people themselves. By December 1826, no longer able to subsidize Nashoba with her own money, Wright gave legal ownership to, among others, Maclure, Lafayette, and Cadwallader D. Colden, a member of the New-York Manumission Society and a powerful New York politician. She then invited Maclure to Nashoba, who was "astonished at the order and good conduct of the Negroes," he told Fretageot, calling it "a master piece of Logical reasoning." Wright was equally taken with Maclure: "McClure . . . possesses one of the soundest heads I have met with," she wrote to Garnett on December 8, 1826. Though no longer the legal owner, Wright still ran the plantation, and she immediately revised its organization, putting a school at its center. A "school for coloured children" would be a "principal part of the plan," she wrote on February 24, 1827, in the abolitionist periodical *Genius of Universal Emancipation*, where a young William Lloyd Garrison would soon find work. She did not explicitly name Maclure's school as her model, nor did she detail its curriculum, but Maclure's influence was clear. Most important, Wright insisted that any new enslaved laborers come without parents. Nashoba would now be "an asylum for . . . negro children," and "the formation of the school" its "most interesting and useful object." "Children sent without parents," she wrote, would be educated in the school and held in bondage until age twenty-one, at which point the profits from their labor would help fund both their "colonization" and "the expenses incurred for their education." Though it is unclear how far she got with the school, Wright clearly relied on the idea of education, and the science behind it, to publicly promote her colonization agenda.[47]

In addition to the new educational focus, Wright continued to emphasize the larger scientific principles on which Nashoba would be based. Echoing Owen, who argued that societies could be scientifically engineered to maximize happiness and equality, Wright contended that her new plan would prove that human nature could be treated like any other branch of science: "Were human life studied as a science . . . it would soon appear that we are only happy in due and well proportioned exercise of all our powers, physical, intellectual, moral." In the same essay, she ridiculed other abolitionists, particularly ones motivated by religion, arguing that their ideas were nothing more than "theories . . . unsupported by experiment." It was a shame, she continued, that these moralists believed "our own species" was "unbecoming of experiment." Nashoba would be different; it would base emancipation on scientific principles, not religious ones. Even if it failed, she wrote in a separate essay earlier that year, "an experiment, that has such an end in view[,] is surely worthy of a trial."[48]

Wright's scientific management and educational ideas certainly made her plan seem rational. But she also may have alienated potential allies with her outspoken secularism. Indeed, for Wright, part of the appeal of Maclure's educational ideas was that they replaced religious instruction with a secular, scientific one. Wright believed religion was a smoke screen that blinded believers from seeing the true causes of their oppression, especially the most vulnerable. If given the proper education, however, they could see the source of their problems. In her revised plan, she prohibited religious instruction and rejected the formal aid of established antislavery organizations, whether the PAS or branches of the ACS, in large part because their religious moralizing "tended rather to irritate than convince" antislavery's opponents. Religion did nothing to prepare the enslaved for emancipation, she argued, adding that it was far more important to give them a scientific education. "Liberty of the mind . . . can alone constitute a free man."[49] Yet Wright's rejection of religion may have also contributed to Nashoba's eventual demise. By rejecting religion, she may have alienated free Black supporters, for whom the church had emerged as a bedrock of Black activism. In addition, by attacking religion, she also risked the support of many other scientific allies, a fair number of whom, like Silliman, saw no fundamental conflict between faith and reason.

The one man of science Wright did not alienate was Maclure. Indeed, no one played a larger role in turning him into an outspoken abolitionist than Wright. Before their meeting, Maclure had attacked slavery only obliquely, as he had in his 1817 geological observations, but after meeting Wright, his antislavery activism grew more intense. Shortly after visiting Nashoba in December 1826, Maclure promised to start his own free Black colony. In 1827 he wrote in his will that one-third of all his wealth would be given to "establishing, maintaining, and supporting a colony of free coloured People on the lands in the vicinity of New Harmony." Skeptical of the ability of men to run the settlement, he named "said Marie Duclos Fretageot and Frances Wright" its future directors.[50] The colony, dubbed New Hope, never came into being, but it signaled Maclure's willingness to distance himself from the more conservative ACS, as well as a new receptiveness to Black colonies generally still shunned by the nation's elite whites.

From 1826 onward, Maclure wrote several antislavery essays for the *New Harmony Gazette*, and in them he often applied his ideas about working-class empowerment to enslaved people, seeing both groups in similar terms. His basic theory was that all societies could be divided into "two great classes"—the productive class and the nonproductive class—and that, through education, the laboring or "productive" classes could equalize the playing field. "The true original sin" that kept all laboring classes in their oppressed condition, he wrote in "Fear, a Concomitant of Slavery," was the denial of "free schools for the education of the laborer." Like the first generation of antislavery men of science, he emphasized that slavery was just as destructive to white people as it was to Black people. Anywhere white children were raised by Black nannies, he wrote in "Effects of Slavery on the Education of Free Children" in 1828, they developed all the bad habits of enslaved people: "abject submission, cringing flattery, and low, artful cunning."[51]

Equally noteworthy was how he turned to geology—the soil and climate in particular—to make an antislavery case. In an essay published on March 1, 1826, he argued that America's temperate climate made it naturally conducive to free labor and democracy. Unlike that in tropical climates, the soil of temperate regions could produce only so much food, he argued. The natural limits on production prevented tyrants from arising, whether politicians or slave masters, since they had no financial incentive. By contrast, tropical climates were more likely to promote tyranny—as well as "slavery with all its most degrading, and horrible features." That was because the soil was inexhaustible, which enticed greedy "tyrants and masters" to create a permanent laboring class to till the land and maximize their own wealth. Maclure's climate-based theory of labor was not new: Benjamin Rush had made a similar case decades earlier, and in 1825 Wright argued that Haiti and Mexico were particularly strong candidates for Black resettlements because, unlike the temperate United States, they had "a climate suited to the complexion of the negro race."[52] But for Maclure, the essay showed a new penchant for fusing science with antislavery. No longer tied to slaveholder patrons, he could now remake himself into an unbridled antislavery man of science.

Maclure's antislavery views may have become bolder in New Harmony, but like Wright's, his ideas still had more in common with the thinking of first-wave elite white abolitionists than with the more radical Garrisonian movement to come. Wright and Maclure may have rejected the ACS and taken more radical positions on other political questions, such as working-class rights and universal suffrage, but they still endorsed gradual emancipation and colonization, both positions fundamentally at odds with the rising new generation of antislavery radicals. Much like Rush, Franklin, and Wedgwood, Maclure and Wright believed that "benevolent" planters could be convinced of liberating their captives, especially if presented with rational, scientific arguments. And like that of those first-generation abolitionists, Maclure's and Wright's embrace of colonization paralleled their predecessors' tacit support for a racially exclusive definition of citizenship.

By the summer of 1827, Nashoba's future looked grim. The scientific management and educational ideas that Wright implemented seemed to make its enslaved captives only more recalcitrant. Wright had hired white overseers to act as both managers and teachers, but to the enslaved laborers it probably felt like a reimposition of slavery's racial hierarchy. Meanwhile, Wright drew the ire of enslaved parents and children when she insisted on separating them, in keeping with Maclure's educational ideas. Then, even though Wright banned corporeal punishment, she learned that while she was on a trip to Europe, one of the enslaved men was "to be punished by flogging." But the final blow came in the summer of 1827. Newspapers reported that Wright's sister, Camilla, had approved of an interracial nonmarital relationship, which caused a national scandal. Wright, it appeared, had gone too far, breeching the taboo against interracial sexual relationships, or *amalgamation*, as it was called.[53]

Yet Wright would not go down without a fight. In her final public defense of Nashoba, published in installments in February and March 1828, she denounced the nation's racial caste system while also using Maclure's educational theories to mask her own racial biases. Wright defended the interracial relationship at Nashoba, arguing that the real problem was white hostility to Black people: "The aristocracy of color is the particular vice of the country," she wrote. Indeed, her defense of interracial marriage, as well as out-of-wedlock sex, in large part explained the vicious public backlash to Nashoba. Yet Wright did not only blame anti-Black racism for Nashoba's difficulties. She also blamed her enslaved workers, using Maclure's theory that, having grown up under the negative influence of slavery, they were nearly impossible to reform. "Raised under the benumbing influence of brutal slavery," she wrote, her enslaved laborers were incapable of being "elevated to the level of society," even after being taught the principles "of moral liberty and mutual cooperation." In other words, there was nothing she could do—once a slave, always a slave.[54]

By the end of 1828, Wright was finished with Nashoba. The sex scandal led many sympathizers to distance themselves from the experiment, which prompted Wright to begin making plans to free the roughly thirty enslaved people then in her possession and resettle them in Haiti. Yet she would use the failure of Nashoba to her advantage. The notoriety made her an intellectual celebrity, albeit a controversial one, and educated Northerners and Southerners alike packed lecture halls to see this rare sight—a woman, speaking publicly, and increasingly about women's rights. As the crowds grew and money poured in, she even went about building her own scientific institution. In 1829 she bought a church in Lower Manhattan and turned it into a secular forum for lectures on anatomy, natural history, arithmetic—in short, all matters scientific—and christened it the Hall of Science. She wrote to Maclure from New Orleans, in January 1829, about her "lecturing to the white and master population," letting him know that she was preparing for the "removal of my colored people to Hayti." Whatever sense of failure she may have felt about Nashoba, her lecturing and fund-raising for the Hall of Science provided a useful distraction. She felt that she was "in twenty places at the same," she told Maclure, giddy if exhausted.[55]

Maclure moved on from Nashoba as well. The failure to implement his educational theories at Nashoba and New Harmony disillusioned him. By 1830 he had resettled in Mexico, writing to Marie Fretageot, still at New Harmony, that he had "lost confidence of being any public use to the instruction of the United States." In Mexico, newly independent and, as of 1829, slave-free, he felt "more could be done for the general good." To be sure, Maclure kept close ties with his scientific colleagues in the United States, including Wright. He instructed Thomas Say, who ran the printing press and laboratory he funded at New Harmony, to print his own geological works, and to adorn them with all the expensive scientific trappings — color plates, fine paper, embossing.[56]

Indeed, New Harmony's reputation has been revived in recent years largely on account of its contributions to science. It became the center for scientific research in the expanding western frontier. The printing press run by Say (and, briefly, by Wright) and funded by Maclure published some of the nation's most important scientific texts in the antebellum period. Lesueur went on to become the first curator of the Muséum d'Histoire Naturelle at Le Havre in 1845. Some of the period's leading European men of science visited New Harmony, inspecting its natural history museum and laboratory, much as American politicians visited Silliman's facilities at Yale. Among the visitors were Charles Lyell, the Scottish geologist who mentored Charles Darwin, and Alexander Philipp Maximilian, the Prussian prince and patron of Alexander von Humboldt. But New Harmony's improved reputation must not obscure its scientific failures, the most significant being the inability of scientific management and scientific educational theories to emancipate the common laborer — slave or free — a shortcoming as evident at Nashoba as it was at New Harmony.[57]

Maclure may have left Nashoba, but he continued to publicly denounce slavery. He published several antislavery essays in the 1830s, using New Harmony's printing press as his mouthpiece. He still rejected the ACS's plan of sending freed Black people to Liberia — an idea "too absurd to be reasoned with," he wrote in 1835—but he remained committed to the earlier generation's gradualist thinking. In contrast to radical Garrisonians, he continued to support colonization and "gradual emancipation." Perhaps, he argued, a separate state for free Blacks within the United States could accommodate white people's racism: "there is room enough here for all, either as a distinct race or mixed." Scientific ideas helped him bolster these arguments. The Ohio River Valley's natural waterways, combined with ever-improving steamboat technology, he wrote, would drive down the cost of labor, rendering "slaves a loss instead of yielding a profit." As enslaved people became better educated, they would no longer accept their subservient status: "Diffusion of knowledge brings both the parties"—slave and master—"nearer a par." He then made an argument that would appear with ever more frequency in the decades to come: slavery was fundamentally at odds with science—it was a system "too great a contradiction with the arts and sciences of the present age."[58]

Maclure's evolution from a modest antislavery critic to a more strident antislavery man of science reveals the tensions abolitionists faced when reliant on slaveholders for scientific support. When he began his scientific career, he was dependent on slaveholding patrons to build his scientific reputation, and, like Silliman, he tamed his antislavery views in order to avoid offending them. Colonization provided a reasonable middle ground, allowing him to embrace some form of antislavery without alienating planter patrons. But once Maclure moved to New Harmony, he was no longer constrained. He was now his own scientific patron, funding his own research and newspaper. Wright, more than anyone else, pushed him to take bolder antislavery positions, and in her Maclure found a kindred spirit, someone who believed with equal fervor that the spread of scientific knowledge would help eradicate social inequalities. Like Wright, Maclure had come to believe that slavery could be solved like any other scientific problem: human relations could be managed scientifically. Running a plantation according to scientific principles and providing enslaved people with a free scientific education would ultimately hasten slavery's demise, and at no financial cost to slaveholders. That free Black Americans might need to live out their freedom outside the United States was not, to this way of thinking, too high a price.

FIVE

The Technological Fix

During the first three decades of the nineteenth century, while American abolitionists experimented with colonization, their British counterparts pushed for a host of reforms aimed at lessening slavery's harshness in their Caribbean colonies and ultimately ending slavery in them altogether. One point of focus was the continued promotion of free-labor settlements outside the Caribbean, which, among other things, might provide models for Caribbean planters to adopt. Sierra Leone still occupied abolitionists' attention, but other tropical colonies acquired by the British after the Napoleonic Wars also drew their interest, particularly colonies in Java, Singapore, and Sumatra, all in Southeast Asia. Another initiative entailed lobbying European nations to ban the African slave trade in the belief that tightening the supply of newly enslaved people would encourage slaveholders to take better care of the ones they had. One last campaign, known as *amelioration*, focused on prodding Caribbean planters to improve the conditions of their enslaved laborers—including the use of labor-saving agricultural technologies—which abolitionists argued would ease the transition to emancipation.[1]

In all these efforts, antislavery men of science played a pivotal role. William Allen, a British chemist, Quaker, and pharmaceutical manufacturer, became deeply involved with the African Institution and the Anti-Slavery Society, the leading British antislavery organizations in the 1810s through the 1830s. Working closely with free Black leaders in the United States and West Africa, Allen supported a litany of scientific agricultural reforms in Sierra Leone, from providing settlers with new farming technologies and experimental seeds to lending advice in the emerging field of agricultural chemistry. In Southeast Asia, Thomas Raffles, a British botanist, abolitionist, and Royal Society fellow, penned an influential natural history of Java that presented the region's climate and soil as perfectly suited to free-labor plantations. Yet even as these men of science lent the antislavery movement credibility, they struggled to explain why their chosen free-labor colonies underperformed; in the case of Sierra Leone, abolitionist elites in London ignored plausible explanations given by Black scientific aides in the colony, ultimately contributing to the failure of their own reforms.

By the 1820s, as Caribbean slaveholders resisted abolitionists' push for gradual emancipation, enslaved laborers revolted, and a new generation of abolitionists demanded bolder action, Britain's aging abolitionist elites shifted emphasis. Rather than encourage planters to adopt new technologies in order to hasten the day to emancipation, they blamed planters' alleged refusal to do so on the institution of slavery itself. Echoing their Lunar Society predecessors, they argued that slavery was a disincentive to scientific and technological innovation—that slavery was, in a word, backward, a check on scientific progress. "In no one instance had slavery and the use of mechanical inventions existed together," wrote the Anti-Slavery Society in 1825, adding that the very "use of slaves prevented the introduction of machinery."[2] The truth was, of course, more complicated. But the idea that science might solve the problem of slavery—that there might be a technological fix—proved seductive, and it would permeate antislavery discourse on both sides of the Atlantic for decades to come.

William Allen was born in London in 1770 to a wealthy Quaker silk manufacturer and began to teach himself chemistry when he was fourteen. Seeing his enthusiasm for the subject, Allen's father apprenticed him to a pharmaceutical manufacturer, Joseph Bevan, in 1792, and Allen's business and scientific careers quickly took off. The Linnaean Society named him a fellow in 1801, the Royal Society followed suit six years later, and the Royal Institution hired him to give public chemistry lectures. Meanwhile, in 1795 Bevan retired and named Allen his successor, and in no time Allen turned the company into one of England's most prominent drug manufacturers. Long before his financial success, however, he immersed himself in the antislavery cause. In 1788 he wrote in his private diary that he hoped England would "put a stop to a traffic which is disgraceful to human nature." Six years later, he met Thomas Clarkson, who became a lifelong friend. Allen helped edit some of Clarkson's antislavery writings, and in his diary Allen cheered the passage of the 1807 slave-trade bill: "A glorious triumph!" he wrote. After the SEAST disbanded in 1807, its goal of abolishing the slave trade accomplished, Allen became a leader of its successor organizations: the African Institution (AI) and the Anti-Slavery Society (ASS), the latter the dominant society from the 1820s onward.[3]

Allen was morally invested in abolitionism, but he also had financial and scientific motives. Public acts of goodwill might reflect well on his pharmaceutical business, but even more important, West Africa stood to become both a potential market for and source of pharmaceutical products. Several Black settlers in Sierra Leone became critical aides to Allen in his search for new medicines. In 1811 a group of Black settlers founded a corresponding society, called the Friendly Society, whose goal was to help strengthen Sierra Leone's commercial ties to Britain and the United States, and it was these men who would act as Allen's on-the-ground aides. "I have sent you three specimens of medicine which you will accept as from a Friend," wrote John Kizell, the Friendly Society's president, in July 1812. He went on to describe in exacting detail how the indigenous population used each medicine: a plant called shargby had to be first "ground to powder," then added to wine or water to control for fevers; mixed with gruel, it "gives good appetites." The locals called another plant lee and used it to help pregnant women combat nausea. A local bark, after being seeped in brandy for "4 or 5 days," helped cure consumption, a term used for a number of illnesses. Allen immediately recognized Kizell's skill. "I have so much confidence in thy integrity and abilities," he wrote to Kizell, in October 1812, "that I shall write to thee with the same freedom as to an old Friend."[4]

Kizell was more than a passive conduit of indigenous medical knowledge, however. His ability to discern the subtle differences between mixtures, illnesses, and cures may have stemmed from an intimate knowledge of African healing traditions. Kizell was born in the Sherbro region of Sierra Leone in 1760, and when he was about thirteen years old, he was enslaved, shipped across the Atlantic, and purchased by a South Carolina planter in 1773. During the War of Independence, Kizell escaped to British lines, and as an emancipated Black Loyalist, he was ultimately resettled in Sierra Leone in 1792. Significantly, Kizell claimed that he was originally enslaved for practicing "witchcraft." If true, then he might have actually been a practitioner of an indigenous healing tradition, which African captors increasingly used as an excuse to enslave people in order to feed the growing European demand for labor. Therefore, as someone versed in both African and European cultures, Kizell was probably not simply relaying indigenous knowledge to Allen, but actively translating it into a language Allen could understand. By omitting the spiritual dimensions of these cures and describing only what he knew Allen felt was legitimate—that is, the naturalistic elements—Kizell revealed himself to be an adept producer of scientific knowledge himself.[5]

The scientific partnership Allen formed with Friendly Society members extended beyond searching for medicines. Allen was also interested in corresponding about all matters of natural history: "I should like to know whether any person in the Colony capable of corresponding has a taste for Natural History," he wrote to James Wise, a Friendly Society member, in March 1814. "My Philosophical Friend, T. F. Forster, a member of the Linnaean Society, request[ed] me to get one of my correspondents at Sierra Leone to procure for him a Box or two of the bulbous roots of some of your flowering plants." Wise responded enthusiastically, "I am very willing to learn something of it," asking Allen to send him "the 1st Vol. of Grove's Dictionary of Arts and Sciences."[6] Here the natural history commodities Allen requested were for ornamental gardens with little commercial value. But their importance lay elsewhere. Exotic objects could help sustain the interest of the AI's elite members, many of whom had a taste for natural history. Their patronage would strengthen the link between Allen and the settlers and in turn allow the settlers to gain a measure of control over their fate.

Allen suggested as much when he wrote directly to the Friendly Society, asking it for natural history objects for the AI's president, the Duke of

Gloucester, a nephew of King George III. The duke had requested "a specimen of African ingenuity or natural curiosities," Allen wrote. The society's members wasted no time fulfilling the request, sending Allen "8 leopard skins," a two-pound "Tortoise shell," and pearls, among other objects. In choosing items without explicit instructions, the Friendly Society's members, all of whom had lived for long periods in either the United States or England, demonstrated a knowledge of natural history collecting. They may not have been considered reputable men of science at the time, but their ability to spot the valued item, to distinguish between the worthy scientific curiosity and mere junk, evinced a scientific eye every bit as discriminating as those of the patrons for whom they collected.[7]

Other scientific exchanges had clearer commercial value. For instance, Kizell sent Allen several different mineral ores, which could be mined and sold to build factories, bridges, and canals. After examining a sample of iron, Allen replied: "If you can find sufficient quantities of the magnetic Iron, it would perhaps turn out a profitable thing for you." Allen's correspondence with another Black settler, Henry Warren, followed a similar pattern: "What is the nature of the soil? Is it rocky, or stony?"; and when he dug down through the soil, "What comes next? Clay, Chalk, gravel or what?" Allen also asked settlers to go on expeditions into the African interior and describe specific climates, ethnic groups, soils, medicines, crops, and trading routes, the details of which could facilitate trade. Henry Savage, a Black settler born in England and trained as a teacher, began to write an ethnographic "account of the Timmanees," which would, he assured Allen, provide a more honest view of "the true state of this country." Savage stressed the "great pains in the collection of facts" that he went through so that Allen "may firmly rely on my observations," demonstrating a subtle mastery of natural history writing's rhetorical tropes. Savage hoped his "short account" would so impress Allen that he would fund a more elaborate expedition "to explore the interior of this country & examine its real extent, position, & local situation."[8]

Unfortunately for Savage, Allen had another settler in mind. In October 1812 Allen sent Kizell explicit instructions and "a little writing paper" for him to use on his own expedition. Like all expeditions into the interior, it was extremely dangerous, but it was a testament to Kizell's commitment to the mission—and, by extension, to science—that he agreed to it. He might "run the risk of [his] life," Allen warned Kizell, but if he were to "keep a regular Journal," he promised Kizell that he would publish it and forward him the profits. Kizell's journal has not been found. But the one known letter he left about the trip makes clear that, like most abolitionists', his motives were not confined to promoting antislavery. Just before the journey, on March 15, 1813, Kizell revealed to Allen that he had a personal motive: his uncle, he wrote, "wants me to go with him to my Country to see my Mother's brothers and friends," whom he had not seen since he was enslaved. "I do not wish that it may trouble you or any of your friends," Kizell added; "That is my reason for informing you of it." Allen understood. Kizell went.[9]

When Sierra Leone's Black settlers agreed to help Allen with these scientific missions, they were not simply doing him a favor. As Black men, they may not have been accepted as the intellectual equals of whites, but they were no less capable of appreciating and valuing the things Europeans did—scientific knowledge chief among them. Like Allen, Sierra Leone's Black settlers understood that knowledge about the region's plants, mineral ores, geography, and indigenous inhabitants might help the colony prosper. Their racial identity may have prevented them from being formally recognized as men of science, but their eagerness to work with Allen evinced a tacit belief in science's value. Mistrust and deception, paternalism and resistance, may have also characterized their relationship, but a shared sense of intellectual values and even mutual respect was in evidence too. Certainly, when it came to science's importance to the colony's success, Kizell, Warren, Savage, and Wise thought in terms similar to Allen's: the colony's future depended on it.

There were other Black men who understood science's utility to Sierra Leone. Perhaps most critical was Paul Cuffe, who was born free in Massachusetts in 1759 to a Native American mother and formerly enslaved African father. As a sailor and merchant, Cuffe made himself into one of the nation's wealthiest Black men, and in the 1810s he became a key ally to Allen and the AI in London. Cuffe has sometimes been depicted as a proto–Black nationalist on account of his support for emigration to Sierra Leone, but casting him in these terms obscures the intellectual world that partly shaped him: the Enlightenment world of white antislavery reformers. Cuffe supported emigration to Sierra Leone not as a kind of racially exclusive separatist project, but in the hope that it would integrate people of African descent into the "civilized" European world, the world of Christianity, commerce, science, and reason. "I see no Reason," Cuffe wrote to Allen from Sierra Leone in 1811, "why they"—meaning indigenous Africans—"may not become a Nation to be Numbered among the historians' nations of the world."[10]

Like white antislavery elites, Cuffe believed that free Black colonies in Africa would enable Black people to demonstrate their equality. He believed, like white reformers, that slavery "degraded" Africans, preventing them from acquiring the Christian faith and secular knowledge they needed to show themselves the equals of whites. "Come, my African brethren and fellow countrymen," Cuffe said upon first greeting Sierra Leone's settlers in Freetown in 1811, "let us walk together in the light of the Lord," and we "may become a people." When he returned to the United States to attract potential settlers, he implored Black audiences to "act worthily of the rank you have acquired as freemen." Celebrating the Franklin-esque virtues of industry, frugality, and education, he instructed Black parents to teach their children "reading, writing, and the first principles of education." He was convinced that if Black people acquired the enlightened, industrious habits that he had himself, whites would come to accept them. "By your good conduct alone," he said, "you can refute the objections which have been made against you as rational and moral creatures."[11]

White abolitionists in the United States recognized a kindred spirit. In 1808 James Pemberton, the former president of the Pennsylvania Abolition Society—who, years earlier, had solicited Banneker's almanacs—asked Cuffe if he would work with the AI to recruit free African Americans to resettle in Sierra Leone. Cuffe had once distanced himself from Black causes, but he now took them up with gusto. And yet neither the protection of his white patrons, his own wealth, nor his growing public stature could prevent him from being treated like any other Black man. As he sailed back to the United States from his first trip to Sierra Leone, in 1811, a Royal Navy ship stopped him. The political context was the embargo between Britain and the United States, but Cuffe knew the real issue was race: "The pretense," Cuffe told Allen, was that on board with him were four "Africans . . . which I did not take away without the Governor's leave." The naval officer assumed they were enslaved on account of their color, and he ordered Cuffe to sail back to the colony so that the governor could sign off on the trip. No amount of fame could guard Black men from these routine indignities.[12]

Cuffe brooked these humiliations with equanimity, however, and they only deepened his commitment to Black uplift. The problem was that he had few models to turn to for what it meant to be Black. A collective sense of Black identity, or even Black identities, was still taking shape, and a key institution that would help facilitate a collective sense of identity—an independent Black press—would not emerge until the late 1820s. In consequence, Cuffe relied on descriptions of Africa written mostly by white people, ones that filtered depictions of Africans through a European lens. Interestingly enough, the accounts he probably read were written by naturalists. Cuffe may have read Henry Smeathman's *Plan of a Settlement to Be Made Near Sierra Leone* (1786), which could be found on the shelves of Newport's African Union Society library, whose members Cuffe befriended. And he certainly had read Mungo Park's *Travels in the Interior Districts of Africa*, re-edited by the AI in 1815; one year later, Sarah Howard, a free Black woman in the United States, wrote to Cuffe to ask if he would open a Black school in Massachusetts and requested a book on geography. Cuffe sent her the "History of Parks Travels."[13]

Cuffe never pretended to be a naturalist. But he knew that becoming conversant in the idioms of natural history—how to describe the soil, climate, and geography; how to collect the enticing natural curiosity—would help secure the support of white patrons. In the summer of 1811, after his first brief stay in Sierra Leone, Cuffe sailed to England to meet Allen and other AI leaders. He made a point of describing the region's natural environment in terms Allen would appreciate. Allen recorded in his diary that Cuffe told him that "the country about Sierra Leone is remarkably fertile, and that sugar cane would grow there as well as in the West Indies." The "coffee grows wild," and, Cuffe added, "cotton, indigo, and rice are indigenous." It was as if Cuffe had taken a page from Smeathman's *Plan of a Settlement*. Indeed, Cuffe's *Brief Account of the Settlement and Present Situation of the Colony of Sierra Leone*, published shortly after his return to the United States, reads much like Smeathman's pamphlet. "Its soil is generally productive," Cuffe wrote, emphasizing how, along with the temperate climate, Sierra Leone was "well calculated for the cultivation of West-India and other tropical productions."[14]

Like the Friendly Society's members, Cuffe also presented the AI's members with natural curiosities. "I made the Duke a present of an African Robe, a letter box and a Daggar," Cuffe recorded in his private journal, "to Shew that the Africans" were capable of "mental Endowments." Cuffe's choice of objects was intentional, as he noted exactly what he wanted these objects to demonstrate: innate African intelligence. But the items he chose also suggested that he knew his hosts' intellectual preferences. Thomas Clarkson, who, with Allen, hosted Cuffe throughout the summer of 1811, had famously relied on natural history objects to prove West Africa's economic potential. During the first slave-trade debates, between 1788 and 1792, Clarkson carried around an "African box" replete with woven textiles, daggers, a cotton loom, and seeds, all signs of the region's capacity for legitimate trade. Importantly, Clarkson's African box was rooted in his own fascination with natural history and science in general. He studied mathematics at Cambridge, wrote anonymous reviews of natural history books, and, in his 1808 history of the anti–slave trade crusade, used the geographic image of a river, with its many rivulets branching outward, to illustrate the growth and spread of the antislavery movement.[15]

Clarkson would have therefore been impressed with Cuffe's selection of items. But when it came to the ways science might help Sierra Leone, Cuffe did not limit himself to curiosity collecting. He also invoked the word *science* as often as he could. When Cuffe spoke to the AI of his method of teaching sailing to Sierra Leone's settlers, for instance, he repeatedly called it a science. Cuffe probably knew that many non-Western cultures had a sophisticated knowledge of nautical navigation. On both sides of his family— his Wampanoag mother, his Akan father—navigation had a rich history. But when writing to Allen, he chose to talk about navigation in a specifically European, *scientific* way. He routinely referred to the skills he was teaching the colony's Black settlers as the "science of Navigation," emphasizing how his student-sailors had become "well versed in arithmetic." The implicit message for Allen was that his students were receiving the best, most "scientific" training around. Cuffe's efforts paid off. After spending several weeks with Cuffe in 1811, Allen wrote in his diary, "Clarkson and I are both of the mind" that "through the means of Paul Cuffee [*sic*]" the chance "for promoting the civilization of Africa . . . should not be lost." The AI even began describing Cuffe in the press as an "enlightened African"—a phrase that evoked a certain learned, scientific worldliness.[16]

Cuffe's embrace of scientific practices and idioms was not simply for show. He seems to have genuinely believed in the utility of science and technology to Sierra Leone's success. Cuffe's own wealth stemmed from his mastery of mathematical skills as well as his investment in modern agricultural technologies. His sailing logbook is filled with precise calculations of the wind's strength, its speed in knots, and the flow of the ocean's currents. As he sailed from Sierra Leone to England in May 1811, he recorded wind variations of "1 ½ point westerly"; he calculated that the distance left to travel, measured by using the "1/2 k[not] current setting W.W 120 hours give 60 miles S. West," was 231 miles; by June he estimated a latitude of 15°49', and longitude of 23°10'. He also invested in an expensive gristmill.[17] Cuffe did not have the privilege of a formal education; everything he knew he acquired through sheer will and everyday experience. But rather than spurn science and the newest machinery, Cuffe came to admire these modern tools. He did so not only to impress his patrons, but also out of what appears to have been a genuine belief that Africans would become "civilized" if they embraced Western modes of thought, science included, as their own.

One of the AI's highest priorities was to implement scientific agricultural reforms in Sierra Leone, whether that meant encouraging the use of labor-saving machinery, using insights from agricultural chemistry to reduce soil exhaustion and maximize yields, or experimenting with new seed varieties. In all these efforts, Cuffe proved an enthusiastic aide. In August 1811 Allen told Kizell to be on the lookout for experimental seeds he had sent with

Cuffe. (He also asked Kizell to send back plants related to "any other subject of Natural History.") In 1813 Cuffe petitioned Congress for a license to send agricultural "machines" to the colony: any day, Allen wrote to Wise, Cuffe would be sending "a Rice & Saw Mill" and "also a Waggon." Allen himself was a serious student of agricultural chemistry, an emerging scientific field aimed at making agriculture more efficient—and one that would become a key battleground in the antebellum debates over slavery.[18]

Allen spent years corresponding with the Friendly Society and imperial botanists about ways to improve Sierra Leone's agricultural output. In 1814 he asked the Friendly Society for a list of "agricultural instruments" they needed, and he inquired about the progress of the sawmills and rice mills they were constructing. He received "seed & roots from this part of India" from Captain T. P. Thompson, a naval captain with a taste for natural history who described India's climate as "not very different from that of Sierra Leone." William Roxburgh, a physician and botanist living in Calcutta, sent yet more seeds to Allen, who passed them on to Kizell. Allen eventually published these exchanges in the AI's annual reports, using them as a kind of scientific propaganda—evidence that scientific know-how was being used to improve the colony's productivity. In 1809, for instance, the AI printed a lengthy letter from Roxburgh that highlighted his shipment of coconut seeds, the "justly famed *Teak*," and the "Cajaputta oil-tree," among other tropical commodities, to Sierra Leone. To underscore the scientific nature of this work, the letter noted that the results of these agricultural experiments—planting Indian crops in West Africa—would soon be published in the *Transactions of the Society for the Encouragement of Arts.*[19]

Though the scientific contributions of Cuffe and the Friendly Society were seldom acknowledged publicly, they were crucial to the scientific work being done in Sierra Leone. When they participated, they were not simply appeasing the whims of their white patrons, either. They seem to have believed, as their white patrons did, that implementing scientific agricultural reforms, that exploring West Africa's interior, and that learning the "science of Navigation" would improve Sierra Leone's chances. White society may have refused to see these men as scientific equals, but that did not stop them from behaving like ones. And on occasion, their white scientific allies privately afforded them a degree of scientific respect. Just before Cuffe departed England in 1811, Allen "presented him with a telescope." It was as if Allen was saying, through the coded language of scientific instruments, that Cuffe was one of them: an enlightened man of science.[20]

The image abolitionists began to craft of Sierra Leone—that it was being re-made into a modern, scientific colony—contrasted with the reality. For one thing, many of the new technologies Cuffe and Allen tried to ship to Sierra Leone were held up by Congress. Materials for the mills and wagons were sent at precisely the moment that the United States and Britain began fighting what became known as the War of 1812. Cuffe sent several petitions to Congress asking for special permission to continue working with the British colony, but he was repeatedly rebuffed: in June 1812 Cuffe wrote to Allen, "It is growing very hazardous for me to keep this path open." Another obstacle was the AI's deep mistrust of Sierra Leone's Black settlers. The Friendly Society tried to explain to Allen that the colony's paltry agricultural output was due not to the settlers' laziness, as several AI members seemed to believe, but instead to the prejudices of the colony's white merchants and officials. "They care not a copper for the colony nor for no Black man," Kizell wrote to Allen on February 14, 1814. "They endeavor to keep us down below the rank of freemen." Allen was not convinced. "We can assure you that the complaint here," Allen wrote to the Friendly Society on August 11, 1815, "still is [that] these people"—Black settlers—"are not at all disposed to cultivate their land." Clarkson echoed him: in 1819 he terminated the AI's relationship with the Friendly Society "because the Friendly Society," he wrote privately, has *"never attended to that which was intended to be the great object of the intercourse between us."*[21]

The implementation of scientific agricultural reforms was also impeded by an influx of liberated Africans to the colony. In 1808 the British Royal Navy began to resettle hundreds of freed Africans—captured and emancipated as part of the British effort to enforce the Atlantic slave-trade ban—in Sierra Leone. Between 1808 and 1811, the number of liberated Africans in the colony—1,991—surpassed Sierra Leone's entire population in 1802—1,917—and the colony's rapid growth made it difficult to maintain stable agriculture.

Yet another factor limiting the colony's agricultural output was Allen's erratic advice. Every few months he encouraged the settlers to grow a different commodity on the basis of its price in English markets, which fluctuated constantly. First it was coffee, then tobacco, then, as war broke out with the United States, cotton. "Its price under the present circumstances with America is extravagantly high," Allen giddily wrote the Friendly Society on November 23, 1814. Yet Allen behaved as if the seasonal cycles of crops did not exist, as if the natural rhythms of hoeing, planting, watering, and waiting could immediately be switched to match the vicissitudes of global markets.[22]

Cuffe also overestimated African American support for emigration. The best he could do was attract 38 African American settlers to the colony in 1816 out of a free Black U.S. population that had numbered 186,000 in 1810. Cuffe's difficulties may have stemmed in part from his willingness to work with the American Colonization Society, which Philadelphia's free Black community famously rejected in January 1817, the year Cuffe died. Many free Black Americans also had family members who were still enslaved, and they did not want to leave them behind. For other Black Americans, the United States was the only home they knew, and still others had not yet given up on the idea of a multiracial republic. And while it is impossible to prove, some may have detected in Cuffe's underlying philosophy too heavy an emphasis on assimilation. After all, Cuffe routinely argued that Black people needed to embrace white culture in order for whites to accept them—and that included an embrace of science. By emphasizing what Black people had to change about themselves, rather than what white people had to change about how they viewed Black people, Cuffe might very well have contributed to emigration's failure.[23]

In the 1820s, as the Royal Navy intensified enforcement of the slave-trade ban, more and more military and government officials poured into Sierra Leone. As a result, British officials increasingly took over the job that the AI had been doing throughout the 1810s—promoting scientific expeditions. Like abolitionists, British officials hoped these expeditions would facilitate Sierra Leone's transformation into a lucrative trading post and open up new markets for the colony's agricultural goods. Regardless of whether the colony's new bureaucratic class was genuinely invested in abolitionism, it presented the slave trade as an impediment to Africa's commercial development, and abolitionists were therefore eager to support its members' scientific endeavors, at least to a point. If nothing else, promoting Sierra Leone as a base for expeditions could turn public attention away from the colony's economic shortcomings and offer hope for a more promising future.[24]

The AI's most important effort to promote the colony as a prime site for expeditions occurred in the 1810s. In 1815 the organization published the unfinished journals of Mungo Park, a revered Scottish explorer and physician who led a heralded African expedition in 1798, but who died on his second mission seven years later. The AI obtained his journals and published excerpts in its annual reports that looked favorably on abolitionism; the AI was keenly aware of the boost Park's scientific reputation would bring to the colony. An AI report from 1813 reprinted quotes from Park's journals, ones that extolled the "fertility of the soil of Africa," and called the Gold Coast "quite favorable to the production of coffee." It even highlighted how, according to Park, Africans in the interior were far more civilized than the "inhabitants of the *coast*." The reason, Park explained, according to the report, was that the "baneful influence of the Slave Trade" corrupted coastal Africans, and that the "advantages derived . . . from the sea and navigable rivers" provided an escape for at least some of them into the *"interior* parts of Africa," where they remained untouched by slavery's degrading influence. The fully edited Park journal that the AI published separately in 1815 expanded on this theme. The deeper one traveled into the interior, it read, the more civilized Africans became. (The AI editors referred to this phenomenon, ingeniously, as Africa's "moral geography.")[25]

Equally noteworthy were the lengths the AI went to to discredit the editor of Park's earlier journal: Bryan Edwards. A distinguished man of science himself, Edwards was also a Jamaican slaveholder who published Park's 1798 journal in large part to undermine the antislavery cause. The Edwards version of Park's journal suggested that European slave traders were neither the cause nor the chief instigators of the slave trade. Rather, the journal argued that European enslavers were saving Africans from an even worse fate—life on their own barbarous continent. Edwards claimed to be quoting Park directly when his version of the journal read, "My opinion is, the effect would neither be as extensive nor beneficial, as many wise and worthy persons fondly expect."[26]

The 1815 journal published by the AI tried to counter the image of Africa shaped by Edwards's edition of Park's journal. The AI edition printed interviews with Park's relatives that attested to his antislavery bona fides and cast doubt on Edwards's integrity. Edwards "came forward on every occasion as an advocate of the planters," the introduction notes; "his first object must have naturally been to gain the services of Park in the direct support of the Slave Trade." It also assured readers that the new journal would not be subjected to the same biases that tarnished the Edwards edition. It would include none of Park's alleged "opinions," only his "facts," which, it added, supported slavery's "opponents"—the abolitionists. Ultimately, the AI hoped the new journal would encourage more expeditions by publicizing tantalizing discoveries that Park's premature death prevented from being known. For instance, the AI edition includes evidence of Africans' sophisticated dyeing and cloth-making skills, which suggested that, with further research, Africans in the interior might make viable trading partners. The AI's separate annual report for 1815 also noted that, in the wake of its newly published Park journal, patrons were rushing to fund their own expeditions, in the hope of finding "further discoveries in the interior of Africa."[27]

By using Park's journals to encourage further expeditions, the AI was taking a risk. Successful expeditions might be a useful distraction from the colony's economic troubles, but they also might attract people to the colony who had little interest in abolitionism. British officials could exploit explorers' discoveries to promote an imperial agenda in Africa that had little to do with ending slavery; explorers could undertake expeditions simply for fame. In other words, science might momentarily conceal Sierra Leone's problems but eventually overtake its core mission. Whatever the risks, the AI took them. In 1816 it celebrated the government's decision to fund several new expeditions, ones that would, it wrote proudly, "follow up the important discoveries of Mr. Park." Many of these expeditions were led by military officials based in Sierra Leone, and the AI touted their findings in part because several explorers attacked the slave trade in their work. While these expeditions lent scientific legitimacy to the antislavery movement, they simultaneously encouraged Britain's imperial expansion in Africa. After all, explorers were trying to determine whether the African interior was a potential source of wealth and, by implication, worth colonizing. Rather quickly, then, the line between promoting abolition and promoting empire became blurred, and scientific expeditions helped muddy the line.[28]

The first government-sponsored expeditions ended in failure: several explorers died from disease and others failed to gain access to the interior. But events began to turn around once private patrons, encouraged by the AI's reissued Park journal, started funding expeditions of their own. Most revealing were the expeditions funded by the African Company. A former British slave-trading firm, the company began to take an interest in expeditions after the ban of the African slave trade, and perhaps its most important expedition was Thomas Edward Bowdich's in 1817. Bowdich's narrative, titled *Mission from the Cape Coast Castle to Ashantee* (1819), portrayed Africans in the interior as highly sophisticated, the surrounding region comparably healthy, and its geography favorable to trade. The implicit purpose of Bowdich's narrative was to entice British officials to turn the Gold Coast, the African Company's base and the region Bowdich explored, into the colonial seat of Britain's nascent West African empire—a direct challenge to British abolitionists, who hoped Sierra Leone would serve that function. Yet the rivalry with abolitionists did not prevent Bowdich from championing the anti–slave trade cause. On the contrary, he understood that fighting the slave trade in Africa would flatter Britons' collective sense of virtue and, more practically, potentially endear indigenous Africans to British officials. The ethnic groups Bowdich met in the interior had a "disagreeable impression of us," Bowdich wrote, adding that they believed all that Europeans wanted was enslaved Africans. Making matters worse was the "inconsistent and selfish" behavior of "the different European powers" that continued the slave trade. This made it more difficult to trade in anything but slaves, he argued, the money simply being too good.[29]

Bowdich's expedition begat others. In 1822 Alexander Gordon Laing, a British army officer in Sierra Leone who was previously stationed in the West Indies, received a commission from the colonial governor to explore the interior. Like Bowdich's narrative, Laing's work, published in 1825, emphasized the importance of enforcing the slave-trade ban in order to promote a "legitimate and honest trade." It also underscored how Britain's implementation of the slave-trade ban ingratiated Africans to the British: British enforcement, Laing wrote, was "rapidly obtaining for Great Britain an influence in this vast continent." He even included a lengthy exchange he had had with the king of the Soolima people, Assana Yeera, to illustrate the point. "Ah . . . you English are good people," Laing recorded Yeera as telling him. "You do not wish to see Black men in trouble . . . you do not sell them; you put them down at Sierra Leone, give them plenty to eat, plenty of cloth, and you teach them to know God."[30]

Explorers often highlighted the scientific benefits of their expeditions, all in an attempt to make them appear more noble, regardless of whether they brought material gain. Bowdich wrote that his expedition shed "much light" on "Natural History and Physical Science." In 1821 the *Royal Gazette and Sierra Leone Advertiser*, Sierra Leone's official government newspaper, noted that another recent expedition was "of the highest interest to science," and that Laing's mission would "benefit . . . geographical science." In 1822 the *Royal Gazette* touted the astronomical observations of Edward Sabine: "The results will unquestionably prove beneficial to the interests of science." Scientific accomplishments in Sierra Leone also helped burnish résumés. Sabine became president of the Royal Society. Dixon Denham, leader of an 1821 expedition, was inducted into the Royal Society in 1825 and appointed governor of Sierra Leone in 1828. The collective effect was to make Sierra Leone seem not only morally, but *intellectually*, virtuous—a gift to science as much as to humanity.[31]

Occasionally, abolitionists attacked explorers for disingenuously co-opting the antislavery cause. One antislavery writer took Bowdich to task for claiming to support slave-trade abolition yet obscuring the fact that his expedition was funded by former slave traders, the "African Company." Kenneth Macaulay, an abolitionist and governor of Sierra Leone in the mid-1820s, cast doubt on Laing's work after proslavery writers cherry-picked parts of it to discredit Sierra Leone. Laing was "brought up in the West Indies," Macaulay wrote, so "may [he] therefore not unreasonably be supposed to have had prejudices against the colony?" For abolitionists, the

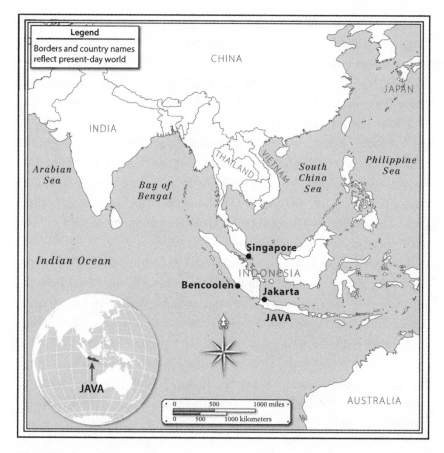

Map of Southeast Asia.
Designed by Gerry Krieg.

problem was not that the slave-trade ban had become unpopular, but that it had become *too* popular: even those with little interest in full emancipation found something to gain in championing slave-trade abolition. British officials in Sierra Leone were even using the ban as an imperial bargaining chip, offering African nations help in suppressing the slave trade in exchange for more territory. Though a diverse set of nations—the Igbo, Ashantee, Fante, and Yoruba—successfully stalled colonization for decades, abolitionists' early support for scientific expeditions provided British imperialists with just the right amount of moral and intellectual cover.[32]

Much as Britain's abolitionist elites contributed to the colonization of Africa, they played a similar role in the East Indies. Abolitionists were particularly vital to the colonization of modern-day Indonesia, Singapore, and Malaysia—part of an archipelago anchored halfway between Australia and India that came under British control in the 1810s. Elite abolitionists hoped that these regions could, like Sierra Leone, become an alternative location for free-labor plantations—once, that is, the region's indigenous form of slavery was eradicated. The challenge for abolitionists was to limit the damage that public knowledge of a preexisting form of slavery might do to their cause. In this regard, men of science stationed in the East Indies became exceedingly helpful—no one more so than Thomas Stamford Raffles, a botanist, colonial official, and antislavery advocate.[33]

The British East India Company named Raffles its lieutenant governor of Java in 1811, the year the company took control of the island from the Dutch and became Britain's de facto government. Raffles held the position for six years, and though the British exchanged Java for other East Indian settlements in 1817, the scientific research he conducted on the island, and the anti–slave trade measures he enacted while there, became touchstones for abolitionists throughout the 1820s. In 1817 Raffles published a natural history of Java that included a robust defense of Britain's humanitarian policies—abolishing slavery above all—and in turn made Britain's presence in the region seem less like a colonization project and more like a humanitarian intervention. In the text, titled A *History of Java*, Raffles wrote that his mission as lieutenant governor was "to uphold the weak, to put down lawless forces, to lighten the chain of the slave," and to promote "humane institutions." When touting the antislavery reforms he enacted while governing, he thanked abolitionists for their support: his own antislavery policy, he wrote, "took for its basis the principles of the African Institution." Raffles in fact worked closely with London abolitionists. In a private 1819 letter to Wilberforce, an AI founder, Raffles reiterated his support for "the cause of the slave," and in another letter he described how the spread of Islam in the region seemed to extend slavery's reach—"May not therefore the spread of the Gospel go hand in hand with the Abolition of the slave trade?" When Raffles returned to London in 1824, Wilberforce, Raffles, and their wives,

Sophia Raffles and Barbara Wilberforce, spent many evenings together, and Wilberforce continued to praise Raffles's "public spirit & Philanthropy" in their private letters.[34]

To be sure, Raffles often exaggerated his antislavery accomplishments. When he arrived in Batavia, the capital of Java, in 1811, he discovered a massive slave system. In 1814, the nearest date when numbers are available, there were 14,239 enslaved people in the city, which had a total population of 47,217. Ten years later, in 1824, the city's enslaved population had declined by 1,753, partly on account of Raffles's reforms, which included an annual tax on all enslaved people over the age of eight, a complete slave-trade ban, and a registry for all enslaved people on the island. And yet that drop paled in comparison to the 20,392 decline between 1788 and 1814. In other words, slavery had been declining at a much faster rate under Dutch rule than it did under Raffles's reign. Raffles's private correspondence also revealed him to be a highly cautious abolitionist. In 1812, for instance, he advised a Royal Navy official not to aggressively enforce his own slave-trade ban, lest he alienate the local slaveholding elites. If Parliament showed great concern for West Indian planters' property rights, Raffles wrote to the officer in explanation, "How much more delicately situated must not the government of a newly captured island be?"[35]

Enforcing antislavery policies required more delicacy than Raffles let on, but he relied on a host of scientific arguments to promote those policies to the broader public. Among the most critical was his reworking of a common explanation for slavery rooted in natural history. Hot climates, naturalists had often argued, made indigenous populations lazy, which in turn made them vulnerable to enslavement by a small elite. In A History of Java, however, Raffles "shifted the blame" away from the tropical climate, as one scholar recently argued, and ascribed slavery's existence to the islands' previous Dutch and Chinese rulers. The native population had not known slavery until the Dutch and Chinese arrived, Raffles argued, adding that "the Europeans and Chinese alone" held slaves in the region.[36]

In truth, slavery was indigenous to Southeast Asia, but it differed markedly from the enslavement of Africans in the Americas. In traditional Southeast Asian societies, master and enslaved had mutual obligations to one another, which was not the case in the New World. In addition, Southeast Asians did not treat enslavement as a permanent, hereditary status, conferred only on outsiders, but as a temporary status reserved for their own communities. And perhaps most important, slavery often functioned as a social safety net in the wake of natural disasters. Earthquakes, tsunamis, and volcanoes frequently left Southeast Asians without homes, families, or food, and thus, when natural disasters struck, the most vulnerable willingly sold themselves into slavery to obtain these basic needs. The very structure of Southeast Asia slavery, in other words, stemmed partly from the natural environment—the exact opposite case Raffles was trying to make.[37]

To be sure, Raffles did not entirely fabricate the idea of slavery being in some way "foreign." When he arrived in Java, the Dutch were, proportionally, the largest slaveholding class (while the local Chinese elites owned the greatest number of slaves). Most slaves were from neighboring Southeast Asian islands, not Java itself. Moreover, by the turn of the nineteenth century, Dutch and Chinese slaveholders had begun to transform slavery from a status that could be shed within one's lifetime to one that was intergenerational and treated as another form of property—it was becoming, in short, less indigenous and more European. Thus, when Raffles arrived, he would have seen a sizable Southeast Asian population being enslaved by "foreign" elites, and in a way that mimicked European-style slavery. But his insistence on Britain's benevolent intentions—to say nothing of the embellishments of his own antislavery record—belied whatever claims he might have made to objectivity.[38]

Divorcing slavery from its environmental origins was not the only way Raffles used natural history to help abolitionists: presenting the region's environment as ideal for agricultural production was another. The soil was Java's "grand source of wealth," he wrote. The "sugar-cane, the cotton tree, and the coffee plants, here flourish in great luxuriance. . . . No area of land . . . in any other quarter of the globe, could exceed it." Fears of natural disasters were also exaggerated: volcanoes, far from posing a risk, accounted for the soil's "extraordinary fertility." He did not deny the frequency of earthquakes, but he downplayed their danger: European settlements "have never sustained any serious injury from them." Raffles also went to great lengths to discount negative perceptions of the disease environment. Most tropical environments, from the West Indies to West Africa, had been notoriously known in Europe as a "white man's grave." But Raffles insisted that Java's mountains and extensive coastline created constant "sea breezes," which blew away the hot, stagnant air that was so conducive to tropical diseases. He not only included exact temperature measurements to prove Java's comparatively cooler climate; he also relied on the best "professional men"— physicians, natural historians, botanists—to further legitimate his claims. "In point of salubrity," these men asserted, according to Raffles, Java was equal to "the healthiest parts of British India, or of any tropical country in the world."[39]

One of the men of science Raffles relied on was Thomas Horsfield, a former student of Benjamin Rush. Born in Pennsylvania, Horsfield was a physician and botanist who left the United States for Java in 1800, for unclear reasons; once there, he worked as a physician for the Dutch East India Company, then for Raffles and the British East India Company when they took over. Though Horsfield does not appear to have been directly involved in abolitionist politics, he embraced the general humanitarian ethos that suffused Rush's teaching—the idea that science should serve humanitarian ends. He dedicated his University of Pennsylvania medical dissertation, a chemical analysis of poison sumac completed in 1798, to the physicians who trained him, including Rush. Medical research was a quest not for "wealth or grandeur," Horsfield wrote on the dedication page, but for the "more solid and durable pillars of *science* and *humanity.*" Raffles, not long after meeting Horsfield in Java, published his research in *Transactions of the Batavian Society of Arts and Sciences*, a scientific journal he created as lieutenant governor of Java. He also sent copies to scientific journals in Europe: "The papers of Dr. Horsfield are highly interesting to science," wrote the editors of the Royal Institution's journal, of whom William Allen was a member, in 1817.[40]

Raffles used Horsfield's research to discredit a long-held rumor: that the East Indies was infested with a poisonous tree that made the islands virtually uninhabitable. (Ironically enough, Erasmus Darwin helped popularize the myth—"the HYDRA-TREE of Death," he called it in 1789.) Raffles tried to refute "the very extraordinary tales told of this tree" by citing Horsfield's research on the plant in *A History of Java*. Having conducted more than twenty experiments with the plant's sap on various animals, Horsfield was confident, Raffles wrote, that all the fantastical claims were "founded in fiction." In addition, Raffles used Horsfield's research to back up his claims about the soil's fertility and the potential discovery of new medicines: several plants Horsfield discovered, Raffles wrote, would probably "become most valuable articles in general medicinal practice." Horsfield thus added scientific credibility to Raffles's research, and in the process he helped Southeast Asia seem like a viable region for free-labor alternatives to Britain's Caribbean plantations.[41]

In *A History of Java*, Raffles repeatedly argued that East Indian free-labor plantations could replace slave plantations in "the West Indies," but he pressed the point even more aggressively in private letters with AI leaders. Some of the most engaged abolitionists were naturalists themselves, like Edward St. Maur, the Duke of Somerset, a member of both the AI and the Royal Society and later a president of the Linnaean Society. "Our advantages over the West Indies are not only in soil, climate and labour, but also in constant markets," Raffles wrote to Somerset on August 20, 1820. He then highlighted the promise of his own scientific experiments with tropical agriculture: "I am attempting to introduce the cultivation and manufacture on the same principle as in the West Indies," he wrote, adding, "I find that sugar-work may be established here at less than one-sixth the expense which must be incurred at Jamaica." No longer lieutenant governor of Java, Raffles was writing from Bencoolen, an old British outpost in nearby Sumatra. But his arguments echoed the ones he had made from Java—that if Britain colonized Sumatra, officials could both prevent slavery's expansion and use the local population as paid labor. "I cannot help regretting that the public attention is not turned to the advantages which may result from colonizing this part of Sumatra," Raffles wrote to Somerset. "The Chinese and natives would be the manual labourers," and the "abolition of the slave trade" would "effectively destroy" the ability of slavery to take root.[42]

Britain's abolitionist elites embraced Raffles as one of their own. In 1817 the AI proudly reported on Raffles's efforts to abolish the slave trade, collect seeds, "promote the cultivation of the soil," and "introduce amongst the Inhabitants [in Java] beneficial medical discoveries." In 1823 Thomas Clarkson pointed to Raffles's experiments with sugar cultivation in the East Indies as proof of profitable, humane alternatives to Caribbean slavery. Indeed, Clarkson wrote, colonizing and employing the indigenous population would be "a blessing to the Natives." In 1828 the *Anti-Slavery Reporter* again referred to Raffles's *History of Java* to highlight the many viable alternatives to West Indian slavery.[43] And yet in doing so, abolitionists enmeshed themselves ever more deeply in Britain's imperial project. The British never seized Sumatra, but in 1822 they took over Singapore, naming Raffles governor. British abolitionists did not view these colonization efforts as anathema to their humanitarian mission, but rather as essential to it: by encouraging free labor in the East, they could undermine slave labor in the West. Men of science were essential to this mission. Men like Raffles and Horsfield provided the scientific evidence abolitionists needed to make free-labor plantations in Southeast Asia seem like viable replacements for Caribbean slave plantations—plantations that abolitionists would increasingly depict as scientifically backward.

Promoting free-labor colonies in the East Indies and West Africa was a vital component of British abolitionism in the early nineteenth century, but no less critical was the promotion of amelioration reforms—that is, attempts to improve the conditions for enslaved people in the Caribbean itself. The most well-known reforms included the outlawing of whippings, creating a slave registry, mandating a weekly day of rest, and providing enslaved people with Bible instruction, but one understudied reform was abolitionists' insistence that planters adopt scientific agriculture, especially the use of labor-saving technologies. In the first two decades of the nineteenth century, abolitionists argued that adopting modern agricultural technologies would make slavery not only more efficient but also more humane. Sympathetic men of science aided them by publicly testifying to modern technology's benevolent influence, but amelioration stalled as planters learned to co-opt reforms to serve their own interests. Over time, abolitionists shifted emphasis, arguing not that planters needed to adopt technological reforms, but that slavery itself was an impediment to technological innovation.[44]

Map of British Caribbean colonies, ca. 1830.
Designed by Gerry Krieg.

Of all the technological improvements that abolitionists promoted, plows received the most attention. Throughout the Caribbean, enslaved people typically used hand-held hoes to plant sugarcane, rather than relying on animal-drafted plows. Abolitionists tried to convince planters that using plows would greatly expedite the first step in sugarcane cultivation—breaking up the soil—while also reducing the physical burden on enslaved laborers. Though plows were hardly a new implement, their widespread use throughout Europe had come to symbolize the technological advancement of European agriculture, particularly when writers contrasted it with non-European hoe-based agriculture. In 1814 the AI commissioned a widely read two-volume study, *Mitigation of Slavery*, that featured nearly twenty planter testimonials advocating adoption of the plow. The second volume was edited by William Dickson, a former Barbadian slaveholder turned abolitionist with a scientific reputation (he once taught mathematics in the island). In 1786 Dickson had moved back to London, eventually joined the SEAST and the AI, and helped edit many antislavery works, including Wadström's *Essay on Colonization* (1794). In *Mitigation of Slavery*, Dickson argued that if planters adopted plows, along with other scientific agricultural reforms, slavery would die away so gradually "as not to be perceived." Evidence of other planters who had successfully experimented with the plow would "prove . . . by a method established by the mathematicians, and followed by all skillful calculators," that gradual emancipation could be both lucrative and humane.[45]

Perhaps the most respected planter Dickson cited was Edward Long, a Jamaican slave owner and naturalist. In *History of Jamaica* (1774), Long encouraged planters to apply insights from agricultural science to their plantations, from labor-saving technologies to "the aid of natural philosophy, chemistry, and some other branches of polite science," lest it stay in the hands of "ignorant clowns." Dickson singled out Long's praise of one "machine" in particular: the plow. With two sets of horses, "a plough could do the same quantity of work in a given time, that one hundred Negroes could do in the same time," Long wrote, in a passage Dickson often quoted. Other planters Dickson cited came to similar conclusions. "Mr. Ashley" said that he nearly "double[d] his crop . . . by means of the plough *alone*." The anonymously written "Treatise on Planting" noted how plows gave "more ease to their slaves," and cited Linnaeus to prove it. "Mr. Fitzmaurice," a Jamaican planter, and George Woodward, of Barbados, also asserted that the plow "*eases the Negroes*." Testimonies that enabled Dickson to prove, mathematically, that plows were cheaper than enslaved workers proved equally useful. From the returns of one planter, Dickson calculated that most planters could save "above 75 per cent" of input costs per acre by adopting the plow in lieu of enslaved laborers. A French slaveholder from Saint Domingue (Haiti) conducted his own comparison, complete with a graphic table, which Dickson reprinted: "Here, says our author," was proof that a single plow could produce profits equal to forty enslaved workers hoeing "280 working days in the year."[46]

Many of the planters Dickson cited had either died or supported slavery—or both. Edward Long died a year before *Mitigation* was published, and he never endorsed emancipation. Joshua Steele, a Barbadian planter who spoke "very favourably" of the plow, died in 1791, and despite experimenting with a host of amelioration reforms, never freed his workforce. Bryan Edwards, who rhapsodized about the plow in *History of the West Indies* (1793) and whom Dickson excerpted, died in 1800, and he was no fan of abolitionists. Dickson did not mention their deaths, which helped: after all, they could not retaliate. But Dickson did highlight their proslavery views whenever he could. Monsieur St. Venant, for instance, "is a decided enemy of *negrophilisme*," he wrote, and supported the plow purely for "oeconomical doctrines." The advantage was that a writer's proslavery politics could preempt charges of bias. After all, though Dickson may have been an abolitionist, what would skeptics say to planters who made the same point?[47]

Of all the "fair trial[s]" with plows Dickson cited, his own may have been the most persuasive. Years earlier, in 1789, he had tested the plow against the hoe on a friend's plantation, setting aside "between 150 and 200" acres for plowed cane, and roughly the same amount for hoed cane. If the experiment failed, Dickson wrote, it "would probably have condemned the plough to a long oblivion, and might even have been held up as *proof* . . . that Abolition itself, from which the experiment emanated, was a wild and pernicious project." The results, to perhaps no one's surprise, clearly favored the plow. His enslaved laborers "were unspeakably eased" by the tool, Dickson wrote; the quality of the plowed sugar far surpassed that of the hoed crop; and profits from the plowed cane "more than paid all the expence" of the initial experiment. Dickson even suggested that the technology made his enslaved workforce happier and less likely to rebel, citing the fact that none had revolted during the Haitian Revolution. Why, then, Dickson wondered, had "the plough . . . not been long ago generally introduced"?[48]

Dickson included several other scientific arguments that countered common planter critiques. To planters who said that the plow killed off too many caterpillars, which helped break down the soil, he argued that, if true, the plow should have also killed off unwanted insects like the termite. But it did not, he said, citing Smeathman's work as evidence. To those who contended that the rocky soil easily broke plows, Dickson noted that plows were now made of iron, not wood, which ought to eliminate that concern. Did plows dry out the soil? Not at all, he said, pointing to the success of his own experiment. Were Africans too dimwitted to learn the science of plowing? Ridiculous: the enslaved people he tested his plows on quickly learned how to use them, and if anything it was the hot climate that "stunts and stupefies" *both* the white overseer and the enslaved African. What, then, was the *"real"* reason planters did not adopt the plow? Dickson asked. His evidence led his readers to one conclusion: the cruelty, if not the ignorance, the *backwardness*, of slaveholders themselves.[49]

Dickson's emphasis on the plow obscured several important truths. For one thing, many of the reasons planters gave for not adopting the plow were valid. In many cases, plowing did dry out the soil, the stony soil did break plows, and the Caribbean's mountainous terrain did make plowing difficult. More important, the focus on plows distracted from the many scientific innovations planters did embrace. In 1789, for instance, Jamaican planters famously imported Tahitian breadfruit from Joseph Banks, the Royal Society president, in a successful effort to find a low-cost, high-calorie foodstuff for enslaved laborers. In 1796 Caribbean planters imported a new breed of Indian sugarcane from William Roxburgh, the same botanist who sent seeds to Sierra Leone. Jamaican slaveholders also hired London chemists and purchased expensive scientific instruments — hydrometers, microscopes — all to increase efficiency. More broadly, by the turn of the nineteenth century British planters increasingly recast the wider amelioration narrative to suit their own ends. Rather than outright rejecting amelioration, many planters selectively implemented ameliorative reforms as a way to improve the image of slavery and ultimately ensure the institution's survival.[50]

The plow argument also concealed another reality. Increased plantation productivity resulted from more than just European-derived scientific and technological innovations: it also resulted from the knowledge of enslaved laborers. Many enslaved Africans came to the Caribbean with a deep knowledge of agriculture. They already knew how to grow rice, yams, plantains — crops indigenous to West Africa — and that meant less time learning how to grow them upon their arrival, which in turn meant greater productivity. Enslaved

Africans also had a long history of using the hoe in Africa and were probably quite efficient with it. Thus, when abolitionists promoted the plow over the hoe, they were taking the very technology that may have contributed to greater efficiency out of enslaved Africans' hands. In any event, plantation hoes were hardly the backward tool abolitionists made them out to be. Planters invested heavily in plow innovations, undoubtedly consulting with their enslaved laborers about which designs worked best. Scholars might debate how "African" the Caribbean hoe was, but it is critical to remember that, wherever the knowledge came from, it was honed and perfected by enslaved Afro-Caribbean people themselves, through the daily backbreaking work of plantation labor, and through their own ingenuity.[51]

By the mid-1820s, the abolitionist elites had begun to take a more confrontational tone. The changed tenor resulted as much from planter resistance to emancipation as from pressure from a younger, bolder, more diverse set of abolitionists—women, enslaved laborers, factory workers, Baptist and Methodist preachers. Though none of these groups was new to abolitionism, many were increasingly visible. In 1824 Elizabeth Heyrick became the first white abolitionist to publicly call for the *immediate* end to slavery; a year later, Lucy Townsend, barred from joining the exclusively male abolitionist societies, established the first female antislavery organization, in Birmingham, England. Enslaved people also asserted their voices with carefully timed revolts. The most famous occurred in August 1823, when roughly 10,000 enslaved people rebelled in Demerara, a British colony on the northern coast of South America. Testimonies revealed that the rebels were not rebelling out of blind rage, or even demanding immediate emancipation; they simply wanted planters to enact the amelioration reforms Parliament had recently recommended, such as Sundays off. Some rebels even pushed those boundaries, calling for "three days in the week for themselves, besides Sundays," one eyewitness reported.[52]

Under these pressures, elite abolitionists in London began to focus less on encouraging planters to adopt modern technologies, and more on how planters' alleged refusal to do so demonstrated slavery's backwardness. In 1823 several founders of the AI established a new antislavery organization,

the Anti-Slavery Society (ASS), which immediately became Britain's preem-
inent abolitionist society. Though much of its core membership remained
the same as the AI's—Allen, Clarkson, Wilberforce—emancipation was its
explicit goal, not simply amelioration (though it still insisted on gradual
emancipation). In Clarkson's inaugural essay announcing the ASS's objectives, published in 1823, he mocked planters for refusing to adopt the plow.
Though "the introduction of the plough has been opposed in the West Indies," he wrote, citing Long, "it has been proven that *one plough* . . . would
turn up as much land *in a day, as one hundred Negroes* could with their
hoes!" In 1825 the ASS's journal, the *Anti-Slavery Monthly Reporter,* wrote
that "in no one instance had slavery and the use of mechanical inventions
existed together." Indeed, the very "use of slaves prevented the introduction
of machinery."[53]

Often abolitionists drew on political arithmetic, an antecedent to modern
demographics, to drive home the point. Their data showed that, between
1808 and the early 1830s, the enslaved Caribbean population was declining
at an average annual rate of .4 percent, and abolitionists linked that rate to
the failure to adopt modern technologies, plows especially. In 1827 the ASS
pointed to the "quite conclusive" fact of a *"decrease in population,"* and
blamed it in part on sugarcane's still being "executed not by ploughs and cat-
tle . . . but by men and women" wielding hoes. The *Anti-Slavery Monthly
Reporter* quoted Raffles's scientific agricultural reforms in the East Indies to
prove that new technologies could literally save lives: in Java, the *Reporter*
noted, the rate of reproduction increased when each farmer had "his own
plough, [and] his own buffaloes and oxen." These "unquestionable statistical facts" meant that modern technologies were "to be infinitely preferred
. . . to the incessant compulsory toil of the Demerara slaves, which is no less
rapidly wearing them down and wasting their numbers."[54]

Occasionally, proslavery men of science offered counterarguments. In
1825 Major Thomas Moody, a proslavery military engineer who had served
in the Caribbean, published a dissenting report from one that Parliament
had commissioned to investigate Caribbean plantations. Moody insisted
that all the scientific agricultural reforms abolitionists promoted had already
been tried and had either failed or did little to improve enslaved people's

lives. He played up his scientific background: as "an officer of engineers" and "active member of agricultural societies" he had no "self-interest" to "bias [his] judgment"; he would stick only to *the most faithful exposition of the facts.*" One of Moody's most extensive critiques focused on why Caribbean planters did not use plows and similar "machinery." In some regions, he argued, the soil made plows impractical, citing his own "fair experiment" in Guiana. In other places, planters did use plows, but their use did little to lengthen enslaved people's lives, and it took twice as long to produce the same amount of sugar. Anyway, he claimed, enslaved people had poor work habits and often broke their tools. They had "no interest in the preservation of the tools they use," Moody wrote, so why should planters trust them with expensive new machinery?[55]

Abolitionists had a response, but rather than back away from technological arguments, they deepened and expanded them. They noted that Moody had mentioned that "steam engines of great power" were erected throughout the West Indies—*but* that plows were used only sporadically. From these two facts, they reasoned that, while steam engines increased the speed of sugar processing, they did not increase the speed of sugar cultivation, and without plows to till the soil, enslaved laborers could not keep up with the new processing machines. "The introduction of steam engines may increase the demand for slave labor" when it was not coupled with other tools that expedited the planting process, the *Anti-Slavery Monthly Reporter* wrote. Though it was a logical argument, it downplayed an important truth. Planters were right: over the long term and in general, technological innovations helped fuel slavery's expansion.[56]

The focus on planters' alleged refusal to adopt new technologies was linked to another argument rooted in agricultural science: that slavery exhausted the soil by encouraging monoculture, or the planting of only one crop. Advocates of agricultural science argued, correctly, that planting a diversity of crops mitigated soil exhaustion, which made slaveholders, who often grew one cash crop exclusively, easy targets. "It seems, indeed, to be an inevitable law of Slavery, to curse the soil on which it is exhausted," wrote the abolitionist James Cropper in 1824, reprinting an article by an American abolitionist. "No soil, however fertile, can resist the deteriorating effects of slave cultivation," the *Anti-Slavery Monthly Reporter* wrote in March 1827. Abolitionists linked soil exhaustion to the lack of new technologies, arguing that if planters used animal-drafted plows, they would have more "manure" to "renew the fertility of the soil." They even likened the soil to a machine that slavery inevitably broke: in the quest for quick profits, sugar planters overtaxed the soil in the same way that a "proprietor of a cotton mill," interested only in fast money, would "infallibly abridge the duration of the machinery."[57]

The overall impression was clear: slavery was at odds with agricultural science. And increasingly, abolitionists made that exact argument. In 1823 Clarkson scoffed at planters, who, he argued, would rather have "Negroes carry baskets of dung upon their head" than purchase "mules, or oxens, and carts." He was done trying to persuade planters to adopt new technologies, and he insisted that the very nature of slavery prevented it: "From when does such a system arise?" he asked, referring to the lack of modern tools on Caribbean plantations; "It has its origins in *slavery* alone."[58]

By the early 1830s, abolitionists in Britain, old and new, young and old, elite and not, had given up on planters' promises to reform slavery themselves. Emancipation, not amelioration, had become abolitionists' central goal. In 1832 Parliament passed the Reform Bill, which expanded the number of working- and middle-class male voters, who in turn voted into Parliament more than one hundred antislavery officials. British women, uniting under their own antislavery societies, forty in all, put a combined 350,000 signatures on scores of antislavery petitions in 1833. Methodist preachers matched them, spearheading 1,900 of the 5,000 petition drives the same year. No less central were enslaved people. On Christmas Day 1831, roughly 25,000 enslaved Jamaicans staged another well-organized revolt, one that came amid rumors that planters were stifling an emancipation bill working its way through Parliament. Many newcomers to abolitionism rejected gradualism in favor of immediatism—the immediate end to slavery—which the abolitionist elites, still in charge of the legislative process, increasingly accepted. And yet in the lead-up to the final emancipation bill, the abolitionist elites made two critical amendments: first, enslaved people would need to serve in a years-long unpaid "apprenticeship" before being fully free; and second, planters, not the enslaved, would be compensated.[59]

Given the clout of the abolitionist elites, it is not surprising that many of their arguments about slavery's scientific backwardness made their way into Parliament. In 1832 Parliament convened its final committee to investigate Caribbean slavery. Repeatedly, the abolitionist-friendly committee members asked planters why they did not "substitute machinery for manual labor." Sometimes they asked why "steam-engines" were not more widely used. But most often they focused on plows. "With reference to cane hole digging," the committee asked, "why is not it now done by the plough?" "Because," one planter responded sheepishly, slaveholders "ha[ve] not sufficient confidence in the superior benefits of machinery to manual labor." A South Carolina slave owner testified that, in the South, the naturally increasing enslaved population was partly due to "the introduction of machinery," which "tended to the preservation of the lives of the slaves." Another person examined by the committee said that some Caribbean islands had "introduced the plough," but "generally speaking, it is manual labor" that did all the work. Yet another attributed the lack of plows "to the prejudices of the planters."[60]

On July 31, 1833, Parliament passed the Slave Emancipation Act, which went into effect on August 1, 1834. The bill required emancipated laborers to work for years in unpaid apprenticeships, though a revised bill abolished the apprenticeship system in 1837, and all enslaved laborers were finally set free on July 31, 1838, at midnight. The original bill also guaranteed compensation to slaveholders, not the enslaved, giving them the remarkable sum of £20 million, roughly 40 percent of Britain's gross domestic product. Emancipation undoubtedly marked a sea change in the lives of the 775,000 women and men who took their first breaths of freedom on that pitch Black August 1 morning in 1838. But few of them doubted how much work lay ahead. Most of them were forced into labor contracts over which they had little control, and almost every aspect of their lives was closely regulated by white officials. What's more, millions of Black people remained enslaved throughout the Atlantic world. William Gibson, a recently freed Jamaican man, conveyed the mix of emotions felt by many liberated Black people when, during an Emancipation Day celebration in 1838, he said, "Let us pray that our brothers and sisters in other lands may be made free"; he added, "And let us look for better freedom."[61]

In the end, the abolitionist elites remained in control of the legislative process and, perhaps more important, in control of slavery's image. By the 1830s slaveholders were increasingly seen not just as cruel and inhumane, but as the enemies of scientific progress. The arguments antislavery men of science made about the environments of West Africa and Southeast Asia, coupled with slavery's alleged incompatibility with "mechanical inventions," cast slavery as an archaic institution fundamentally at odds with modernity. To be sure, the British Caribbean's post-emancipation decline in sugar productivity made certain kinds of social science arguments—particularly ones about the economic superiority of free labor—difficult to maintain.[62] But abolitionists had always been most successful not when retrospectively debating the economic success of emancipation, but when arguing about what science and technology could do to eradicate slavery in the future. In the decades to come, Black and white abolitionists in the United States would expand on the scientific arguments British abolitionists perfected in their push toward emancipation. And in an age increasingly enthralled with science and technology, antebellum abolitionists would find it extremely useful to have science on their side.

SIX

Antislavery in an Age of Science

In February 1861 Jonathan Baldwin Turner, a naturalist, inventor, and Republican Party organizer in Illinois, printed a lecture on racial science that he had been tinkering with for years. The "ESSENTIALS OF CHARACTER" of each race, he argued, were shaped by the natural environment. Black people, he claimed, could thrive only in climates closest to the equator, from where they emerged, whereas white people, having originated in climates nearest the poles, could thrive only in cooler temperatures. The problem, he argued, was that slavery was now preventing both groups from realizing their racial destinies—as separate and equal races on two different continents: "the democratic whites on the North, and the monarchial Blacks on the South." In essence, Turner was making a scientific case against slavery using the theory of polygenism, which held that races were created separately and is often assumed to be a proslavery theory. Moreover, Turner argued that a new chemical technology would soon make it possible to transition from slave-grown cotton to free-labor, white-grown flax. To secessionists who clamored, "COTTON IS KING," Turner countered: science would soon make a "QUEEN [of] FLAX."[1]

Turner saw all the scientific evidence as pointing in the same direction: emancipate and colonize African Americans somewhere outside the United States. Turner also claimed, not without merit, to be speaking as an "abolitionist . . . the true friend of the Black man." In the early 1840s, he hid runaways in Illinois and was fired from his professorship at Illinois College in part for promoting antislavery views. In 1854 he helped write parts of the antislavery platform for what became Illinois's Republican Party, and on September 19, 1862, he had "a long talk" with Abraham Lincoln, an old friend, about the "Proclamation of Emancipation." Turner's antislavery views were not unique. Like many white antislavery men of science in antebellum America, he promoted a moderate form of antislavery that ultimately congealed in the Republican Party platform. Much like abolitionism's Revolutionary era founders, the majority of antebellum white Northerners opposed to slavery believed that slave owners, not enslaved people, should be compensated. They were willing to experiment with voluntary Black colonization. They were skeptical of, and often hostile to, Black citizenship, and they promised to tolerate slavery where it existed and only stop its expansion westward. Ultimately, they hoped that a mix of political, economic—and scientific— solutions to slavery could be found before the issue tore the nation apart.[2]

The popularity of these moderate antislavery views is sometimes overshadowed by the period's most radical abolitionist voices. Whether it is William Lloyd Garrison or Sojourner Truth, Abigail Kelley Foster or Frederick Douglass, radical activists have come to define the meaning of abolitionism in the antebellum period, despite being a small minority; in this chapter, following academic custom, the term *abolitionism* is reserved for radical activists, whereas *antislavery* is used for more moderate antislavery supporters as well as both groups together. Following the free Black communities' lead, Garrison famously rejected the early movement's gradualism and its toleration of colonization, and in 1833 he founded the American Anti-Slavery Society (AASS), the abolitionist movement's largest organization before the Civil War. Unlike the first wave's leadership, these new leaders created racially integrated organizations and embraced women as equal members, however uneasily. During the 1840s and 1850s, abolitionists splintered over tactical questions, such as whether to support political parties and political violence; but despite their divisions, collectively their activism spawned a powerful political party based on a fundamental opposition to slavery—the Republican Party—and turned slavery into the nation's defining political issue.[3]

When the Civil War broke out, most abolitionists supported the Republican Party, though its modest antislavery platform made them "critical allies."[4] But not all Republicans, even those staunchly opposed to slavery, supported abolitionists. Men of science, at least in the North and Midwest, often shared the Republican Party's cautious disposition: they could be ruthless critics of slavery, but they distanced themselves from calls for immediate, uncompensated emancipation and Black citizenship. Often moderate Republicans and like-minded men of science claimed that *they* were the true abolitionists, the ones who best embodied the spirit of the movement's earliest leaders, whether Franklin or Rush. And they were not wrong.

Antislavery men of science kept their distance from the new abolitionist societies for a number of reasons. For one thing, objectivity was emerging as a hallmark of scientific credibility, and to lay bare the political implications of their work risked undermining the basis of their authority. In addition, the American scientific community was beginning to professionalize: scientific journals and societies were proliferating, colleges were establishing new scientific departments, and the federal government was significantly increasing funding for scientific research.[5] All this encouraged men of science to uphold an image of disinterestedness. Another reason for distancing themselves from abolitionists was that many northern men of science still maintained close ties to their southern colleagues. As had happened with Silliman, northern men of science learned that expanding the nation's scientific institutions often required them to soften their antislavery views. To be sure, antislavery ideas did not disappear from their work, but they were framed in ways that made it seem like adopting new scientific technologies would make slavery fade away without any struggle.

Abolitionists, Black and white, saw how men of science were using science to promote an antislavery agenda. But the cautious antislavery politics of most men of science posed a problem: if abolitionists accepted their arguments, they might appear to be endorsing their tepid politics. Yet if they rejected them, they might limit the movement's appeal. It was an "age of science," Frederick Douglass said in 1854, and abolitionists could either turn away from science and risk being seen as against modernity—exactly the opposite image antislavery leaders had been generating since the movement's inception—or they could engage with scientific ideas and shape them to their own ends.[6] The difficulty was that abolitionists had so few men of science in their ranks, or at least ones accepted by the white scientific establishment. The few who did attain a measure of scientific legitimacy saw white men of science not even critique their work, but worse, ignore it. Moreover, Black men of science, even while challenging the anti-Black claims of the era's racial science, sometimes replicated the very notions of biological determinism they tried to undermine.[7]

When it came to agricultural science, another prominent scientific field bearing on the debates over slavery, abolitionists—with a few notable exceptions—also chose to engage with rather than reject it. Black abolitionists were particularly conspicuous supporters. Some believed that Black people's adoption of science and technology would help counter claims of intellectual inferiority, while others believed that adopting scientific agriculture would help free Black communities achieve economic independence. For still others, engaging in agricultural science would prove Black people's commitment to scientific progress and, by implication, to the nation, which was increasingly devoted to its own scientific and industrial advancement. But perhaps most important, Black abolitionists of all kinds knew that most white Northerners, far larger in number than they were, increasingly looked to science as a savior. Black leaders may have had humbler hopes for science, but they were willing to sing its praises if it would broaden the movement and bring King Cotton to its knees.

The scientific study of race had taken a new turn in antebellum America. The earlier environmentalist framework had explained racial difference as a result of changes in the natural and social environment. It had stressed race's malleability: the possibility that each race, if moved to a different climate and raised under proper social conditions, could become like any other. In addition, most environmentalists believed that all human beings descended from a single human pair, a theory known as monogenism. But during the antebellum period a more fixed view of racial difference took hold, one that was less beholden to monogenism and its biblical underpinnings. New scientific subfields, like geology and comparative anatomy, facilitated the shift to a more fixed understanding of race by lending greater weight to physical differences. Changes in society also heightened the attractiveness of racial fixity. Slavery had grown dramatically in the first half of the nineteenth century: nearly four million Black Americans were enslaved on the eve of the Civil War, which made the United States the largest slaveholding nation in the world. The free Black population had also risen sharply, increasing 450 percent between 1800 and 1860, from 108,395 to 487,970. The rapid growth of the enslaved and free Black populations made debates over emancipation and Black citizenship—and, by extension, debates over Black people's innate capacity for freedom—suddenly urgent.[8]

Ethnology, a new scientific subfield devoted to race, promised answers. Ethnologists combined elements of geology, linguistics, natural history, anatomy, and craniology to limn alleged biological differences among racial groups. They generally split into two camps: monogenists, who believed that all races descended from a single human pair, and polygenists, who argued that the races were created separately, each suited to the particular natural environment of their origins. Polygenism is often remembered as a proslavery theory, and many proslavery thinkers certainly claimed it as their own. But what is sometimes overlooked is that northern men of science, some of them strident antislavery advocates, were its most important theorists.[9]

Polygenism's earliest champion was not a proslavery Southerner. It was a Philadelphia Quaker and physician named Samuel George Morton—who, it turns out, relied heavily on the work of the first generation of antislavery men of science. Benjamin Rush's medical lectures, Morton wrote, informed his ethnological research and had given him "such delight." William Maclure helped finance his research: "Few cabinets of Natural History," Morton wrote in July 1841, "have not been augmented from his stores." Silliman sent Morton a skull, for which he felt "much obliged." These antislavery allies enabled Morton to write *Crania Americana* (1839), a foundational text for the American school of ethnology, which rested on the premise that skull size was a proxy for intelligence. Morton's chief claim was that Caucasians (white people) had, on average, the largest skulls and highest intellect, and that Ethiopians (Black people) had, on average, the smallest skulls and were by every measure "the lowest grade of humanity." But his assertion of Black inferiority did not make his work unique, even among antislavery sympathizers: everyone from Rush to Wadström had depicted Black people as in some way deficient. What was unique was his insistence that these ra-cial capacities were long-standing and unchanging. "Each Race was adapted from the beginning to its peculiar local destination," he wrote; racial differ-ences were, in a word, fixed, "independent of external causes." It would be another decade before Morton openly backed polygenism, but by making racial differences impervious to change, he established a line that would separate antislavery racial theorists—whether polygenist or monogenist—from their proslavery adversaries.[10]

Morton never publicly denounced slavery, but throughout the 1840s he went out of his way to align his research with men of science who held well-known antislavery views. He dedicated *Crania Americana* to James Cowles Prichard, a leading British ethnologist with recognized antislavery sympa-thies. He also highlighted his ties to George Combe, a Scottish anatomist who wrote a "phrenological table" for *Crania Americana* and who called slavery "one of the most grievous wrongs ever perpetrated on humanity." In associating his work with British men of science publicly opposed to slavery, Morton could protect himself from accusations of proslavery bias.

Ultimately, however, it may have been the validation his work received from American scientific journals in the North that secured its legitimacy.[11] No one was more important in this regard than Benjamin Silliman. Combe lobbied Silliman to review *Crania Americana* in the *American Journal of Science*, the nation's leading scientific journal, which praised the book as "the most extensive and valuable contribution to the natural history of man." In a private letter Morton profusely thanked Silliman for "having given so much space to the review," and he gently urged him not to publish another author's work that contradicted his own. (Silliman obliged.) Silli-man continued to publish Morton's research into the late 1840s, a time when, in the wake of the Mexican-American War, slavery's westward expan-sion emerged as the nation's defining political issue. In 1847 the *American Journal of Science* printed Morton's response to critics who questioned his assertion that two different races could not produce fertile offspring. Three years later, Silliman published Morton's newest skull measurements, in which Morton claimed to "confirmatively establish" that the difference in size between Black and white skulls was the greatest of all races, a difference of "at least nine cubic inches." In the same essay, Morton undercut the abolitionist argument that Black people descended from ancient Egyptians, which abolitionists often used as proof of Black potential. No, Morton claimed: ancient Egyptians were white.[12]

Morton's ideas gave polygenism a respectable and apolitical gloss. But even more influential in this regard was Louis Agassiz. A Harvard professor born in Switzerland, Agassiz was trained by the era's most revered men of science, including Alexander von Humboldt, the German geologist, and Georges Cuvier, the French paleontologist. In 1847 Harvard offered him a professorship in zoology and geology, lending U.S. science a measure of international prestige. Before arriving in the United States, Agassiz was a monogenist, though that began to change after he met Morton, as well as African Americans, in the United States. Indeed, Agassiz's interest in polygenism seems indebted less to his support for slavery than to his aversion to free African Americans in the North. Upon arriving in Philadelphia in 1846, he wrote to his mother about his disgust at being served by a Black waitstaff: their "Black faces with their fat lips and grimacing teeth, the wool of their heads, their bent knees, their elongated hands" made him want to "leave in order to eat a piece of bread apart rather than dine with such service." In the same letter, he began to question "the fraternity of humankind and the unique origin of our species."[13]

By 1850 Agassiz had begun to argue unequivocally that Black people were a separate species. Like all flora and fauna, he argued, God had created each race for a particular "zoological province," meaning that they could survive only in climates similar to ones in which they were created. Like Morton, Agassiz went out of his way to preempt criticism from antislavery Northerners. Writing in the liberal Boston journal the *Christian Examiner,* he patiently explained that even though the races were created separately, they were all part of the same human family. Therefore, Black people should still be treated decently: "It is because men feel thus related to each other," he wrote, "that they acknowledge those obligations of kindness and moral responsibility which rest upon them in their mutual relations." And yet treating Black people decently was not the same thing as treating them equally. Agassiz, even more emphatically than Morton, insisted that Black people were a distinctly inferior race that had no potential for improvement. "It will not do to assume their equality," he wrote, "so long as actual differences are observed." Rather than "treating them on terms of equality," he concluded, it was wiser to promote only the kinds of changes in their condition that fit their natural constitutions. Agassiz seemed to intuit that many white anti-slavery Northerners, even if they disdained slavery, were skeptical of Black citizenship. What Agassiz offered, then, was a scientific rationale that could simultaneously justify emancipation and segregation.[14]

The legitimacy that Agassiz conferred on polygenism in the North enabled proslavery southern racial theorists to use his work for their own purposes. Perhaps obviously, Agassiz's insistence that Black people not only were created separately, but also were irreversibly inferior, provided proslavery ideologues with a potent scientific claim. Agassiz himself notoriously contributed an essay to *Types of Mankind* (1854), a widely read polygenist text that barely concealed its proslavery bias. Written by Josiah C. Nott, an Alabama physician, and George Gliddon, an ethnologist and American ambassador based in Egypt, *Types of Mankind* also included an essay by Morton, who had died in 1851, that appeared to support polygenism. Nott's writings made the book's proslavery agenda clear, but it was Morton's and Agassiz's appended essays, which judiciously avoided any mention of slavery, that gave the work legitimacy.[15]

Morton's and Agassiz's contributions to *Types of Mankind* made polygenism difficult to refute. While most northern scientific journals (and even a few proslavery southern ones) harshly reviewed *Types of Mankind*, they either defended the integrity of Morton and Agassiz, or let certain problematic claims go unchallenged. For instance, the Philadelphia-based *Medical Examiner* wrote that Nott and Gliddon had "failed to establish any one point they have ventured to assert"—but it added that Morton's "conclusions are entitled to profound respect." *Scientific American*, based in New York, critiqued Agassiz's logic yet insisted on his work's integrity, something "we cannot say so much for that of either Dr. Nott or Mr. Gliddon." When Nott and Gliddon published a sequel three years later, which included another Agassiz essay, the *Buffalo Medical Journal* called it "villainous." But it reprinted without comment the claim that the "highest forms of monkey and lowest forms of negroes" were nearer to each other than Black people were to white people. When northern journals did critique polygenism, they tended to do so on religious grounds, not secular-scientific ones. But in doing so, they offered readers a stark choice—either accept science or accept religion—a distinction many at the time did not actually make. The result was to leave readers to judge the new science of race not on the evidence, but on the reputations of the men of science that provided it.[16]

Morton's and Agassiz's endorsements allowed polygenism to stay within the bounds of scientific respectability. But other northern men of science, ones with deep antislavery convictions, went further, shaping the theory to promote a broader antislavery agenda. Jonathan Baldwin Turner, a student of Silliman's at Yale, a patented inventor, and the founding president of Illinois's natural history society, articulated one of the period's more elaborate antislavery polygenist accounts. Born in western Massachusetts in 1805, Turner had helped found Illinois's Republican Party, and he successfully lobbied Congress to pass the Morrill Land Grant Act of 1862, which helped create public universities nationwide. Turner's interest in the practical sciences, rather than "abstract" fields like physics, was shaped by Silliman. In one of Turner's few surviving student notebooks, he noted how Silliman emphasized the importance of "practical knowledge" over the kind that could only be "acquired from books," which was "useless in the most common and daily interactions of life." It was this kind of science—a "practical" science devoted to agriculture, manufacturing, and transportation—that Turner would eventually champion. And it was this kind of science that would allow him to discredit the image of science as an effete, elitist pursuit and instead align it with the era's democratizing spirit, an ally to the era's "American free laborer."[17]

It was also at Yale that Turner began to engage in antislavery activism. In addition to his notebooks, Turner carefully preserved a signed copy of a pamphlet written by Ibrahima Abd al Rahman, an enslaved Guinean-born prince. In 1788, when Ibrahima was twenty-two, British enslavers captured him in Guinea, which led to his twenty-year enslavement on a Mississippi plantation. Fluent in four languages, Ibrahima snuck out letters to abolitionists, and by 1828 antislavery societies took notice, working with the

American Colonization Society to free and resettle him in Liberia. Turner appears never to have joined the ACS, but he did attend a New Haven event for Ibrahima hosted by the "Managers of the Colonization Society," which would have included Silliman. He also seems to have absorbed the ACS's central message: emancipation without colonization was a nonstarter.[18]

Turner engaged in more radical antislavery activities when he moved to Illinois. In 1833 Illinois College offered him a job teaching rhetoric and science courses, but he quickly gained a reputation for his brazen antislavery lectures. He also helped fugitive enslaved people escape on the Underground Railroad—in the late 1830s, he aided "three colored women, who had escaped from the St. Louis slave market." It was dangerous, thankless work. In 1842 a Kentuckian, looking out for Turner, warned him of a lynch mob that had pledged "the secret death of every abolitionist they can find." In 1843, when he briefly worked as an editor for a local antislavery newspaper, an outraged reader castigated him for his sympathetic coverage of abolitionists who had been helping fugitives: how dare he give "sanction to a shameful assault"? In 1848 Illinois College fired him in part for these antislavery lectures.[19]

Yet even as Turner partook in and defended radical antislavery activities, he did so reluctantly. He avoided openly associating with abolitionists, and he never clearly answered the question that a Yale classmate asked him in a private letter in 1837: "Have you turned Abolitionist?" Most often, he equivocated. In an 1843 lecture he argued that "though one hates slavery," the true abolitionist "will not be of a party"—that is, will not join an antislavery society or political party. In the same lecture, he distanced himself from abolitionists' emphasis on Black citizenship. "Equality in society, in an absolute sense, is impossible," he said, though he kept his subject vague, avoiding any mention of racial groups. Turner ultimately defies the simple distinction between abolitionist and moderate antislavery reformer, demonstrating that at times many people showed shades of both.[20]

By the mid-1850s he had given up his opposition to joining abolitionist organizations. In 1857 he joined the newly formed National Compensation Emancipation Society (NCES), a short-lived antislavery society that revived the gradualist agenda of the earlier generation and had named Benjamin

Silliman its president.[21] The NCES attracted few members, but its significance lay elsewhere. At root, its moderate antislavery agenda mirrored the prewar Republican Party's—which is to say, the most politically palatable, most mainstream form of antislavery. Like the Republican Party, the NCES committed itself only to slavery's nonextension, not to full-scale emancipation, and it encouraged slaveholders to free their enslaved laborers themselves rather than waiting for federal intervention. It prioritized national union over immediate emancipation, stayed silent on the question of free Black citizenship, and supported efforts to voluntarily colonize freed Black Americans outside the United States.

Perhaps unsurprisingly, many abolitionists pilloried the NCES: "When you ask us to give up our principles and subscribe to a falsehood, a libel against God and humanity, you cut off every true anti-slavery man from a participation in your work," read an article in *Frederick Douglass' Paper* in 1859. Yet other abolitionists tolerated it. Gerrit Smith, a leading white abolitionist, spoke at its first convention, in Cleveland; in 1859, when a local Delaware society held a convention, Frederick Douglass covered it favorably, not because he supported its agenda but because it was a promising sign: "the first Emancipation Convention ever held in a slave state."[22]

The NCES also embodied the high hopes many antislavery Northerners had for a scientific solution to slavery. The organization's 1857 manifesto—which Turner, scribbling on his copy, noted should be "carefully preserved"—made it seem that scientific advances were destined to keep the Union together, and only slavery was preventing the nation from becoming a unified, free white republic. "Steam and electricity have brought the extremest States nearer together," it read, making the "necessity of each other's productions . . . ten times greater." The nation's most visionary scientific goals—"like the Pacific Railway, a Ship-Canal across the Isthmus of Panama, or a Trans-Atlantic Telegraph"—could be achieved only through national union—and "SLAVERY alone stands in the way of this glorious consummation."[23]

Whereas the NCES focused on slavery's being an obstacle to scientific progress, Turner promoted scientific education as a means of defeating slavery. Echoing Frances Wright and Joseph Priestley, Turner argued that access to a free scientific education, embodied in the land-grant college movement, would provide working-class Americans with the "scientific practical instruction" needed to outcompete and ultimately destroy slave labor. "If the Union is saved," he said in 1857, at an agricultural fair promoting land-grant colleges, "it will not be by . . . demagogues at Washington," or by radicals on either side of the slavery debate. It will be by the "forgotten and unknown" white laboring classes, who, through the "power of science," will bring about the "POLITICAL salvation of man." The cultivation of "science—mind—knowledge" would teach white workers how to use "the great labor-saving machines patented in the high court of heaven"; these machines would in turn drive slaveholders out of business. He depicted southern resistance to these colleges as the ultimate proof of slaveholders' aversion to science. If southern schools taught science at all, he wrote, they focused only on impractical, elite, or abstract fields like "mathematics," rather than the practical scientific subjects that were the "very foundations of all that knowledge [that] befits and ennobles a free man."[24]

Turner's ability to alloy the image of science to his antislavery views served a number of purposes. For one thing, it helped him dismiss more radical abolitionist positions. If science supported *his* agenda, one of gradual emancipation and voluntary colonization, then all the alternatives could be seen as going against scientific truth. Using science to buttress his ideas also made his antislavery views look more rational, the reasonable middle ground between two extremes. In 1857 he told a group of farmers that if they supported "professorships, endowments, and universities" devoted to "agriculture, the most elevating science and the most noble study," they would avoid the warpath on which abolitionists and fire-eating secessionists were placing the nation. Offering white laborers a free public education also made antislavery an easier pill to swallow. For many free white workers— farmers, mechanics, artisans, especially in the Midwest—opposition to slavery was rooted less in their moral discomfort than in the threat slavery's expansion posed to their ability to acquire western land.[25]

Turner's political allies in Illinois understood these dynamics. In 1856 Owen Lovejoy, the brother of the murdered abolitionist printer Elijah Lovejoy and a Republican congressman from Illinois, wrote to Turner that a free and practical education would make the common white laborer "more loyal to his country"—that is, less likely to side with secessionists. One year later, Lyman Trumbull, Illinois's antislavery Republican senator who helped write the Thirteenth Amendment, fretted over the "spread and perpetuation of negro slavery and the degradation of white labor," as he wrote in a private letter to Turner, and let him know that he "re-read [Turner's] pamphlet in regard to Industrial Universities" and felt it was "a good one." In effect, Turner and his antislavery allies attracted white laborers' support not only with promises of free government land, or "homesteads," and not only with moral arguments, but also with promises of a free scientific education, one that would help them till the land more efficiently and without slave labor. In the process, the alliance between science and antislavery was updated for an age of free labor.[26]

Turner often appealed to a white-solidarity ethic and used the language of racial science to do so. You are the "representatives of a new race," he told a group of white farmers at an Illinois agricultural fair in September 1857 while promoting land-grant colleges. "You have MIND, which they had not," he had said to a similar group four years earlier, in explaining how white people had conquered the continent while Native Americans were now, he claimed, "an extinct race." Therefore, it was not surprising that he turned to ethnology to promote his antislavery views. In *The Three Great Races of Men* (1861), Turner adapted several polygenist arguments made by Morton and Agassiz to fit his own antislavery agenda. Like Morton and Agassiz, he insisted that the differences between the races were so "extreme" and "hereditary" that the two could not possibly live as equals in the same environment. Like them, he argued that Black and white people were both fully human and God's children, even if they were created separately. But unlike them, he argued that Black people were in no way inferior and were simply suited to a climate similar to that in their place of origin. And unlike them, he insisted that Black people could improve.[27]

Turner's essay began with a central premise: Black and white people were "two extreme races . . . two extreme civilizations" now living in a climate suited only to one. Each of the "three great races" (the third being Asian) was created by God to thrive only in its climate of origin. Black people's "natural home," he argued, was nearest the equator, preferably Brazil. Moreover, he claimed that the alleged essential characteristics of each race, their "*natural tendencies,*" could be explained by each group's environmental origins. Black people, having originated in tropical environments where their wants were few, naturally exhibited "indolence, ease, quiet, grace and repose." They were innately emotive, a people of the heart, which accounted for their intense loyalty and fervent religiosity. White people, by contrast, originated in northern "polar" climates, which explained their intellectual, pioneering, and conquering nature. Coming from climates where nature provided them with little, white people excelled in "mechanics, science, metaphysics," all of which they needed to survive in their unforgiving, primordial natural environment.[28]

Turner's theory provided a perfect scientific rationale for his antislavery views. By emphasizing Black people's alleged indolence and loyalty, he could condone slavery, but only momentarily and only where it already existed. Yet by stressing the ability of Black people to improve, and insisting on their full humanity, he could argue that emancipation was, sooner or later, a necessity. "The negroes . . . are actually rising in the scale of being," and "therefore demanding for themselves a broader sphere of freedom, just as we and all others have done," he wrote. Moreover, his contention that Black people were created separately but still fully human allowed him to support their removal to equatorial climates, where they could build free Black monarchies that complemented, rather than replicated, natural "white" democracies. Their loyal nature made them natural monarchists, he argued, while their comparative lack of intellect explained why, in the temperate United States, they could never be made "a SAFE democrat," the "political machinery" of democracy being simply too complicated for Black people to operate. His claim that Black people were ill-suited to democracy also allowed him to support emancipation without citizenship, or what he called "practical freedom." After all, he argued, Black people had every right to deny citizenship to white people in their own equatorial monarchies, where whites were ill-suited, just as white people had every right to deny Black people citizenship in the temperate United States: "For the white race could not survive the one, nor could the Blacks endure the other."[29]

Turner's lecture was not an outlier. It was in high enough demand that he delivered it "to a great variety of audiences," beginning in 1857. Some white abolitionists even asked him for copies. In 1863 Robert Collyer, a British-born white abolitionist preacher in Chicago, said he was "very anxious to get it, *immediately*," according to John Kennicott, a wealthy antislavery botanist and close friend of Turner's. Collyer was about to give one of his "*sledge hammer* sermons on the character of the negro," Kennicott wrote to Turner: "*He* is an abolitionist, like the rest of us, and believes we shall yet have to ask aid from the Negro to end the war."[30]

Other prominent men of science published variations on Turner's theme. Thomas Ewbank, a British-born founder of the American Ethnological Society and head of the U.S. Patent Office from 1849 to 1852, offered his own version in 1860. Like Turner, he wrote that the question of slavery "is as much a question of natural philosophy"—that is, of science—"as of moral and political economy." While he was less willing to grant Black people any comparable advantages over white people, his core argument was essentially the same: each race was "constituted to flourish best in climates akin to its native one"; the only solution was to emancipate and colonize them in a tropical climate. Ewbank also shared Turner's belief that science in general held the key to emancipation. As head of the U.S. Patent Office, Ewbank helped Turner publish his "State Agricultural University" plan (an outline of what the curriculum and mission of an agricultural university should look like), and Ewbank's own ethnological essay argued that if the nation invested in new forms of energy, like steam power, electricity, and coal, it could be used to power new machines that would replace slave labor. The real "friends of the negro," he claimed, were men of science, not abolitionists; for it was men of science who could work on "*promoting the application of inanimate force*"—energy—to new slavery-eradicating machines. Enslaved people should not look to radical abolitionists "for 'signs' of emancipation," he scoffed: they should look to "the agency of physical science to hasten its approach."[31]

Ewbank's essay, published in 1860, was a last-ditch effort to stave off war. He pointed to science as a way to soothe sectional anxieties, emphasizing that slavery would neither need to end immediately, nor last long if scientific ingenuity was put to work. Yet even after Lincoln issued the Emancipation Proclamation, on January 1, 1863, other prominent men of science published similar works. Edward Bissell Hunt, a Yale-educated chemist employed by the Army Corps of Engineers and a frequent contributor to Silliman's *American Journal of Science*, published his own account just a few weeks after the proclamation was announced. He was thrilled that *Harper's*, based in New York and opposed to slavery, published an excerpt.[32]

Titled *Union Foundations: A Study of American Nationality as a Fact of Science*, Hunt's essay cheered slavery's imminent demise, but it put rather plainly what many white Northerners opposed to slavery had long felt: "Towering far above the social problem of slavery or freedom for the negro rises this momentous question of races"—meaning, for white people, what to do with Black people once they were free? He would answer the question, like Ewbank, not with moral suasion, the Garrisonian approach that emphasized slavery's sinfulness, but by using "the cold light of Science." Ethnology clearly indicated that "North America belongs to the Caucasian race," Hunt wrote, and it "is not a natural home for the negro. . . . He belongs within the tropics, whence he came." Like Turner, he played to slaveholders' sympathies, conceding that slavery had civilized Black people to a point, but it was now getting in the way of enslaved people's progress. Slaveholders had done what they could, in other words, and it was now time to emancipate black people and let them create "a destined African empire" somewhere in South America. The alternative was stark: Black people would "inevitably be eliminated" living in a climate unfit for their constitution, and in so close proximity to a white race that was in "every way the strongest." With science as his witness, he claimed it was a "fatal error" to push Black people "away from the tropics into regions consecrated to white labor." Therefore, he advocated "friendly deportation." For Hunt and many like him, there was "no other solution for our race problem."[33]

The antislavery agenda all these men supported—gradual emancipation coupled with voluntary colonization—seems like an anomaly in hindsight. But it was in fact a continuation of the early abolitionist agenda, and it was what many moderate Republicans, particularly those in power, initially had in mind. In the first year of the Civil War, President Lincoln experimented with colonizing Black Americans in Panama, on a piece of land the indigenous population named Chiriquí. A patented inventor himself, Lincoln also believed, like Turner and Ewbank, that science would facilitate the effort, specifically by using chemists to search for coal in the region, in the hope of establishing a viable industry for future Black settlers. Lincoln eventually abandoned colonization in the face of abolitionist pressure, wartime necessity, African Americans' refusal to emigrate, and the brazen escape of tens of thousands of enslaved men and women during the war. But long before the conflict, and well into it, Lincoln shared colonizationists' ambivalence about Black citizenship. On April 14, 1862, Lincoln tried to convince free Black leaders in Washington, D.C., to support colonization, and he sounded much like Turner: "We have between us a broader difference than exists between almost any other two races," he told Black leaders summoned to the White House. "I think your race suffer[s] very greatly . . . by living among us, while ours suffer from your presence."[34]

Five months later, Turner spoke with Lincoln about emancipation. Turner was in Washington visiting his son, who was recovering from wounds he received while serving in the Union army. While there, Turner gave a sermon to African Americans who had freed themselves by running to Union lines, and he tried to convince "them myself of colonization." Turner then visited Lincoln, an old political ally. According to a letter Turner wrote to his wife, Rhodolphia, on September 19, 1862, Lincoln made a joke about abolitionists, saying that a few of them had recently tried to put the fear of God in him. The war was God's will, the abolitionists said, and would not end "without the freeing of negroes." Lincoln replied, "Is it not a little strange that the Lord should tell this to you, who have so little to do with it, and not tell it to me, who have a great deal to do with it!" After a good laugh, Turner wrote, the "sly old coon" pulled the Emancipation Proclamation out of his pocket, as if to acknowledge that slavery's time had come. In mocking abolitionists, both men underestimated their role in the emancipation process. But they were quite right in suggesting that, if Black people were to be free, it would ultimately happen on terms set by white politicians, people whose politics proved far less visionary than many abolitionists had hoped.[35]

Abolitionists, Black and white, clearly had a problem on their hands. The respect polygenism received in northern scientific circles, to say nothing of southern ones, threatened to undermine a central part of their agenda: full Black citizenship. But rejecting scientific arguments simply because they supported a more conservative agenda risked making themselves appear out of touch. "This is, you know, an age of science," Frederick Douglass told a graduating class of seniors at Western Reserve College in Ohio in 1854, in a lecture about ethnology. Like everyone else, he was in awe of what science had recently achieved: railroads, steamboats, telegraphs, which together had "almost annihilated . . . time and space." He had hoped that these magnificent new technologies would bring humanity closer: oceans, instead of borders, would become "bridges—the earth a magnificent hall—under which a common humanity can meet in friendly conclave." So enthralled was Douglass with science that, after the war, he planned to write "a lecture on Banneker," and he encouraged an early biographer to do the same, emphasizing that "such examples of mental industry and success" would be eagerly consumed by "newly emancipated people." Given his hopes for science, he was therefore troubled, deeply troubled, that "a phalanx of learned men—speaking in the name of *science*" was now arguing that Black people had separate origins.[36]

Many abolitionists shared Douglass's disappointment. In contrast to the religious fanatics Lincoln made them out to be, abolitionists believed science would help advance their agenda. Many hoped that racial science would help prove that Black people were, without qualification, the same as white people, descended from the same human pair. "Next to our Maker do we revere Science as the clearest manifestation of his law which he has vouchsafed for us," wrote the Black abolitionist newspaper the *Colored American* in 1839. It was "therefore almost with the anguish that springs from blasted hope" that the editors had to report on the "flimsy attempt to demean" Black people with racial science. Abolitionists soon began to say what few within the scientific community would: far from being immune to the politics of slavery, science was being subsumed by it. In 1841 James McCune Smith, the nation's first Black physician, lambasted ethnologists for "bend[ing] science to provoke popular prejudice." In 1849 James Russell Lowell, a white abolitionist, wrote in the AASS's newspaper, the *National Anti-Slavery Standard*, that polygenism's sudden popularity proved that "slavery demands the expurgation of science."[37]

But even as abolitionist newspapers denounced polygenism, they did not reject the larger science of which it was a part: ethnology. Nor did they question the integrity of prominent polygenist theorists like Agassiz. In 1850 the *National Anti-Slavery Standard* covered Agassiz's theory respectfully. Reprinting an article from a Boston newspaper, the *Standard* allowed Agassiz's central claim to go unchallenged: that the "unity of the human race and the diversity of their origin . . . were two separate questions"—in other words, that even if Black people were created separately, they could still be fully human. It also allowed him to assert, without comment, that his theory had no bearing on the "political question of the Negroes." The article ultimately concluded that Agassiz's theory posed no danger to abolitionism, since it comported with the biblical account of Creation, and thus kept intact the abolitionist premise that Black people, because they were human, should be treated just like any other human being. Even when Agassiz contributed to *Types of Mankind*, Garrison's *Liberator* shielded Agassiz from the vitriol it heaped on the book's main authors, Nott and Gliddon. "With the exception of Agassiz's contribution," read the *Liberator*'s review, "the whole book is saturated with colorphobia of the most virulent kind." It also printed a letter defending Agassiz from accusations that he was selling himself out to slaveholders: "I see nothing in his essay [*Types of Mankind*] to implicate him, in any way, in the pro-slavery aims and proclivities of the book."[38]

That two of the nation's most prominent abolitionist papers—the *Liberator* and the *National Anti-Slavery Standard*—tolerated, if not quite accepted, Agassiz's work suggests, as historians have long known, that many white abolitionists either were blind to their own anti-Black biases or had a high tolerance for others'. The editors of both papers, Garrison and Sydney Howard Gay, were white, but their radical politics did not make them immune to the racial prejudices that saturated their world. Indeed, it was their paternalistic treatment of Black abolitionists that led Black leaders to organize independently in the 1850s. James McCune Smith, despite a close friendship with Gerrit Smith, the white abolitionist who spoke at the NCES, did not hide his disdain for the AASS's official silence on the question of full Black equality: "It is a strange omission in the Constitution of the American Anti-Slavery Society," he wrote to Gerrit Smith on March 31, 1855, "that no mention is made of Social Equality, either of slaves or Free Blacks, as the aim of that Society."[39]

But white abolitionists were not the only ones to keep Agassiz's work within the bounds of legitimate discourse. Douglass's 1854 commencement speech, despite otherwise attacking polygenism, ultimately allowed, as a thought experiment, the possibility that his own case for monogenism was wrong. What if, he asked, "they are able to show very good reasons for believing the negro to have been created precisely as we find him in the Gold Coast? . . . Does it follow, that the negro should be held in contempt?" He answered with an unequivocal no: after all, "a diverse origin does not disprove a common nature."[40] In essence, he affirmed what Agassiz, Morton, and Turner had long been arguing—that separate origins did not mean Black people were not human. It would be easy to dismiss Douglass's entertaining of polygenism as mere rhetoric, but a more intriguing possibility is that Douglass, understanding polygenism's legitimacy in the antislavery North, could not afford to reject the theory entirely. To do so risked alienating too many white people, people he knew who, whatever their anti-Black biases, would be essential to emancipation. Faced with a wrenching choice, Douglass, ever the pragmatist, allowed white Northerners to believe what they wanted to believe—that Black people were somehow fundamentally different—even if he still refused to deny Black people their humanity. It was a brutal choice, but one that arose from a context outside his control.

Douglass admitted that he was not "scientific," and his comments could carry only so much weight. But abolitionists did have some bona fide men of science in their ranks. The most prominent was James McCune Smith, who, like Banneker, was that rare thing: a Black man of science whom white people acknowledged as such. Smith was born into slavery in New York City in 1813, the child of an enslaved woman named Lavinia and her owner, Samuel Smith, a white merchant. New York State's 1827 Emancipation Act freed Smith, and as a youth he attended Manhattan's African Free-School, where his classmates included other future Black abolitionist elites: Henry Highland Garnet, Samuel Ringgold Ward, and Philip Bell. Despite Smith's being a standout student, Columbia College rejected him because he was Black. But he was accepted to the University of Glasgow, where James Watt, inventor of the steam engine, had studied, and where Smith received a medical degree. New York City's Black community raised the money for Smith's tuition, and when he returned to the city in 1837, he plunged into a career as a physician and editor, becoming one of the nation's most prominent Black abolitionists.[41]

To a significant degree, Smith overcame the barriers preventing Black men from receiving scientific recognition. In the 1840s he published two articles in prestigious medical journals edited by white men, and in 1856 he was inducted into the nearly all-white American Geological Society. Yet when it came to his actual *ideas*—and specifically, ones bearing on racial science—white scientific authorities rarely cited his work.[42] Moreover, he often felt the need to cite white authorities to legitimate his own claims, as if he knew that the broader public would not take his work seriously otherwise. In 1844, for instance, he quoted Edward Jarvis, a white Harvard-trained physician sympathetic to abolitionism, who came to the same conclusions that he did about the notorious 1840 U.S. census. The census suggested that Black people living in free states suffered from insanity in far greater numbers than those living in slave states, the implication being that freedom made Black people mentally ill. Jarvis's article cast doubt on the census (a "bearer of falsehood," he called it), but Smith went further, trying to prove that freedom was in fact beneficial to Black health. According to Smith, New York City records showed that Black mortality rates dropped significantly after emancipation. The data "cannot be charged with fanaticism," he concluded: they "give cold, silent evidence" that "freedom has not made us mad; it has strengthened our minds by throwing us on our own resources and has bound us to American institutions with a tenacity which nothing but death can overcome."[43]

The 1840 census was not directly about polygenism, but Smith proved equally adept at challenging that theory as well. In the late 1850s he wrote a pair of essays that dismantled polygenism's central claims and offered an alternative monogenist account that promoted the abolitionist aim of freedom and Black citizenship. When Smith focused on critiquing polygenism, his work was necessary and poignant, offering Black readers, always his most receptive audience, a sophisticated rebuke to the day's vicious new form of scientific racism. Smith's critique rested on two core premises: first, that all races derived from a single human pair, and, second, that physical differences emerged only on account of changes in the natural and social environment. Like eighteenth-century environmentalists, he believed that race was rooted in the external environment and was malleable, not fixed. He also emphasized the ability of all racial groups to improve. Given the potential for each group's racial progress, it was "incumbent on every American citizen who holds dear the cause of Human Progress" to overthrow the "caste which slavery has thrown in our midst."[44]

The strength of Smith's critique lay in his decision to attack polygenism on its own terms: the terms of ethnology. In an 1859 essay, for instance, he challenged the notion that Black people's skulls were closer in size to apes' than to white people's by showing craniologists' idiosyncratic use of evidence. He quoted James Cowles Prichard, the widely respected white British monogenist, who argued that the "skulls of the African and the white" shared a "uniform resemblance" and therefore represented "the sublime argument of the unity of the human race." To counter the claim that differences in skin color were proof of separate origins, Smith highlighted the works of Morton and Edward Wilson, the latter a white British physician. Even a polygenist like Morton acknowledged that the skin tones of Native Americans, allegedly a separate species, ranged from white to Black, Smith wrote, which suggested that color was not a reliable indicator of separate origins. Meanwhile, Wilson's "microscopic science" demonstrated that all human bodies, regardless of outward appearance, were "made up of very minute cells" that looked identical under a microscope—proof that all races were more similar than different. In another 1859 essay Smith pointed out the absurdity of the polygenist claim that Anglo-Saxons were racially "pure": Did not history prove that they were a mix of Celtic, Dutch, Asian, and Germanic peoples? he asked. "Far from being a distinct race of mankind, endowed as a race with a superior genius," Smith concluded, "this Anglo-Saxon race is an admixture of all the Indo-European races."[45]

But Smith did not limit himself to a critique of polygenism. Instead, he retooled ethnology into a radical abolitionist science, one that promoted both emancipation and Black citizenship. In both his 1859 ethnological essays, he argued that all human races improved by intermixing. The most advanced civilizations in history, he claimed, were ones where the natural geography encouraged interracial mixing. The achievements of Greco-Roman, Asian, and Egyptian civilizations could all be explained by their proximity to water and their relatively navigable terrain. By contrast, the deeper one traveled into any one of these continents, the more racially "pure" its inhabitants became, and the less civilized. "Not only is the dwelling and assembling together of men an essential condition of civilization," he concluded, but "the larger the dwelling together, the greater is their advancement."[46]

A favorable climate also encouraged racial progress, he argued. The temperate climate of the United States was ideal for all racial groups, and it was already showing that Black people, if moved from tropical Africa to the United States, would gradually improve. Black Americans had become "not only far superior, in physical symmetry and development, to pure Africans now found on the coast, but actually equal in these respects [to] the white race of the Old Dominion [Virginia]." If for Smith climate no longer stood in the way of Black people's racial progress, neither did geography. Technology could solve whatever challenges the vast and uneven topography of the United States presented: "having, in the Steam Engine, a means of keeping alive the intercourse between the various sections of this territory," Black people could mix more easily not only with each other, but, most important, with white people too. On occasion Smith inverted the racial hierarchy: whiteness, not Blackness, he wrote, was the "color of *defect*." But Smith's larger point was not that white people were innately inferior: it was that both races needed each other to improve, and anyone who defended either slavery or colonization was defying the logic of science.[47]

If the scientific scaffolding was removed, Smith's racial theory promoted a political agenda that was visionary. He was arguing for immediate emancipation and full Black rights at a time when most Americans thought both were impossible. He was making the case that diversity was the nation's strength, not its weakness: "As our motto indicates, E PLURIBUS UNUM," he wrote, "From many nations . . . are made up the unity of the American People." The problem was that in using racial science to make these arguments, he sometimes replicated the racial stereotypes he tried to undermine. In an

1849 article, he examined an African-born teenager living in New York, named Henry, and he wrote that he came from "the lowest grade of Barbarism to which the human family can be sunk." Smith was trying to show that Henry's embrace of Christianity and American mores proved that beneath outward appearances all humans could change, and, thus, all were the same. Nevertheless, it led him to denigrate indigenous African societies and hold up Euro-American societies as the ideal to which all other cultures ought to aspire. Other social groups—women, Jews, Catholics—did not come off any better. By repeatedly praising the "manhood" of Toussaint-Louverture, the Haitian revolutionary, and using him as prime evidence of Black humanity, he perpetuated limiting, if widely held, gender stereotypes. Similarly, by praising Jewish wealth and Catholic political power and claiming they were rooted in nature, he provided scientific cover to equally disturbing myths.[48]

The deeper issue was not the stereotypes Smith perpetuated, but the assumption that racial science forced all its practitioners to accept: that race was a fact of nature, and perhaps immutable to change. Smith himself was aware of the dangers of thinking this way. In an 1843 essay he tried to counter the claim that he was being biologically deterministic, reemphasizing that the destiny of all racial groups was rooted in the external environment, not in biology or bloodlines, and therefore that all racial groups were in essence the same.[49] But even this form of extreme environmentalism could be deeply deterministic. After all, whether one explained racial characteristics in terms of innate biology or the external environment, one was still accepting the premise that race was rooted in nature and could be changed only through natural processes. Whether race was malleable or fixed, rooted in the natural environment or in the blood, to adopt scientific ways of thinking about race was to downplay the social foundations of all group identities, and to ignore the role that agency, contingency, and history played in shaping one's future.

Smith's critique of polygenism was not the only route taken. John S. Rock, another Black man of science and an abolitionist, chose another tack. Born free in Salem, Massachusetts, Rock received a medical degree from the American Medical College in Philadelphia in 1852, and he gained a reputation for challenging ethnologists' anti-Black claims in the lead-up to the Civil War. In his only surviving lecture, printed in the *Liberator* on March 12, 1858, Rock wasted little energy trying to prove Black people's sameness in scientific terms. Instead, he undermined claims of white superiority by focusing on the historical record, showing that, if placed under similar social conditions, white people would act no differently from how Black people were then acting; thus, any emphasis on biological or even environmentally conditioned racial differences were, for all intents and purposes, meaningless. Rock homed in on the claim of Black people's allegedly innate servility. Theodore Parker, a prominent white abolitionist in Boston, had recently tried to ease white northern anxieties about abolitionists, whose agitation they feared would provoke a slave revolt. Not to worry, Parker argued in a widely circulated lecture: Black people were content in slavery, for if they were naturally more rebellious, they would have ended slavery decades ago, with "the stroke of an axe."[50]

Rock was not pleased. He responded with his own essay, which argued that the only reason Black people did not revolt more frequently was that they were vastly outnumbered. In any event, if Black people were naturally content in slavery, how did Parker explain "the history of the bloody struggles for freedom in Hayti?" What of the runaways, "hundreds annually," who escaped north: was that not resistance? And if the love of liberty was proof of white superiority, why did so many white people "not dare open [their] mouth" against slavery? Rock went on to argue that "sooner or later" Black people's military service in an inevitable war against slavery would put to rest any claim of servility. But unlike Smith or even Douglass, Rock did not bother paying lip service to the accomplishments of white civilization: "I do not envy the white man the little liberty he enjoys," he wrote. "It is his right, and he ought to have it. I wish him success, though I do not think he deserves it. But I would have all men free."[51]

Rock may have rejected scientific theories of race, but he refused to deny the importance of race as a social identity. "White men may despise, ridicule, slander and abuse us," he said, "but no man shall cause me to turn my back on my race. With it, I will sink or swim." He was under no illusions about the need for white people to help Black people end slavery and achieve full citizenship—"but we must not rely on them. They cannot elevate us. Whenever the colored man is elevated it will be by his own exertions." He argued that Black people would gain full equality only when they took control of their own affairs—economically, politically, educationally: when they did so, they would "wield a power that cannot be misunderstood."[52] Rock may have underestimated the extent to which that process was already under way, and the extent to which forces were arrayed against Black Americans. But he understood that race, however meaningless in terms of science, carried immense social significance. The nation would become a true democracy not by denying that race existed, he suggested, but by learning how to live with it.

Rock ultimately challenged racial science not with more science, but with logic and with history. In doing so, he offered another way of understanding race—one rooted in history, and subject to change—that provided readers with a plausible counter-narrative to the powerful claims of ethnology. Yet for Rock, much like Smith and Douglass, the alternative—to refuse to engage with ethnology—was no option at all. As Douglass said in his 1854 speech, to stay silent on ethnology's claims was to leave the realm of science open to its exploitation by anti-Black intellectuals. "The neutral scholar is an ignoble man," he said. "The lukewarm or the cowardly will be rejected by earnest men on either side of the controversy." Perhaps it was because Douglass was not a man of science himself that he was more willing to highlight the flaws in scientific thinking: "Scientific writers, not less than others, write to please," he said. Like anyone else, they might "sacrifice what is true to what is popular." Douglass understood, in other words, that science was shaped by society and was perhaps forever constrained by the blinkered worldviews of the people who create it.[53]

Ethnology was not the only scientific discipline that mattered to antebellum abolitionists. Equally important, and far less well studied, was agricultural science. Agricultural science was a loosely defined discipline based on the idea that the scientific study of soil, crops, husbandry, and agricultural machinery could make agriculture more efficient. British abolitionists used it remarkably well in the lead-up to Caribbean emancipation, and antislavery men of science in antebellum America followed in their tracks. In the United States, the discipline had roots in the Revolutionary period, when Benjamin Rush founded one of the nation's first scientific agricultural societies, the Philadelphia Society for Promoting Agriculture, in 1785. But the field came into its own in the antebellum period. Scientific agricultural journals flooded bookstores. Advocates such as Turner lobbied Congress to create an agency devoted to agricultural science—the origins of the U.S. Department of Agriculture, established by antislavery Republicans in 1862—and pushed Washington to support land-grant colleges in the hope that these schools would conduct cutting-edge agricultural research.[54]

Throughout the antebellum period, northern antislavery men of science routinely portrayed slavery as fundamentally at odds with agricultural science. Some presented slavery as hindering the implementation of scientific reforms. In 1839 the Boston botanist William Kenrick published a letter in the *New England Farmer*, a prominent agricultural journal, that argued that if only Virginians would emancipate their enslaved laborers, white farmers would "flock" to the South and, armed with their knowledge of scientific agriculture, replenish its depleted, slave-worn soils. In no time, Kenrick wrote, the dilapidated Virginian countryside would turn into "the garden of all the Atlantic States." In 1861 Turner argued that anywhere slavery spread, "all industry and mechanical skill [were] penalized." Four years earlier, the *New York Tribune*, the nation's largest newspaper and a leading advocate of both scientific agriculture and antislavery, wrote that the only thing preventing slaveholders from adopting labor-saving technologies was the "stumbling block of slavery."[55]

Others argued that new technologies could peacefully wean the nation off slave-grown cotton. Edward Everett, a Harvard University president who established the Lawrence Scientific School in 1847, the precursor to Harvard's engineering school, and who brought Agassiz to the university, presented agricultural inventions as the best, most peaceful means of ending slavery. "Agricultural machinery," he argued in 1854, as a U.S. senator from Massachusetts, would be the "most likely means by which the emancipation of the slave may be brought about." In 1860 Thomas Ewbank, the former head of the U.S. Patent Office, wrote that steam-powered plows and other unforeseen technologies were the last best "hope for the negro"; he added, "The ultimate extinction of human slavery depends on it."[56]

Everywhere slavery was depicted as exhausting the soil and, in consequence, feeding slavery's westward expansion. The "ruinous system" of slavery was squarely to blame for "the list of '*Virginia worn-out lands*,'" read an 1849 letter in the *Cultivator*, an agricultural journal based in Albany, New York. The *New York Tribune* routinely portrayed cotton planters as scientifically ignorant. In an 1851 article, it argued that slaveholders plainly defied agricultural science by insisting on planting one profitable crop alone—cotton—rather than rotating several crops, as advocates of scientific agriculture urged. "Not only is the soil thereby exhausted," it wrote, but the "population [is] led to scatter." In his *Three Great Races* pamphlet, Turner argued that nature would do what God had ordained if slavery expanded northward—"for he smites every Northern soil with barrenness and with curses, on which the slave is permitted too long to linger." In 1863 Edward Bissell Hunt, the Yale-educated chemist, wrote that slavery leached the "carbonaceous and nitrogenized ingredients of its soil," concluding that the "worn-out land of Virginia, and the desolate circuit of abandoned plantations . . . are monumental protests against such frauds on the soil."[57]

But, crucially, many of these men, though they publicly opposed slavery, disassociated themselves from abolitionists. Science was objective, they stressed, impervious to political influence. Indeed, it was on account of their work's disinterestedness, they claimed, that a scientific solution to slavery would be "effectual, without being violent . . . injurious to no class, but advantageous to all interests," wrote Ewbank. In 1852 Augustus Hascall, an agricultural reformer and antislavery congressman from upstate New York, argued that creating a national agricultural department "would do more to cure slaveholders of their slaveholding monomania than all the lectures that Abolitionists have ever read or written." One year earlier, the *New York Tribune* wrote something similar: "Science has not generally paid much regard to the sacredness of vested interest"; therefore, "if a new discovery or invention is brought forward which promises benefit to the nation and to mankind, we shall not suppress it nor oppose it because Mr. A. or Mr. B. will be a loser."[58]

These scientific claims were potentially useful to abolitionists, but they also had the potential to undermine their radical agenda. As some abolitionists would eventually point out, any attempt to promote a scientific solution to slavery might weaken public support for more aggressive political action. Yet it was also true, as other radicals would suggest, that scientific arguments might keep moderate white Northerners within the broader antislavery fold. If scientific arguments made these Northerners feel that they did not have to sacrifice their sense of reason, of objectivity, to a cause many saw as being hijacked by intemperate radicals, then so be it. It was an age of science, as Douglass might have put it, and to neglect a scientific argument because of its co-option by more moderate political forces would be to let too good an opportunity go to waste.

In part, northern men of science advocated scientific solutions to slavery because they felt that abolitionists were needlessly pushing the nation toward war. Scientific and technological solutions offered a soothing alternative, one that fit neatly within the enduring Enlightenment belief in scientific progress as the means toward social improvement.[59] Yet another reason northern men of science supported scientific solutions was that some of them, even ones with staunch abolitionist credentials, had lucrative ties to slave owners.

Nowhere was this clearer than with John Pitkin Norton. Born in Albany in 1822, he was the nation's first professor of agricultural chemistry, a position Yale created for him in 1846, and the author of *Elements of Scientific Agriculture* (1851), one of the nation's first textbooks in the field. When Norton began studying chemistry at Yale, under Silliman, in 1840, he regularly attended the Connecticut Anti-Slavery Society's meetings. He found Black abolitionist speakers particularly captivating. In a March 20, 1841, entry in his unpublished diary, he wrote enthusiastically about an electrifying speech by James W. C. Pennington, "a colored man from Hartford" and prominent abolitionist. In May 1841 he heard Henry Highland Garnet, another leading Black abolitionist, speak of the "the horrors of a servile war," and he found it no less thrilling. The antislavery activism of his parents—his father, a wealthy industrialist, helped found the Connecticut Anti-Slavery Society—

was a great influence on him. But perhaps most significant was the intimate relationship he formed with several *Amistad* rebels. In 1839 fifty-three enslaved West Africans seized control of their slave ship, *La Amistad*, originally bound for Cuba, and redirected it to Long Island. Norton's father helped fund their successful Supreme Court freedom suit two years later, a victory that momentarily united a splintering abolitionist movement.[60]

The Norton family housed several liberated rebels until they were able to return to Sierra Leone, so the younger Norton had the opportunity to form a bond with them. In his diary he seethed at white neighbors who feared their presence: "People will in a few days learn that these thirty Africans are not about to murder every inhabitant," he wrote in March 1841. But not everyone was convinced. In September a white mob assaulted one of the captives, an event that Cinque, the group's famous leader, privately recounted to Norton. Though he did not go into details, Cinque told Norton that he *"put his hand on him"* — the assaulter — *"& he fell down,"* Norton's underline evincing his private excitement. Other rebels were more forthcoming. Kenna "was very communicative," Norton wrote, "and told me the whole story of his capture & conveyance to Cuba" and "interesting facts about the Mendi country." Norton seems to have got along with the rebels, but there was no shortage of suspicion. In one instance, a group whom Norton had been teaching English quizzed *him* on the meaning of a word. They already knew its meaning, but they wanted to be sure he was not deceiving them.[61]

Norton remained involved in antislavery activities through the mid-1840s. But like Turner's, his antislavery views defy simple categorization. He rejected colonization yet attended at least one meeting of the local ACS chapter (and "hope[d] they will do well"). He joined racially integrated abolitionist societies and opposed essentialist racial thinking—finding particularly appalling an anti-Black writer who thought "barbarism & stupidity" were "an essential of their character." But when it came to women's equality, he was no radical. In his diary he wrote that he opposed the inclusion of "women's rights" in the Connecticut Anti-Slavery Society platform, and he probably supported the American and Foreign Anti-Slavery Society (AFASS), founded in 1840 by abolitionists, including his father, who rejected the AASS's support for women's equality. But unlike the AFASS, which opposed antislavery political parties, Norton wrote approvingly of the Liberty Party, the nation's first official antislavery party, founded in 1840. And though his antislavery activism tapered off after he began teaching chemistry at Yale in 1846, he continued to attack slavery in print. In 1851 he published an article denouncing proslavery congressmen for refusing to offer an official welcome to the antislavery Hungarian revolutionary Lajos Kossuth: "Has it come to this, that we at the North who oppose slavery may not lift up our voice on behalf of suffering humanity elsewhere?"[62]

Yet even while attacking slavery in newspapers, he began to soften the antislavery comments in his scientific work. In 1850 he went out of his way to appease his New York City editors, who worried that the antislavery remarks in a new book that he had edited, *The Farmer's Guide to Scientific and Practical Agriculture*, would prevent its sale in the South. In an unpublished early draft, Norton described slavery as an inherent check on scientific progress: "Southern Agriculture has to contend with the difficulties arising from the inherent defects of slave labour that are unknown to the North," he wrote. "However just," his editors responded in a letter dated February 19, 1850, the remarks "may, in consequence of the extreme sensitivities of the subject of Slavery in the Southern States, prejudice the sale of the Book in that direction." Norton decided to cut the "offensive matter" and ultimately bent over backward to appease slaveholder sensibilities. In the final version he wrote that the slaveholder "does not, it is true, do the work with his own hands; but he can encourage good work among his people." He acknowledged that planters worked under certain constraints—such as the need to "keep the movements of the subordinate in check." But no matter the peculiarities of their labor system, he assured slaveholders that they would find his book useful.[63]

But even as Norton appealed to slaveholders, he included several veiled critiques. He wrote, for instance, that "our Northern states are taught to consider honest labor honorable," which, though not an explicit attack on slavery, implicitly functioned as one. He presented northern agricultural practices as far more scientific than southern ones: "We"—meaning Northerners—"are far in advance of many southern districts." But perhaps most significant was his vision for the nation's agricultural future. He argued that the Union should promote "Indian corn . . . our greatest national crop"; unlike any other crop, especially cotton, it grew everywhere, "from north to south, and from east to west." Conspicuously, Norton ignored the fact that southern cotton was then dominating the nation's economy: in 1850 it accounted for 53 percent of the nation's exports. His claim that corn could replace cotton was rooted not only in a scientific claim—that it could grow in most climates—but also in its nationwide cultural resonance. For both sections of the Union, corn symbolized a founding national myth: that Native Americans graciously welcomed European colonists and passed the torch onto the continent's new inhabitants. By reminding Northerners and Southerners of their shared origins, the passage used agricultural science to offer a solution to their sectional divide.[64]

Norton did not stop at corn, however. In *The Farmer's Guide* he presented a celebrated recent invention, the McCormick reaper, as the embodiment of the nation's idealized values: freedom, openness, technological progress. In contrast to eastern states, whose rocky and hill-pocked topography stymied the machine's use, it was "in the broad, unbroken, smooth fields of the West, [where] these machines find their natural sphere of action." Much as Maclure's 1817 geological map had, Norton was depicting the west or, more specifically, the Midwest, as freedom's natural home; as proof, he pointed to the widespread use of the McCormick reaper, manufactured in Chicago, throughout the Midwest's free-labor wheat fields. Only in a region naturally blessed with good soil, a temperate climate, and relatively flat terrain, and only in a place so implicitly innocent—that is, a place without slavery—could a technological innovation like the McCormick reaper take root. Of course, Norton did not mention that the McCormick family first designed the machine on Virginia plantations. Nor did he mention that it was now helping revive slavery in Virginia and Brazil. Perhaps he did not realize it; perhaps he did, and preferred not to mention it. Either way, his vision of modern machines plowing through free and open western prairies offered a hopeful scientific alternative to the slave-based reality.[65]

The delicacy of Turner's antislavery critiques can in part be explained by his growing reliance on slaveholders for income. When Yale created a professorship for him in its new School of Applied Chemistry in 1847, the position came without a fixed salary. Like many science faculty today, he was working in a soft-money environment—that is, he needed to generate his own income. He found one reliable revenue stream in planters, who paid him handsomely to run soil analyses from his laboratory at Yale. Another source came from writing for southern agricultural journals, most notably the *Soil of the South*. The writing job paid well—ten cents a word, or roughly one hundred dollars an article, which meant that the income he earned from the nine articles he wrote for *Soil of the South* in the fall of 1851 and in 1852 nearly equaled the average annual salary of a Yale professor, which was $1,150. Unsurprisingly, he avoided any mention of slavery in his *Soil of the South* articles, and he often used them to promote his "chemical analysis" business.[66]

The *Soil of the South*'s editor, William Chambers, anticipated Norton's potential unease with writing for a publication marketed to slaveholders. In his initial letter to Norton, Chambers wisely framed the job as an opportunity to advance science and national union, rather than help slave owners. "Practical men" were editing the journal, Chambers assured Norton, subtly distancing himself from nakedly proslavery agricultural science writers like Edmund Ruffin. He also played to Norton's nationalist instincts, writing that "science and practice would 'both be benefited by Union.'" Last, Chambers emphasized how the journal would help Norton promote science to an audience woefully in need of it: "Our planters need the light of science, and we believe that you, better than any other man in our country, are prepared to give it to them." Norton must have been flattered, but he also must have realized that, in taking the position, he would need to soften his antislavery language. As Silliman had learned decades earlier, promoting science in a slaveholding union came at the price of compromising one's antislavery views.[67]

The need for self-censorship did not occur only when writing for southern journals. It was also something Norton had to consider when writing for northern publications. In the fall of 1851 Norton agreed to write for the *Plow*, a New York-based journal edited by Solon Robinson. Abolitionists had recently vilified Robinson for contributing to *De Bow's Review*, a leading proslavery journal, but for Norton the benefits seemed to outweigh the costs. The *Plow*'s editors assured Norton that their journal would have "the *largest circulation* of anything of that kind in the U.S." and would help "both you and your department grow." Moreover, Robinson could help Norton promote a land-grant college he was trying to build in Albany—a college that would, Norton wrote, start "indoctrinating [students] with a love of science." The plans Norton drew up for the college included the "full instruction in scientific agriculture, engineering, mining & chemistry," and Norton had received verbal teaching commitments from some of the nation's leading men of science, including the Princeton geologist Arnold Guyot, the Harvard mathematician Charles Sanders Peirce, and Louis Agassiz. To alienate Robinson, who exchanged several letters with Norton about the college, would be to sacrifice his scientific dreams on the altar of abolitionism.[68]

Norton did not live to see his dream realized: he died of tuberculosis in September 1852, barely thirty years old. But he did write at least one article for the *Plow*. Published in February 1852, it portrayed science as inherently moral, a field of study that combated the darkness of prejudice with the light of reason. The journal's "name is significant in more senses than one," Norton wrote. "In your hands it will come to represent a *moral plow*, driving through the tough sand & tangled roots of error & prejudice." Norton did not mention slavery directly, but he did write that anywhere modern plows were found in "Southern States," there was promise for agriculture to become "something above the mere exertion of muscular strength." In other words, wherever machine technology took root, slave labor would no longer be necessary. He was being deliberately vague, of course, but that was a skill experience had taught him, a skill one needed when promoting science in a slaveholder's union.[69]

Norton may have muted the antislavery messages in his scientific work, but other men of science were less inhibited. In the 1850s, with slavery increasingly under attack, men of science on both sides of the Atlantic began to promote technological innovations as a way to reduce reliance on slave-grown cotton. One of the most talked about inventions, and one that has received minimal scholarly attention, was a newly patented machine that promised to make flax into a cheap and humane alternative to cotton. The advantage of flax was that it grew in temperate climates, ones largely inhabited by white farmers, especially in the Midwest and Ireland. The problem was that raw flax was heavy to transport, and when turned into cloth, it was stiff, scratchy, and brown—far less desirable than soft, white cotton. But those problems seemed to be solved in 1851, when a Brazilian chemist, aided by a London chemistry professor, invented a flax-processing machine and pitched it as a solution to slavery. In no time, antislavery supporters in the United States, from moderate antislavery Republicans to radical abolitionists, began pointing to the invention as proof that slavery's days were numbered.

The flax machine was first unveiled in 1851 at the Great Exhibition in London. The exhibition was a major international event, covered widely in the press, and attracted the world's political and intellectual elites, thousands of common visitors, and even a few Black abolitionists. It was also a showcase for Britain's scientific, industrial—and moral—progress; several of the inventions on display were intended to demonstrate how modernity and morality went hand in hand. To no one's surprise, slavery hung over the entire affair. "The abolition of slavery," wrote London's *Morning Chronicle*, referring to slavery's recent demise in much of the Caribbean and Latin America, was the chief impetus behind many of the agricultural inventions on display. The flax machine's main inventor was Pierre Claussen, who had come to England by way of Brazil. According to one journal, he was "formerly an extensive cotton-grower and slave-owner in Brazil," though journalists more frequently cited his scientific credentials, all in an effort to bolster his invention's legitimacy. The *Morning Chronicle*, for instance, highlighted Claussen's work as a chemist with the Brazilian Institute, a scientific society devoted to practical inventions. It also noted that he was "well-known as the inventor of the circular loom," and that, as a botanist, he had donated numerous plants to natural history museums throughout Europe.[70]

Claussen's flax machine was relatively simple. First, a device mechanically separated the useless flax stalk from the valuable fiber; then, a boiling vat rinsed the fiber in a patented chemical solution. Despite its simplicity, or perhaps because of it, it was a major attraction at the exhibition: Queen Victoria and Prince Albert "inspected [the machine] with considerable attention," wrote the *Chronicle*, and it took home a prize for best new invention. Claussen marketed his machine to fit the temper of the times. By 1850 Britain had taken the lead in the international campaign against the

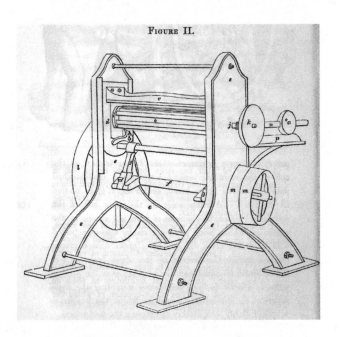

FIGURE II.

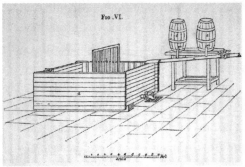

Fig.VI.

Fig. 5a. Claussen's flax machine, straw breaker, 1851.
Fig. 5b. Claussen's flax machine, chemical solution bath, 1851.

Pierre Claussen, a Brazilian chemist, debuted his flax machine at London's Great Exhibition in 1851. Claussen claimed his invention would turn flax into a cheaper alternative to slave-grown cotton and force planters to emancipate their enslaved laborers. Black and white abolitionists in the United States heralded Claussen's machine as the death knell of slavery. Fig. 5a depicts the machine that separated the raw flax straw from the valuable fiber. Fig. 5b shows the chemical solution bath that made the flax fiber as soft as cotton. From *The Preparation of Long-Line Flax-Cotton, and Flax-Wool, by the Claussen Processes* (1852) by Dr. John Ryan. Courtesy of Hathitrust.org / Getty Research Institute.

transatlantic slave trade, and many Britons took great pride in the fact. Between 1807 and 1850, Britain signed anti–slave trade treaties with more than a dozen nations, from Bolivia to Brazil. The treaty system drastically reduced the number of enslaved Africans shipped across the Atlantic: by the 1850s, the annual rate fell to 15,000 people per year, the lowest it had been since the early seventeenth century. Of course, the legislative successes coexisted with darker realities. Signatory nations often refused to enforce the ban, and evasion became routine. Americans were particularly notorious smugglers, helping transport 275,000 enslaved Africans to Brazil in the 1840s alone. Meanwhile, between 1838 and 1860, Britain imported 145,000 indentured servants ("coolies"), nearly half from British East India, to its Caribbean islands to replace enslaved laborers, and while not slavery, the indentured servant system bore many similarities to it.[71]

Even if slavery's slow death concealed new forms of coercive labor taking its place, abolitionists on both sides of the Atlantic—and shrewd marketers like Claussen—gladly hitched their sails to the antislavery winds. In an 1851 pamphlet aimed at investors in the United States and Britain, Claussen appealed directly to antislavery sympathies: "The abolition of slavery will be accelerated by the success of the Company," he wrote, referring to the business he formed to sell the technology. He cited two American newspapers that highlighted his invention's antislavery potential, adding that "the natural sagacity of the Americans is directed to this subject." For Britons there were additional benefits. Most important, it would help with the "employment of the poor" in Ireland, where many farmers had recently been devastated by the Great Famine of 1845–49. "When it is considered that British cotton is the produce of free instead of slave labour," Claussen wrote, there was no reason not to invest in his company.[72]

Claussen was careful to balance moral arguments with economic ones, but not all men of science were so cautious. John Ryan, a chemistry professor at the Royal Polytechnic Institute who seems to have had a financial stake in Claussen's company, unabashedly claimed that the flax machine would lead to the total "extinction of human slavery, with all its cruel and debasing horrors." When Ryan touted the machine's emancipatory potential during a lecture at the Polytechnic Institute in January 1852, the audience erupted in applause. Critics accused Ryan of gaming a righteous cause, but he countered that the flax machine had been studied by "most distinguished members of the scientific and manufacturing world," and all had given it their approval.[73]

Many men of science did in fact examine Claussen's machine, but they often distanced themselves from explicit political claims. James Thomas Way, the official chemist for the Royal Agricultural Society of England, published a lengthy chemical analysis of a finished flax fiber, but he made no mention of slavery; he alluded to it only indirectly, noting that the flax invention would decrease Britain's dependence "upon the United States for cotton." *Scientific American*, reporting from New York in March 1851, referred to the antislavery implications of Claussen's invention—it "causes so much talk in England and in this country"—but presented its analysis as purely scientific, apolitical, not mere "talk." Yet in the end it gave it the periodical's approval: it "is certainly a philosophic idea." The tempered tone of these reports did not mean that their authors were, in reality, disinterested. Quite the opposite: it was precisely because so many newspapers, politicians, and abolitionists highlighted the antislavery implications of Claussen's invention that they did not have to make their own politics explicit. By endorsing the technology, men of science and science journal editors could maintain an image of neutrality while tacitly promoting an antislavery agenda.[74]

Abolitionists of all kinds sensed an opportunity. One of the first abolitionist papers to endorse Claussen's machine was Frederick Douglass's *North Star* in January 1851. Ten months later, the editors of the *Anti-Slavery Reporter*, the journal of the British and Foreign Anti-Slavery Society (BFASS), appealed to cotton manufacturers to invest in "flax cotton, discovered by M. Claussen." The same article cited James Way's chemical report, as well as another scientific study proving that flax was not as "exhaustive" of the soil as some critics supposed. In 1855, after Claussen's invention proved a failure, the BFASS promoted another, improved flax machine that immediately appeared in its wake. "Chemical solvents are making of flax fibre a more than rival for the Mississippi fibre which underlies the Southern system of labour," it wrote, arguing that the technology spelled "THE DOOM OF THE COTTON-HAND."[75]

But few were as enthusiastic as Horace Greeley. The editor of the *New York Tribune*, Greeley became an outspoken antislavery Republican in the 1850s, and he had long been a champion of scientific agriculture. He hired Solon Robinson to write an agricultural column for the *Tribune* and sent a reporter to London's Great Exhibition to report on the new agricultural technologies, including Claussen's flax machine. Greeley was always quick to connect his antislavery politics to scientific agriculture, and nearly all his articles on the machine tied it to the antislavery cause. The fiber Claussen produced was "scarcely distinguishable from the Sea Island sample," a *Tribune* reporter wrote enthusiastically on June 23, 1851, referring to the South's main cotton fiber. On account of Claussen's invention, "slavery will lose one of its principal supports," the *Tribune* wrote later that summer. It also informed readers that a sample of "fabrics manufactured . . . by Claussen's process" could be "inspected at the Anti-Slavery office" on Beekman Street, in lower Manhattan. Greeley ultimately portrayed slaveholders as competing not only against abolitionists and politicians, but against science itself. As the *Tribune* put it: "Perhaps South-Carolina would secede from Chemistry as well as from the Union of Common Sense."[76]

Northern interest in flax technologies increased on the eve of, and during, the Civil War. In March 1860, Greeley's *Tribune* pointed to the latest U.S. Patent Office report that referred to a new flax machine, concluding that "FLAX [IS] TO SUPERCEDE COTTON" and that, since "every State can grow flax," it would "weaken the only prop which American Slavery possesses." The *New England Farmer*, Massachusetts' leading agricultural journal, sounded a similar note: "SCIENCE is king—not *cotton*," the editors wrote in January 1861, referring to yet another flax machine. By August 1862, the Civil War now raging, the *New England Farmer* wrote that "the invention of machinery for spinning flax satisfactorily and cheaply would at once depose and overwhelm King Cotton." Jonathan Baldwin Turner, in his *Three Great Races* lecture, was unambiguous in his praise of "QUEEN FLAX," as was Edward Everett, the Harvard president and founder of its Lawrence Scientific School. On September 12, 1861, Everett told farmers in upstate New York that growing flax would aid the Union effort. He noted that London's Association for the Advancement of Science had recently reported "the most promising experiments" on a different flax machine—yet more evidence that "the fibre of flax" could substitute "the tissues for which cotton has hitherto been employed."[77]

For abolitionists, one of the dangers of the flax machine was that many of its loudest champions did not have the same goals or strategies as they did. Greeley, for instance, stayed silent on the question of Black citizenship, and when the war began, he argued that the South should be allowed to leave the Union peacefully, even if it meant keeping slavery intact. He often denounced radical abolitionists, and he tended to speak of flax technologies as a substitute for, rather than a complement to, their moral and political program. In 1853 the *Tribune* argued that " 'moral suasion' has been found to have very little effect upon the slaveholder"; it added, "They will give up slavery only when they can't keep them any longer with positive and palpable loss." The paper went on to argue that, unlike activists or politicians, science cared not a whit about the fate of slavery, and it was therefore the "most legitimate and inoffensive" way of "acting against" it. The *Boston Journal*, also skeptical of abolitionists yet opposed to slavery, sounded a similar note. In October 1861 it wrote that the "genius [who] shall perfect machinery for the manufacture of flax into cotton at a low cost will do more for the abolition of slavery than a thousand Senators who confine themselves to making speeches against the institution."[78]

Abolitionists, Black and white, men and women, saw how Northerners framed new inventions like the flax machine as an alternative, rather than a complement, to their agenda. Yet with a few notable exceptions, abolitionists generally supported technological arguments anyway. They did so not because they naively believed science or technology could single-handedly solve the problem of slavery, but because technological solutions could aid their own political programs. Abolitionists' support for the flax machine was in part premised on the belief that it would help their larger boycott of slave-grown cotton. Since the 1790s abolitionists had been creating "free produce associations," which, like the modern fair-trade movement, bought only free-grown products and were usually organized by middle-class women. Confined to the home and responsible for household purchases, women found that these boycott organizations could provide them with a way to engage in political activism. In the 1830s, as cotton production soared, Black and white women abolitionists in the United States revived slave-grown boycotts. In 1836 the interracial Philadelphia Female Anti-Slavery Society (PFASS) reissued an essay by the British abolitionist Elizabeth Heyrick, which called for an earlier boycott against slave-grown sugar. A few years earlier, Lydia White, a white antislavery Quaker in Philadelphia, opened the nation's first free-produce store, and by the 1850s there were dozens more like them.[79]

Black men also supported boycotts. Alexander Crummell and Henry Highland Garnet promoted the free-produce movement in lectures throughout England in the early 1850s. In 1852 Jacob C. White Jr., a prominent abolitionist in Philadelphia, gave a rousing lecture to his largely Black audience urging them to embrace the free-produce movement. If more free-produce stores existed, he said, "I doubt that there could be found one colored person in fifty that would put themselves to the least trouble in order to patronize the store that sold the free produce." He was clear that boycotts alone would not cause "abolition of Slavery," but at the very least, he argued, they would guard against charges that free Black people were not acting "with consistency."[80]

Less radical, more mainstream white antislavery supporters were happy to associate flax with the free-produce movement—but only when it served their purposes. In 1853 Greeley's *Tribune* published a letter from a member of the American Free Produce Association applauding Greeley's "Agricultural experiment" with flax, but he did so even while positioning flax—and, by extension, science—as a more viable route to emancipation than what radical abolitionists had on offer. Yet it was a sign of abolitionists' pragmatism that they supported flax technologies anyway. The *National Anti-Slavery Standard*, the AASS's official newspaper, reprinted several *Tribune* articles about Claussen's machine. Even after Claussen's invention failed, the *Standard* continued reporting on subsequent flax experiments. In 1854 it reported that, although "the experiment of Claussen in making poor cotton out of good flax failed," another inventor, with just "a few chemicals," seemed to be succeeding. The *National Era*, the paper of the American and Foreign Anti-Slavery Society, also reprinted several Greeley articles on the "invention from Chevalier Claussen."[81]

Both the *National Era* and the *National Anti-Slavery Standard* were edited by white abolitionists. But Black-controlled abolitionist papers also gave flax machines their support. Douglass's *North Star* had been writing about flax machines well before Greeley. On January 16, 1851, it reported not only on Claussen's machine, but also on a similar new technology that combined "chemical and mechanical means" to spin flax on "cotton machinery." Douglass's subsequent periodical, *Frederick Douglass' Paper*, continued in a similar vein, reprinting several flax machine articles from Britain's *Anti-Slavery Reporter*. Douglass did not limit himself to American cotton either: he touted scientific solutions that might undermine slave-based agriculture in other slave societies, particularly Brazil, Cuba, and Puerto Rico. In 1854 he enthusiastically reported on a chemical paper written by Justus Liebig—a revered German chemist, whom Norton briefly studied under—which suggested that asparagus might be a substitute for coffee. "The illustrious German chemist" had found a strand of asparagus containing a chemical similar to caffeine, Douglass wrote; if free laborers grew it in the United States, it might bankrupt coffee planters throughout the Atlantic world and force them to emancipate their enslaved laborers.[82]

Black abolitionists outside the United States also supported technological solutions to slavery. Henry Bibb, who escaped his Kentucky plantation in 1842 and founded a free Black settlement in Canada, extolled Claussen's machine. In July 1851, writing in his Canada-based newspaper, *Voice of the Fugitive*, he quoted directly from "the pen of Mr. Greeley at the World's Fair"—that is, the Great Exhibition—and urged all formerly enslaved people in Canada to grow flax in light of the invention. "The fugitive slaves should be standing in the front ranks of this experiment," Bibb wrote, "which is destined to work out the emancipation of our chattelized race." Other Black leaders in northern states also encouraged flax cultivation on account of the promising new inventions. In 1853 the Ohio State Convention of Colored Freemen passed a resolution to promote flax cultivation among Black farmers: "*Resolved*," it read, "that to the colored farmers of this State be hereby suggested the propriety of considering the cultivation of flax—and of aiding, thereby, as much as in their power, the cotton flax movement." Black editors in Philadelphia urged free Black inventors to experiment with their own flax machines. "We advise our inventive readers to examine and see if they cannot produce improvements in flax dressing machinery," wrote the *Christian Recorder*, the newspaper of Philadelphia's African Methodist Episcopal (A.M.E.) Church, on September 5, 1853.[83]

The reasons Black and white abolitionists supported scientific solutions were often similar. Both groups believed that scientific solutions would exert economic pressure on slaveholders. Both also believed that scientific solutions might keep northern economic pragmatists on their side, ones who cared little for abolitionists but could be swayed by appeals to the nation's economic interests. But Black abolitionists also had reasons particular to their communities. For many Black leaders, supporting scientific solutions merged with a larger integrationist philosophy. Embracing science signaled a commitment to the nation's economic development and, by extension, to the nation itself. For this reason, Black integrationists, women as well as men, urged Black audiences to engage with scientific knowledge whenever they could. Maria W. Stewart, an early Black abolitionist and teacher, routinely implored Black audiences to master scientific subjects: "Where can we find among ourselves the man of science?" she asked a group of Bostonians in 1833. "Where are our lecturers in natural history, and our critics in useful knowledge?" In the 1850s the abolitionist and educator Sarah Mapps Douglass (no relation to Frederick) made botany, geography, and anatomy central to her curriculum at the Institute for Coloured Youth, a prominent Black school in Philadelphia. Because Black support for scientific education was often tied to a larger uplift strategy—premised on the notion that Black self-improvement would facilitate white acceptance—Black support for science represented a broader claim on citizenship.[84]

The racial barriers in Black people's way, however, made acquiring scientific knowledge difficult. In the North, free Black Americans were barred from scientific societies as they were from most civic institutions. In response, Black communities created their own intellectual societies, ones that made scientific topics central to their intellectual programs. The Banneker Institute, established in 1854 in Philadelphia, was representative. The institute held regular lectures open to paying members on topics ranging from mathematics to botany. In 1855, for instance, Sarah Mapps Douglass lectured on anatomy; four years later, Henry Black spoke about Linnaeus. Jacob C. White Jr., a mathematics teacher and cofounder of the institute, had long argued that attaining scientific knowledge would aid the larger fight for full citizenship. When, on May 24, 1855, he was invited to deliver a lecture at the Institute for Coloured Youth, on the occasion of a visit by the Pennsylvania governor, he highlighted the school's emphasis on practical scientific subjects, particularly "the mechanical arts and agriculture." Though free Black people were "not recognized in the political arrangements of the commonwealth," White said, the school would show that Black people were "nevertheless preparing ourselves for a future day when citizenship" would be possible.[85]

Black leaders supported the creation of their own industrial colleges for the same reason they promoted scientific education in general: they would provide Black laborers with the practical scientific skills they needed to make themselves valued citizens. In 1853 Charles L. Reason, a professor of mathematics at New York Central College and a revered teacher within the Black community, worked with Frederick Douglass to create "The Colored People's Industrial College," though lack of funds prevented it from materializing. Reason's desire for an industrial college was largely the same as Turner's—empowering common laborers—but with one key difference. Turner understood science to be an inherently white enterprise, a form of knowledge white people created themselves, and one that only white people could truly understand. By contrast, Reason saw science as racially inclusive. Everyone could and ought to engage with science, he wrote, "everyone man and everyone woman, equipped by its discipline to do good battle in the arena of active life." In an 1854 essay promoting the college, Reason argued that Black people would never truly be free until they attained the means to support themselves in an increasingly mechanized society—they would remain "in what Geologists call, the 'Transition State,'" no longer enslaved, but far from free. A practical scientific education would not only make Black people self-sufficient, he believed, but also show how they could be "to the assistance of the entire American people."[86]

Of course, not all Black Americans who embraced science were integrationists. Emigrationists—ideological adversaries of integrationists like Douglass—also believed that science could help liberate Black people, but it would not necessarily increase the respect they received from white Americans. Unlike integrationists, emigrationists believed that entrenched anti-Black racism made fighting for Black citizenship a fool's errand. In principle (if not always in practice), they opposed the white-led ACS and its Liberia settlement and hoped instead to create Black-controlled nations that would "regenerate their race," as Martin Delany, a leading emigration-ist, wrote in 1861. But they were no less committed to ending slavery, and they believed that independent Black communities could use their eco-nomic power to undermine slavery's profitability. To that end, emigration-ists championed the free-produce movement and saw science as buttressing its mission. In addition, Delany—who founded a short-lived Black settle-ment in modern-day Nigeria in 1859—argued that if free Black colonies in Africa adopted scientific agriculture to cultivate cotton and convinced Brit-ish manufacturers to buy from them rather than from American cotton planters, it would "furnish an irresistible motive to Emancipation."[87]

Science, in fact, suffused Delany's entire African emigration plan. He briefly attended Harvard Medical School in 1850, leaving only in the face of fierce white opposition. But he continued to engage in scientific activities, writing essays on ethnology and astronomy for the *Anglo-African Magazine* and attending the International Statistical Congress, a prestigious scientific conference in London, in 1860. Like Smeathman, he used his scientific rep-utation to bolster his claims of West Africa's natural fecundity—cotton "grows profusely in all this part of Africa," he wrote—offering it as proof that a free-labor African colony could easily outcompete slave plantations. In ad-dition, he recruited Black men of science exclusively to join him on his ex-pedition to West Africa, in large part to prove, publicly, that Black men were just as capable of conducting the kinds of scientific expeditions that white explorers had long been carrying out on Black people's behalf. Delany's published report on his 1859 expedition, *Official Report of the Niger Valley Exploring Party* (1861), highlighted the scientific backgrounds of the origi-nal all-Black group, including "Dr. Amos Aray, surgeon, a highly intelligent man," and "Dr. [James H.] Wilson . . . a physician, surgeon, and chemist." But limited funds meant that only one actually joined him: Robert Campbell, a Jamaican-born man educated at the Institute for Coloured Youth in Philadelphia, and "a very accomplished Chemist."[88]

Delany also worked closely with at least one white flax technology promoter whose antislavery politics contrasted sharply with his own: Benjamin Coates. As a member of the Pennsylvania Colonization Society, Coates was eager to help Delany promote Black emigration to Africa. Coates funded Delany's scientific expedition to the Niger River Valley and wrote that "this new discovery of the process of preparing flax" would help defeat slavery. Despite their political differences—Coates supporting Black emigration because he did not believe in Black citizenship, Delany because he did not believe Black citizenship was possible—they agreed on one thing: science could help bring down slavery.[89]

To be sure, not all radical abolitionists supported scientific solutions to slavery. The most prominent dissenter was William Lloyd Garrison. Before the war broke out, Garrison routinely denounced flax technologies, steam engines, and scientific solutions of any kind. To him, they were based on an economic understanding of slavery that contrasted sharply with his own. Slaveholders kept their captives not for financial reasons, he argued, but for psychological ones—they craved the power far more than the money. "It's not profit that makes slavery so devilish, but man's love of power," he wrote. Therefore, no amount of economic pressure, whether hastened by science or not, would convince slaveholders to relinquish their enslaved laborers. They needed a moral reckoning.[90]

In the *Liberator*'s issue of October 18, 1861, Garrison attacked the liberal *Boston Journal* for writing that the "genius" who discovered a flax substitute "will do more for the abolition of slavery than a thousand Senators." Nonsense, he replied, and in any event, "the humane [reason] is infinitely more important." Three years earlier, he had railed against the *Atlantic Monthly* for making a similar claim. The magazine's Boston-based editors opposed slavery, but they were wary of starting a war over it and tried to cool the national furor by arguing that technology would solve the problem: "With our prodigious development of mechanical invention," the *Atlantic* wrote, "the time for chattelizing men . . . has passed by." Garrison was all too happy to reprint a slaveholder's reply, which argued that new technologies would only make slavery more profitable: "If we do ever plow by steem [*sic*]," the slave owner wrote, "we'll turn our niggers into Pickers—make more cotton and sell it at less a price." Here Garrison was critiquing white antislavery moderates, but he aimed his fire at more radical Black abolitionists too, and for the same reason. Slaveholders craved "the possession of absolute power" more than money, he scolded Black abolitionists in 1850, after several of them had embraced the free-produce movement. Promoting flax technologies in particular would, he argued, simply make Northerners less reliant on slave-grown cotton and delude them into thinking they were untouched by slavery's "contamination."[91]

In one sense, Garrison was right: slavery was not just about money, it was about power. But he could make that argument because he had a luxury that Black abolitionists lacked: the luxury of moral purity. Many Black abolitionists had experienced the brutality of slavery firsthand, and they had family members still enslaved. Black abolitionists felt the daily ostracism that came with being Black in America, a stigma many of them believed would not lessen as long as slavery remained intact. And never in the nation's history had a political party been founded on a principled opposition to slavery. The Republican Party's antislavery platform may have been less bold than Black abolitionists hoped, but they saw it as a vehicle to push through a more radical agenda. Science functioned in the same way. Black abolitionists did not believe science was the solution to slavery, but many white Northerners did, or wanted to, and Black abolitionists needed their support. If it meant promoting scientific solutions to achieve their goals, so be it. Science was power, and with the power scientific arguments gave them, they would set about making America free.

Scientific arguments featured prominently in antebellum antislavery discourse, just as they had throughout the antislavery movement's long history. But as opposition to slavery became widespread in the North, antislavery scientific arguments increasingly served agendas that abolitionists themselves—a radical minority in the antebellum period—rejected. Mainstream white antislavery advocates turned polygenism, often thought to be a proslavery racial theory, into a racial science that could support gradual emancipation and colonization. Meanwhile, northern men of science depicted slavery as scientifically backward and presented technological solutions to slavery as a more peaceful alternative to radical abolitionist agitation. Radical abolitionists nonetheless embraced scientific and technological arguments. Sometimes their ethnological theories perpetuated dangerous scientific conceptions of race, but more important was their decision to engage with and challenge scientific racism rather than ignore it. Black and white abolitionists' support for technological solutions was similarly prudent. It represented neither a capitulation to political moderation nor a naive belief in technological solutionism, but instead a keen understanding that realizing their radical agenda required serious engagement with, not blithe disregard for, the ideas of the society they sought to change.

Conclusion

Sat
8/14/21

On December 6, 1865, with the ratification of the Thirteenth Amendment, slavery was finally abolished. The Thirteenth Amendment was an extraordinary document, not least because it ended slavery immediately and without owner compensation, an agenda only radical abolitionists had imagined possible before the war. Lincoln ran on an antislavery platform, but his moderation had always been his main appeal. Eighteen months into the war, Lincoln continued to keep emancipation off the table: "My paramount object in the struggle is to save the Union," Lincoln wrote to Horace Greeley on August 22, 1862. "If I could save the Union without freeing any slave I would do it, and if I could save it by freeing some and leaving others alone, I would also do that." Lincoln believed that immediate emancipation would alienate the border states, as well as many of his own supporters; he therefore preferred to quietly experiment with more modest antislavery measures. But not a single slave state entertained his offers, and free Black leaders scoffed at his proposal for emancipation in exchange for their support of colonization. Not until Lincoln issued the Emancipation Proclamation, on New Year's Day, 1863, did the war become an "abolition war," as Douglass called it. Even still, emancipation could not be put into effect until the Union Army regained control of the seceded states. As many Black abolitionists suspected, emancipation would come only through force. "Sooner or later, the clashing of arms will be heard in this country," said John Rock, the Black abolitionist physician, in 1858, "and the Black man's service will be needed."[1]

246

In recent years historians have rightly emphasized the importance of Black agency to emancipation. The fugitives who escaped to Union lines at the start of the war prompted Union generals, and ultimately Lincoln and Congress, to issue limited emancipation decrees as a threat to the South, hoping such actions would bring Confederate leaders to the bargaining table. After the Emancipation Proclamation, Black enlistees, many of them formerly enslaved, became essential foot soldiers in the Union Army, making up one-tenth of its enlistees. As emancipation became interwoven with the war's mission, more and more white Northerners, once skeptical of immediate emancipation, slowly began to accept it.[2]

But modern-day interpretations that highlight Black agency do not necessarily reflect the views of historical subjects. To many white people, both during and after slavery, emancipation came about not because of the agency of Black people, but because of the backwardness of slaveholders themselves. What this book has tried to show is how that perception came into being. To be sure, the image of slaveholders as backward, as the enemies of progress, had many sources, including a broader economic narrative that cast slavery as less efficient than wage labor. But another crucial, and connected, source was the scientific critiques that abolitionists used to discredit the institution. For at least a century before the Civil War, antislavery men of science and their abolitionist allies drew on reams of scientific evidence—medical, agricultural, botanical, chemical, geological, demographic—and proposed numerous technological solutions to slavery, all of which cast slavery as inimical to progress at odds with a world defined by science and technology. Slavery, abolitionists argued, prevented planters from adopting machine technologies that could easily replace slave labor. Physicians showed how slavery caused many of the diseases from which Black people suffered. Botanists, naturalists, and explorers extolled the natural environments of Sierra Leone, Java, and the expanding American frontier, suggesting that these regions would provide ideal alternatives to slave plantations.

Black people were excluded from early abolitionist and scientific societies, but they nonetheless found ways to promote a similar scientific narrative. Whether it was Benjamin Banneker's antislavery almanacs, Paul Cuffe's and John Kizell's botanical research and expeditions in Sierra

Leone, or Maria W. Stewart's, Sarah Mapps Douglass's, and Charles L. Reason's insistence that Black people embrace a scientific education, Black leaders saw science and technology as essential tools in their own struggle for freedom. Yet it was seldom easy: white antislavery elites rarely consulted them about the political ends to which their scientific ideas were being used, and their antislavery racial theories often assumed a degree of Black backwardness, something Black intellectuals were not immune to themselves. All too often, the technological solutions to slavery promoted by white antislavery elites—from steam engines that might replace slave labor to flax machines that might destroy cotton's profitability—functioned as a kind of anti-politics: a way to avoid political action. In moments of crisis, it was convenient to turn to science and technology to solve the problems that society lacked the will to fix.

Though this book has focused on how abolitionists used scientific ideas to cast slaveholders as the enemies of progress during the era of slavery, the image of slavery as scientifically backward hardly went away after emancipation. Almost as soon as the Civil War ended, northern men of science held up the North's victory as proof of the natural affinity between science and emancipation. In April 1865, William P. Atkinson, a professor at the newly founded Massachusetts Institute of Technology, argued that the spread of scientific knowledge in the North before the Civil War had made emancipation possible. By emphasizing fact over theory, evidence over dogma, a scientific worldview, he claimed, had enabled Northerners to see slavery for what it was: cruel, inefficient, backward. By contrast, Britain's neutrality during the war could be explained by its political class's being educated in a university system that "utterly ignores the great body of modern physical science." Only politicians who were averse to science could take "sides with slaveholding assassins in a war waged for republican freedom."[3]

Black abolitionists helped cement the image of slavery as scientifically backward. In Frederick Douglass's final memoir, published in 1881, he depicted the technological sophistication of New Bedford, Massachusetts, the city to which he escaped in 1838, as the mirror opposite of the slave-dependent, technology-starved South. Speaking of New Bedford, he wrote, "Here were sinks, drains, self-shutting gates, washing-machines, wringing-machines, and a hundred other contrivances for saving time and money." He concluded, "In a word, I found everything managed with much more scrupulous regard to economy . . . than in the country from which I had come." Commenting on his return to his former owner's plantation in 1881, he marveled at how the labor that was once completed by sixty enslaved people was now "by the aid of machinery . . . accomplished by ten men."[4]

In the twentieth century, scientists, government officials, and popular science writers updated this narrative with only minor alterations. In 1937 the federal government published a report, authored by a blue-ribbon committee of academic scientists, that studied the social implications of new technologies. The report noted that every epoch in human history had been characterized by "new inventions or discoveries, which increase his comfort, his safety, his intelligence, or his well-being," and went on to explain slavery's demise, in all societies, as a result of technological breakthroughs. "Human slavery has given place in modern civilized lands to mechanical power, which has placed a new valuation on human life." One year later, Silas Bent, a popular science writer, published a book titled *Slaves by the Billion*, which argued that new domestic appliances—refrigerators, gas stoves, washing machines—could emancipate "housewives" from the drudgery of domestic chores, just as inventions had "liberated from a condition of servitude" the enslaved of the past. In 1962 Gerald Piel, publisher of *Scientific American*, repeated the claim: "Slavery was abolished only when the biological energy of man was displaced by mechanical energy in the industrial revolution." He added, "Slavery became immoral when it became technologically obsolete."[5]

These depictions were not confided to scientists and a public beholden to an Enlightenment-era narrative that equated science with progress. Academics more skeptical of Whiggish interpretations also cast slavery and science in oppositional terms. In 1942 the Marxist sociologist Edgar Zilsel argued in a classic essay, titled "The Sociological Roots of Science," that the Scientific Revolution was the product of an emerging capitalist society. As class barriers separating artisans from scholars broke down in early modern Europe, knowledge derived from manual labor—from experimentation—grew in prestige, and the idea of modern science was born. In the same essay,

Zilsel contrasted early modern Europe with the Greco-Roman classical world. The reason the classical world never had a scientific revolution, he argued, was that "machinery and science cannot develop in a civilization based on slave labor." The explanations he offered—that slave labor devalued manual labor; that the availability of slave labor discouraged mechanical invention; that slaves, being unskilled, were unlikely to develop scientific or technological knowledge—could have been written by Joseph Priestley himself.[6]

Perhaps the most prominent modern historian to depict slavery and science as diametric opposites was Eugene Genovese. In his influential studies of slavery, written in the 1960s and early 1970s, Genovese cast slavery as a fundamentally retrograde, antimodern, precapitalistic institution. As such, he argued that scientific knowledge and modern technologies—like the labor-saving plow—had only limited appeal in the South, and he cited Zilsel's work as proof that societies based on slave labor were inhospitable to science. "Negro slavery retarded technological progress," Genovese wrote in *The Political Economy of Slavery* (1965), adding that slavery "encouraged ways of thinking antithetical to the spirit of modern science." Like Zilsel, Genovese was writing from a Marxist perspective. In describing the South's slave-based economy as backward, he did not mean to praise northern capitalism or the scientific and technological advances he believed capitalism made possible. Indeed, the long afterlife of slavery's image as incompatible with science and technology might in part be explained by its wide ideological appeal. Intellectuals who saw science, capitalism, and freedom as of a piece and worth celebrating have been just as likely to depict slavery as scientifically backward as Marxist scholars more critical of capitalism.[7]

Of course, the notion that slavery thwarted scientific progress has been crumbling in the decades since Genovese's work appeared. Since at least the 1970s, scholarship demonstrating how slavery shaped racial science has become so rich that historians and the public alike sometimes assume that proslavery ideologues had a monopoly on scientific ideas. More recent scholarship highlighting slavery's importance to the production of other forms of scientific and medical knowledge, coupled with a growing literature highlighting the technological sophistication of many plantations, threatens to obscure our appreciation of just how well the enemies of slave-holders—the abolitionists—seized the mantle of science for themselves. Even scholars of the antislavery movement have largely left the study of science, medicine, and technology to their slavery-focused colleagues. Whether this results from a belief that racial science was the only realm of science that mattered in the debates over slavery—and therefore that slaveholders "won" the scientific debate—or from a wariness of reviving outdated interpretations that cast abolitionists as enlightened heroes and paragons of modernity, the effect is the same: we simply lack an appreciation of the crucial role that scientific knowledge, in its broadest sense, played in undermining slave regimes.

This study claims to be neither a comprehensive study of all the forms of scientific knowledge abolitionists and their scientific allies mobilized, nor a direct challenge to the recent scholarship showing slavery to be in fact deeply modern—that is, inextricably tied to capitalism, science, medicine, and technology. Rather, it offers a counterpoint to this scholarship, arguing that from the moment the organized abolitionist movement took shape, slaveholders were depicted as antimodern—the enemies of progress—and in no small part because abolitionists successfully cast slaveholders and their institution as scientifically backward. The scientific ideas antislavery men of science and abolitionists drew on were necessarily wide-ranging—geological, medical, technological, chemical, astronomical—in part because, at the time, the disciplinary boundaries of science were more fluid, and in part because focusing on a narrow scientific discipline fails to capture the broader picture abolitionists were painting. Indeed, it has been the singular focus on racial science—on questions of biological difference—that has led scholars to miss this larger story. Abolitionists were most successful in claiming the scientific mantle for themselves not when they attacked anti-Black racial science—though they did that plenty—but when they drew on all the forms of scientific knowledge at their disposal. Unlike popular perceptions of abolitionists today, few at the time would have seen them as making only a moral argument: they were also making a *scientific* one. And it worked.

The success of their scientific narrative—their science of abolition—should also force us to rethink some basic assumptions within the history of science, medicine, and technology today. For the past few decades, scholars have rightly rejected histories that cast scientific knowledge as an unabashed source of human progress, and they have instead highlighted the many ways Western science has sanctioned oppressive social hierarchies. While this study does not attempt to overturn this newer narrative—indeed, it is deeply informed by it—it does hope to offer a generative contrast. The story of the science of abolition suggests that scientific knowledge knowledge did not only bolster oppression, but also helped combat it.

To be sure, this book is not calling for a return to old Whiggish narratives of science. The history told here has been at pains to show how too much faith in science could make large swaths of society susceptible to the fiction that scientific and technological progress would inevitably destroy slavery. In this regard, Black abolitionists offer useful insights. Having experienced science as a source of both oppression and liberation, they were particularly well positioned to understand science's power—and its limitations. They were under no illusions that science was impervious to human influence: "Scientific writers, not less than others," often "sacrifice what is true to what is popular," Douglass wrote in 1854.[8] Nor did they see slavery or the racism that underwrote it as a problem that science and technology could fix—science might offer useful tools in their struggle for emancipation, but it was no substitute for political action. Yet rather than allow the abuses of scientific knowledge to curdle into cynicism, Black abolitionists realized the need to engage with scientific ideas and shape them to their own ends. In doing so, they showed us not only how scientific knowledge could be used to uphold systems of oppression, but also how it could be transformed into a source of liberation.

Forensic Medicine used To oppress & liberate

Introduction

1. See Jeffrey Auerbach, *The Great Exhibition of 1851: A Nation on Display* (New Haven: Yale University Press, 1999), 1, 91–127. For the role of men of science on the organizing committee, see ibid., 32, 70–75. For the percentage of cotton imported, see Sven Beckert, *Empire of Cotton: A Global History* (New York: Alfred A. Knopf, 2014), 243. For the U.S. enslaved population, see *1860 Census: Population of the United States* (1864), ix.

2. "Chevalier Claussen's Preparation of Flax," *Morning Chronicle* (London), Jan. 19, 1852, 7; "Claussen's Improvements in the Preparation and Bleaching of Flax," *Scientific American*, March 29, 1851, 220; Henry Bibb, "Flax Cotton," *Voice of the Fugitive*, July 30, 1851, n.p.; "Flax Cotton," *North Star*, Jan. 16, 1851, 2; "Science Is King," *New England Farmer* 13, no. 1 (Jan. 1861): 24.

3. Mr. Whitmore, "Proceedings of Second General Meeting of the Society," *Anti-Slavery Monthly Reporter* (London) 1, no. 1 (June 1825): 6.

4. A note on terminology: the words *abolitionist* and *antislavery* can carry different meanings to scholars in Britain and the United States, and those meanings are further complicated when focusing only on the antebellum United States. In this book I use both terms interchangeably in the first five chapters, which cover the period before the 1830s. In these chapters, both terms refer to anyone who either joined an antislavery society, or whose political views clearly aligned with the ideology of those societies. Only in the final chapter, on antebellum American abolitionism, do I deliberately use the terms differently. Following other scholars, I reserve *abolitionist* for members, Black or white, of the era's antislavery societies. By contrast, I use *antislavery* when referring to more moderate opponents of slavery in the antebellum era, which includes most of the white men of science in this chapter. When writing in broad terms about the entire antislavery movement, in both

centuries and in both the United States and Britain, I generally use the word *anti-slavery*, which is meant to encompass both radical and more cautious antislavery agendas.

5. For the social identity of the man of science in this general period, see Steven Shapin, "Image of the Man of Science," *Cambridge History of Science*, vol. 4, *Eighteenth-Century Science*, ed. Roy Porter (New York: Cambridge University Press, 2003), 159–83; Paul Lucier, "The Professional and the Scientist in Nineteenth-Century America," *Isis* 100, no. 4 (2009): 699–732; Jan Golinski, *The Experimental Self: Humphry Davy and the Making of a Man of Science* (Chicago: University of Chicago Press, 2016), 3–9.

6. For the institutional structures supporting the sciences in the eighteenth century, see James E. McClellan III, *Science Reorganized: Scientific Societies in the Eighteenth Century* (New York: Columbia University Press, 1985). For the institutionalization of the sciences in the early United States, see George H. Daniels, *American Science in the Age of Jackson* (New York: Columbia University Press, 1968), 6–33; John C. Greene, *American Science in the Age of Jefferson* (Ames: Iowa State University Press, 1984). For the development of scientific disciplines in general, see James Chandler, "Introduction: Doctrines, Disciplines, Discourses, and Departments," *Critical Inquiry* 35 (2009): 729–49. For the "second scientific revolution," see Thomas Kuhn, *The Essential Tension: Selected Studies in Scientific Tradition and Change* (Chicago: University of Chicago Press, 1977), 147, 218–20.

7. For a useful definition of scientific knowledge and production, see Peter Dear, *The Intelligibility of Nature: How Science Makes Sense of the World* (Chicago: University of Chicago Press, 2006), 1–14; Jan Golinski, *Making Natural Knowledge: Constructivism and the History of Science* (Chicago: University of Chicago Press, 1998).

8. For challenges to elite scientific authorities in the early American context, see Andrew Lewis, *A Democracy of Facts: Natural History in the Early Republic* (Philadelphia: University of Pennsylvania Press, 2011). For examples of nonelites participating in scientific knowledge production in the American context, see Cameron Strang, *Frontiers of Science: Imperialism and Natural Knowledge in the Gulf South Borderlands, 1500–1850* (Chapel Hill: University of North Carolina Press, 2018); Susan Scott Parish, *American Curiosity: Cultures of Natural History in the Colonial British Atlantic World* (Chapel Hill: University of North Carolina Press, 2006); James Delbourgo, *A Most Amazing Scene of Wonders: Electricity and Enlightenment in Early America* (Cambridge: Harvard University Press, 2006). For the British context, see Deborah E. Harkness, *The Jewel House: Elizabethan London and the Scientific Revolution* (New Haven: Yale University Press, 2007); Londa Schiebinger, *The Mind Has No Sex? Women in the Origins of Modern Science* (Cambridge: Harvard University Press, 1989); Jan Golinski, *Science as Public Culture: Chemistry and Enlightenment in Britain, 1760–1820* (New York: Cambridge University Press, 1992).

9. For recent scholarship on how Western medicine exploited Black people during slavery, see Deirdre Cooper Owens, *Medical Bondage: Race, Gender, and the Origins of American Gynecology* (Athens: University of Georgia Press, 2017); Rana Hogarth, *Medicalizing Blackness: Making Racial Differences in the Atlantic World, 1780–1840* (Chapel Hill: University of North Carolina Press, 2017); Christopher Willoughby, " 'His Native, Hot Country': Racial Science and Environment in Antebellum American Medical Thought," *Journal of the History of Medicine and Allied Sciences* 72, no. 3 (July 2017): 328–51; Peter McCandless, *Slavery, Disease and Suffering in the Southern Lowcountry* (New York: Cambridge University Press, 2011). For recent work on scientific racism, see Suman Seth, *Difference and Disease: Medicine, Race and the Eighteenth-Century British Empire* (Cambridge: Cambridge University Press, 2018); Andrew S. Curran, *The Anatomy of Blackness: Science & Slavery in an Age of Enlightenment* (Baltimore: Johns Hopkins University Press, 2011); Bruce Dain, *A Hideous Monster of the Mind: American Race Theory in the Early Republic* (Cambridge: Harvard University Press, 2002); Ann Fabian, *The Skull Collectors: Race, Science, and America's Unburied Dead* (Chicago: University of Chicago Press, 2010). For the scientific cultures of enslaved people, see Sharla M. Fett, *Working Cures: Healing, Health, and Power on Southern Slave Plantations* (Chapel Hill: University of North Carolina Press, 2002); Susan Scott Parrish, *American Curiosity: Cultures of Natural History in the Colonial British Atlantic World* (Chapel Hill: University of North Carolina Press, 2006), 259–305; Kathleen Murphy, "Translating the Vernacular: Indigenous and African Knowledge in the Eighteenth-Century British Atlantic," *Atlantic Studies* 8, no. 1 (2011): 29–48; Londa Schiebinger, *Secret Cures of Slaves: People, Plants, and Medicine in the Eighteenth-Century Atlantic World* (Palo Alto: Stanford University Press, 2017); Londa Schiebinger, *Plants and Empire: Colonial Bioprospecting in the Atlantic World* (Cambridge: Harvard University Press, 2004); Karol K. Weaver, *Medical Revolutionaries: The Enslaved Healers of Eighteenth-Century Saint Domingue* (Urbana: University of Illinois Press, 2006); Jerome Handler, "Slave Medicine and Obeah in Barbados, circa 1650 to 1834," *New West Indian Guide* 74, no. 1/2 (2000): 57–90; Sasha Turner, *Contested Bodies: Pregnancy, Childrearing, and Slavery in Jamaica* (Philadelphia: University of Pennsylvania Press, 2017); Katherine Paugh, *The Politics of Reproduction: Race, Medicine, and Fertility in the Age of Abolition* (New York: Oxford University Press, 2017); Judith Carney, *Black Rice: The African Origins of Rice Cultivation in the Americas* (Cambridge: Harvard University Press, 2001); Neil Safier, *Measuring the New World: Enlightenment Science and South America* (Chicago: University of Chicago Press, 2008), 9, 59–64, 268–72. For wholesale alternative epistemological traditions, see James H. Sweet, *Domingos Álvares, African Healing, and the Intellectual History of the Atlantic World* (Chapel Hill: University of North Carolina Press, 2011); Pablo Gomez, *The Experiential Caribbean: Creating Knowledge and Healing in the Early Modern Atlantic* (Chapel Hill:

University of North Carolina Press, 2017); Britt Rusert, *Fugitive Science: Empiricism and Freedom in Early African American Culture* (New York: New York University Press, 2017); Carney, *Black Rice*; Fett, *Working Cures.*

10. Frederick Douglass, *The Claims of the Negro, Ethnologically Considered* (Rochester, N.Y.: Lee, Mann, 1854), 9. For Black abolitionists' use of violence in the U.S. context, see Kellie Carter Jackson, *Force and Freedom: Black Abolitionists and the Politics of Violence* (Philadelphia: University of Pennsylvania Press, 2019). For differences between Black and white abolitionists, see Richard J. M. Blackett, *Building an Antislavery Wall: Black Abolitionists in the Atlantic Abolitionist Movement, 1830–1860* (Baton Rouge: Louisiana State University Press, 1983); Shirley J. Yee, *Black Women Abolitionists: A Study in Activism, 1828–1860* (Knoxville: University of Tennessee Press, 1992); Benjamin Quarles, *Black Abolitionists* (New York: Oxford University Press, 1969).

11. Robin Blackburn, *The American Crucible: Slavery, Emancipation and Human Rights* (New York: Verso, 2011), 146. For abolitionist scholarship emphasizing scientific racism, see Manisha Sinha, *The Slave's Cause: A History of Abolition* (New Haven: Yale University Press, 2016), 35; Patrick Rael, *Eighty-Eight Years: The Long Death of Slavery in the United States, 1777–1865* (Athens: University of Georgia Press, 2015), 19; David Brion Davis, *The Problem of Slavery in the Age of Emancipation* (New York: Alfred A. Knopf, 2014), 33–34; Blackburn, *American Crucible*, 145–50. The notable exception of a study to look beyond racial science is Seymour Drescher's *The Mighty Experiment: Free Labor versus Slavery in British Emancipation* (New York: Oxford University Press, 2002). Drescher focused on the antecedents to modern social sciences, particularly economic and demographic arguments, virtually ignoring the natural sciences—chemistry, botany, geology, natural history—as well as technology, which are central to this book.

12. For slavery's role in shaping early modern science, see Kathleen S. Murphy, "Collecting Slave Traders: James Petiver, Natural History, and the British Slave Trade," *William and Mary Quarterly* 70, no. 4 (Oct. 2013): 637–70; David Lambert, *Mastering the Niger: James MacQueen's African Geography and the Struggle over Atlantic Slavery* (Chicago: University of Chicago Press, 2013); James Delbourgo, *Collecting the World: Hans Sloane and the Origins of the British Museum* (Cambridge: Harvard University Press, 2017), 97–98, 101, 122, 203; Susan Scott Parrish, "Diasporic African Sources of Enlightenment Knowledge," in *Science and Empire in the Atlantic World*, ed. James Delbourgo and Nicholas Dew (London: Routledge, 2008), 281–310; Deirdre Coleman, *Henry Smeathman, the Flycatcher: Natural History, Slavery and Empire in the Late Eighteenth Century* (Liverpool: Liverpool University Press, 2018); Londa Schiebinger, "Prospecting for Drugs: European Naturalists in the West Indies," in *Colonial Botany: Science, Commerce, and Politics in the Early Modern World*, ed. Londa Schiebinger and Claudia Swan (Philadelphia: University of Pennsylvania Press, 2005), 119–33. For slavery's role in shaping early

modern medicine, see Cooper Owens, *Medical Bondage*; Turner, *Contested Bodies*; Hogarth, *Medicalizing Blackness*. For technology, see Daniel Rood, *The Reinvention of Atlantic Slavery: Technology, Labor, Race, and Capitalism in the Greater Caribbean* (New York: Oxford University Press, 2017); Joyce Chaplin, *An Anxious Pursuit: Agricultural Innovation and Modernity in the Lower South, 1730–1815* (Chapel Hill: University of North Carolina Press, 1993); Angela Lakwete, *Inventing the Cotton Gin: Machine and Myth in Antebellum America* (Baltimore: Johns Hopkins University Press, 2003); Robert H. Gudmestad, *Steamboats and the Rise of the Cotton Kingdom* (Baton Rouge: Louisiana State University Press, 2011).

13. The literature on slavery and capitalism has a long history, beginning with Eric Williams's *Capitalism and Slavery* (Chapel Hill: University of North Carolina, 1944). For more recent scholarship, see Caitlin Rosenthal, *Accounting for Slavery: Masters and Management* (Cambridge: Harvard University Press, 2018); Sven Beckert and Seth Rockman, eds., *Slavery's Capitalism: A New History of American Economic Development* (Philadelphia: University of Pennsylvania Press, 2016); Edward Baptist, *The Half Has Never Been Told* (New York: Basic Books, 2014); Daina Ramey Berry, *The Price for Their Pound of Flesh: The Value of the Enslaved, from Womb to Grave, in the Building of a Nation* (Boston: Beacon, 2017); Walter Johnson, *River of Dark Dreams: Slavery and Empire in the Cotton Kingdom* (Cambridge: Harvard University Press, 2013); Calvin Schermerhorn, *The Business of Slavery and the Rise of American Capitalism, 1815–1860* (New Haven: Yale University Press, 2015). For Black studies scholarship underscoring modernity's ties to slavery, see Paul Gilroy, *The Black Atlantic: Modernity and Double-Consciousness* (Cambridge: Harvard University Press, 1993); David Scott, *Conscripts of Modernity: The Tragedy of Colonial Enlightenment* (Durham: Duke University Press, 2004). For an early challenge to Eric Williams's second claim—that the unprofitability of British slavery eased the way for abolitionists—see Seymour Drescher, *Econocide: British Slavery in the Era of Abolition* (Pittsburgh: University of Pittsburgh Press, 1977).

14. The older depiction of slaveholders as backward, or premodern, can be found in the scholarship of Eugene Genovese, among many others of his generation. Genovese also details midcentury planter critiques of wage-based labor and capitalism. See, for example, Eugene Genovese, *The Political Economy of Slavery: Studies in the Economy & Society of the Slave South* (1965; repr., Middletown, Conn.: Wesleyan University Press, 1989). For the four-stage theory of development, known as conjectural history and promoted by Scottish Enlightenment thinkers, see Chaplin, *An Anxious Pursuit*, 26–37. In *Anxious Pursuit*, Chaplin also argues that southern planters—long before the 1850s—depicted slavery as a paragon of modernity. For a similar depiction of British slaveholders as Enlightenment moderns, see Justin Roberts, *Slavery and the Enlightenment in the British Atlantic, 1750–1807* (New York: Cambridge University Press, 2013).

15. See Margaret H. Bacon, appendix to *History of the Pennsylvania Society for Promoting the Abolition of Slavery: The Relief of Negroes Unlawfully Held in Bondage; and for Improving the Condition of the African Race* (Philadelphia: Pennsylvania Abolition Society, 1959), iii.

16. For a small sample of studies of abolitionism, whether in an American, British, or Atlantic world context, see Sinha, *Slave's Cause*; Rael, *Eighty-Eight Years*; Christopher L. Brown, *Moral Capital: Foundations of British Abolitionism* (Chapel Hill: University of North Carolina Press, 2006); Seymour Drescher, *Abolition: A History of Slavery and Antislavery* (New York: Cambridge University Press, 2009); Richard Newman, *The Transformation of American Abolitionism: Fighting Slavery in the Early Republic* (Chapel Hill: University of North Carolina Press, 2002); Paul Polgar, *Standard-Bearers of Equality: America's First Abolition Movement* (Chapel Hill: University of North Carolina Press, 2019); David Brion Davis, *The Problem of Slavery in the Age of Revolution, 1770–1823* (Ithaca: Cornell University Press, 1975); Davis, *The Problem of Slavery in the Age of Emancipation*; David Brion Davis, *Inhuman Bondage: The Rise and Fall of Slavery in the New World* (New York: Oxford University Press, 2006), 141–74, 205–321; Blackett, *Building an Antislavery Wall*; John R. Oldfield, *Transatlantic Abolitionism in the Age of Revolution: An International History of Anti-Slavery, c. 1787–1820* (Cambridge: Cambridge University Press, 2013); Blackburn, *American Crucible*; John L. Brooke, *"There Is a North": Fugitive Slaves, Political Crisis, and Cultural Transformation in the Coming of the Civil War* (Amherst: University of Massachusetts Press, 2019). For Somerset, see Drescher, *Abolition*, 99–102; for the PAS's founding, see Newman, *Transformation of American Abolitionism*, 16–38; Sinha, *Slave's Cause*, 73–76. For the SEAST's founding, see Davis, *Inhuman Bondage*, 234–45.

17. For statistics, see Blackburn, *American Crucible*, x (table 1).

18. For scholarship on technology fueling slavery's expansion, see Rood, *Reinvention of Atlantic Slavery*; Maria Portuondo, "Plantation Factories: Science and Technology in Late-Eighteenth-Century Cuba," *Technology and Culture* 44, no. 2 (April 2003): 231–57. For Georgia's and South Carolina's population increases, see Beckert, *Empire of Cotton*, chap. 5. For steamboat numbers, see Johnson, *River of Dark Dreams*, 6.

19. For studies on amelioration, see Christa Dierksheide, *Amelioration and Empire: Progress and Slavery in the Plantation Americas* (Charlottesville: University of Virginia Press, 2014); J. R. Ward, *British West Indian Slavery, 1750–1834: The Process of Amelioration* (Oxford: Oxford University Press, 1988). For colonization, see Eric Burin, *Slavery and the Peculiar Solution: A History of the American Colonization Society* (Gainesville: University Press of Florida, 2005); Beverly Tomek, *Colonization and Its Discontents: Emancipation, Emigration, and Antislavery in Antebellum Pennsylvania* (New York: New York University Press, 2010); Beverly Tomek and Matthew Hetrick, eds., *New Directions in the Study of African Colonization* (Gainesville: University Press of Florida, 2017); Nicholas Guyatt, " 'The Outskirts of

Our Happiness': Race and the Lure of Colonization in the Early Republic," *Journal of American History* 95, no. 4 (March 2009): 986–1011. For specific examples of the denial of free Blacks' rights, see Rael, *Eighty-Eight Years*, 136.

20. For the term *second wave* see Sinha, *Slave's Cause*, 195–227. For other works detailing the second-wave, or "immediatist," era of abolitionism in the United States, see Newman, *Transformation of American Abolitionism*, 107–75; Rael, *Eighty-Eight Years*, 163–235; Edward Rugemer, *The Problem of Emancipation: The Caribbean Roots of the American Civil War* (Baton Rouge: Louisiana State University Press, 2008); Blackett, *Building an Antislavery Wall*. For British emancipation, see Drescher, *Mighty Experiment*.

1 Stars and Stripes

1. Benjamin Rush, "Observations Intended to Favour a Supposition That the Black Color (As It Is Called) of the Negroes Is Derived from the Leprosy," *Transactions of the American Philosophical Society* 4 (1799): 297. The lecture was published in the American Philosophical Society's main publication, *Transactions*, in 1799, but Rush delivered it on July 14, 1792.

2. For more examples, including a discussion of Rush, see Joanne Pope Melish, "Emancipation and the Embodiment of 'Race': The Strange Case of the White Negro and the Algerian Slaves," in *A Centre of Wonders*, ed. Janet Moore Lindman and Michele Lise Tarter (Ithaca: Cornell University Press, 2001), 223–36; Kariann A. Yokota, "Not Written in Black and White: American National Identity and the Curious Color Transformation of Henry Moss," *Common-place* 4 (2004); Kariann A. Yokota, *Unbecoming British: How Revolutionary America Became a Postcolonial Nation* (New York: Oxford University Press, 2010), 215–17; Charles D. Martin, *The White African American Body: A Cultural and Literary Exploration* (New Brunswick: Rutgers University Press, 2002), 19–48; Rana Hogarth, "The Strange Case of Hannah West: Skin Colour and the Search for Racial Difference," *Social History of Medicine* 29, no. 3 (Aug. 2016): 557–72; Winthrop Jordan, *White over Black: American Attitudes toward the Negro, 1550–1812* (Chapel Hill: University of North Carolina Press, 1968), 512–28.

3. Benjamin Rush, "Diseases of Negroes" (n.d.), Rush Family Papers (hereafter cited as RFP), Library Company of Philadelphia, subseries VII, vol. 125.

4. The presidents of the Pennsylvania Abolition Society during this time span were Benjamin Franklin (1787–90), James Pemberton (1790–1803), Benjamin Rush (1803–13), and Caspar Wistar (1813–18). All but Pemberton were men of science. See Margaret H. Bacon, appendix to *History of the Pennsylvania Society for Promoting the Abolition of Slavery: The Relief of Negroes Unlawfully Held in Bondage; and for Improving the Condition of the African Race* (Philadelphia: Pennsylvania Abolition Society, 1959), iii. For the movement's early Quaker leaders,

see Jean Soderlund, *Quakers & Slavery: A Divided Spirit* (Princeton: Princeton University Press, 1985); David Brion Davis, *The Problem of Slavery in the Age of Revolution, 1770–1823* (Ithaca: Cornell University Press, 1975), 213–54; Patrick Rael, *Eighty-Eight Years: The Long Death of Slavery in the United States, 1777–1865* (Athens: University of Georgia Press, 2015), 46–47; Manisha Sinha, *The Slave's Cause: A History of Abolition* (New Haven: Yale University Press, 2016), 12–14, 18–21.

5. Gary Nash and Jean Soderlund, *Freedom by Degrees: Emancipation in Pennsylvania and Its Aftermath* (New York: Oxford University Press, 1991), 89.

6. "To the Honourable Counsel & House," Jan. 13, 1777, in Herbert Aptheker, ed., *A Documentary History of the Negro People of the United States*, 2 vols. (New York: Citadel, 1951), 1:10.

7. Davis, *Problem of Slavery in the Age of Revolution*, 23–24.

8. Alan Taylor, *The Internal Enemy: Slavery and War in Virginia, 1772–1832* (New York: W. W. Norton, 2014).

9. For the Copley Medal, see Joyce Chaplin, *The First Scientific American: Benjamin Franklin and the Pursuit of Genius* (New York: Basic Books, 2006), 116; for Franklin's political and scientific circles in London, see James Delbourgo, *A Most Amazing Scene of Wonders: Electricity and Enlightenment in Early America* (Cambridge: Harvard University Press, 2006), 59–60. For Franklin's London years in general, see Carla Mulford, *Franklin and the Ends of Empire* (New York: Oxford University Press, 2015), 184–273; George Goodwin, *Benjamin Franklin in London: The British Life of America's Founding Father* (New Haven: Yale University Press, 2016). For Franklin's slave owning, see David Waldstreicher, *Runaway America: Benjamin Franklin, Slavery, and the American Revolution* (New York: Hill and Wang, 2004), 25–26; Claude-Anne Lopez and Eugenia W. Herbert, *The Private Franklin* (New York: W. W. Norton, 1975), 296–307. For his will, see Benjamin Franklin, "Will and Codicil," July 17, 1788 (unpublished), in *The Papers of Benjamin Franklin* online, franklinpapers.org (accessed June 11, 2020) (hereafter cited as *PBF* online).

10. For revenue from slave advertisements, see Waldstreicher, *Runaway America*, 7, 19, 24–25.

11. Ibid., 83. Historians have traditionally argued that Franklin became more genuinely invested in antislavery from the 1770s onward. See, for instance, Gary Nash, "Franklin and Slavery," *Proceedings of the American Philosophical Society* 150, no. 4 (Dec. 2006): 618–35; Gordon S. Wood, *The Americanization of Benjamin Franklin* (New York: Penguin, 2004), 226–29. Waldstreicher counters this narrative in *Runaway America*. For a similarly skeptical view, see Emma Lapsanky-Werner, "At the End, an Abolitionist?" in *Benjamin Franklin: In Search of a Better World*, ed. Page Talbott, Richard Dunn, and John C. Van Horne (New Haven: Yale University Press, 2005), 273–97. My work here mostly aligns with Waldstreicher's view.

12. Benjamin Franklin to John Wright, Nov. 4, 1789 (unpublished), *PBF* online (accessed June 11, 2020).

13. Benjamin Lay, *All Slave-Keepers That Keep the Innocent in Bondage* (Philadelphia, 1737), 18, 56–57. For early Quaker objections to slavery, see David Brion Davis, *The Problem of Slavery in Western Culture* (Ithaca: Cornell University Press, 1966), 291–332; Nash and Soderlund, *Freedom by Degrees*, 41–51; Soderlund, *Quakers & Slavery*. For Franklin not necessarily agreeing with Sandiford and Lay, see Waldstreicher, *Runaway America*, 79–82.

14. Benjamin Franklin, *Benjamin Franklin's Autobiography*, ed. Joyce Chaplin (New York: W. W. Norton, 2011), 109–10; Franklin to Joseph Priestley, Sept. 19, 1772, *Papers of Benjamin Franklin*, 41 vols. (New Haven: Yale University Press, 1954–) (hereafter cited as *PBF*), 19:299. For Franklin's religious views, see John Fea, "Benjamin Franklin and Religion," in *A Companion to Benjamin Franklin*, ed. David Waldstreicher (Malden, Mass.: Wiley-Blackwell, 2011), 129–45. For probability, not certainty, as the Enlightenment science ideal, see Lorraine Daston, *Classical Probability in the Enlightenment* (Princeton: Princeton University Press, 1988).

15. Benjamin Franklin, "Observations concerning the Increase of Mankind," in *PBF*, 4:225–34. Here I draw on the work of Waldstreicher and Paul Conner, both of whom argue that Franklin's racism was, more than anything, strategic and political, rather than personal. See Waldstreicher, *Runaway America*, 139, 271n36; Paul W. Conner, *Poor Richard's Politicks: Benjamin Franklin and the New American Order* (New York: Oxford University Press, 1965), 69–79. For useful discussions of "Observations," see Edmund Morgan, *Benjamin Franklin* (New Haven: Yale University Press, 2002), 75–80; Chaplin, *First Scientific American*, 173–74, 351–52.

16. Deborah Franklin to Benjamin Franklin, Aug. 9, 1759, in *PBF*, 8:425. For Deborah Read Franklin's role, see Jennifer Reed Fry, " 'Extraordinary Freedom and Great Humility': A Reinterpretation of Deborah Franklin," *Pennsylvania Magazine of History and Biography* 127, no. 2 (April 2003): 188; Nash, "Franklin and Slavery," 627.

17. Benjamin Franklin to John Waring, Dec. 17, 1763, in *PBF*, 10:395–96. For Associates of Dr. Bray, see John C. Van Horne, "Impediments to the Christianization and Education of Blacks in Colonial America: The Case of the Associates of Dr. Bray," *Historical Magazine of the Protestant Episcopal Church* 50, no. 3 (1981): 243–69.

18. Waldstreicher, *Runaway America*, 190.

19. [Benjamin Franklin], "A Conversation on Slavery," *Public Advertiser* (London), Jan. 30, 1770, in *PBF*, 17:37–44.

20. Christopher L. Brown, *Moral Capital: Foundations of British Abolitionism* (Chapel Hill: University of North Carolina Press, 2006), 96–98; Davis, *Problem of Slavery in the Age of Revolution*, 471–78.

21. [Benjamin Franklin], "The Sommersett Case and the Slave Trade," *London Chronicle*, June 18–20, 1772, in *PBF*, 19:187–88.

22. Franklin, "Observations concerning the Increase of Mankind," in *PBF*, 4:225–34; Anthony Benezet to Franklin, April 27, 1772, in *PBF*, 19:112–16. For Franklin's reply,

see Franklin to Benezet, Aug. 22, 1772, in *PBF*, 19:269. Drescher briefly mentions Franklin's use of political arithmetic in "Observations" but dismisses their significance to abolitionism; see Seymour Drescher, *The Mighty Experiment: Free Labor versus Slavery in British Emancipation* (New York: Oxford University Press, 2002), 36–37.

23. For Franklin's role in the Treaty of Paris, see Wood, *Americanization of Benjamin Franklin*, 196–200. For enslaved people's escapes to British lines during the War of Independence, see Cassandra Pybus, *Epic Journeys of Freedom: Runaway Slaves of the American Revolution and Their Global Quest for Liberty* (Boston: Beacon, 2006); Sylvia Frey, *Water from the Rock: Black Resistance in a Revolutionary Age* (Princeton: Princeton University Press, 1991); Benjamin Quarles, *The Negro in the American Revolution* (Chapel Hill: University of North Carolina Press, 1961). For Philadelphia free Black population before 1780, see Nash and Soderlund, *Freedom by Degrees*, 3–5, 75–76, 88–89, 95.

24. "Franklin's Thoughts on Privateering and the Sugar Islands, Two Essays," n.d. (after July 10, 1782), in *PBF*, 37:617–20.

25. Waldstreicher, *Runaway America*, 219.

26. Franklin to David Hartley, May 8, 1783 (unpublished), *PBF* online (accessed May 7, 2020). Franklin to Hartley, Oct. 11, 1779, in *PBF*, 30:517–19.

27. Nicolas de Condorcet to Franklin, Dec. 2, 1773, in *PBF*, 20:489–91; Franklin to Condorcet, March 20, 1774, in *PBF*, 21:151–52; Benezet to Franklin, April 27, 1772, in *PBF*, 19:112–16.

28. *Definitive Treaty of Peace and Friendship, between His Britannick Majesty, and the United States of America. Signed at Paris, the 3d of September, 1783* (1783), Article 5:6–7; Benjamin Franklin, "Sundry Maritime Observations," *Transactions of the American Philosophical Society* 2 (1786): 323.

29. Franklin, "Sundry Maritime Observations," 315. For the political significance of the new map, see Chaplin, *First Scientific American*, 321–22.

30. Franklin, "Sundry Maritime Observations," 304. "It is remarkable that the people we consider as savages, have improved the art of sailing- and rowing-boats in several points beyond what we can pretend to," he wrote. For Franklin's habit of disassociating North America from the West Indies in order to defend its image as a beacon of liberty, see Waldstreicher, *Runaway America*, 180–81. For Black sailors, see W. Jeffrey Bolster, *Black Jacks: African American Seamen in the Age of Sail* (Cambridge: Harvard University Press, 1997).

31. For the political reason that the PAS leadership elected Franklin president, see Nash, "Franklin and Slavery," 635; Richard Newman, *The Transformation of American Abolitionism: Fighting Slavery in the Early Republic* (Chapel Hill: University of North Carolina Press, 2002), 41. Joyce Chaplin argued that Franklin's scientific accomplishments gave him a "reputation for dispassionate wisdom [that] made him a coveted spokesman on a variety of topics, including political ones"; Joyce

Chaplin, "Benjamin Franklin's Natural Philosophy," in *The Cambridge Companion to Benjamin Franklin*, ed. Carla Mulford (New York: Cambridge University Press, 2008), 72.

32. "Report on the Abolition of Slavery," March 5, 1790, 1st Cong., 2nd sess., Misc., *American State Papers*, vol. 1, no. 13, p. 12. For Pemberton's role in getting Franklin to sign the petition, see Waldstreicher, *Runaway America*, 236; Newman, *Transformation of American Abolitionism*, 41–42. Also worth noting is that Franklin's final public antislavery remarks came in the form of a letter, published anonymously in the *Federal Gazette*, on March 23, 1790, in which he reacted to Congress's rejection of this petition. For the petition itself, see "Memorial . . .," Dec. 12, 1790, Pennsylvania Abolition Society Papers (hereafter cited as PAS Papers), Historical Society of Pennsylvania, Philadelphia, series V, part 1.

33. Benjamin Rush, *The Autobiography of Benjamin Rush: His "Travels through Life" Together with His Commonplace Book for 1789–1813*, ed. George W. Corner (Philadelphia: American Philosophical Society, 1948; original autobiography written by Rush in 1800), "Commonplace Book, 1789–1791," 183. Rush sent the lock to Richard Price, a British mathematician, antislavery advocate, and prominent Dissenting minister.

34. For recent scholarship discussing the influence of politics on Rush's medical work, see Jacquelyn C. Miller, "The Body Politic and the Body Somatic: Benjamin Rush's Fear of Social Disorder and His Treatment of Yellow Fever," in Lindman and Tarter, *A Centre of Wonders*, 61–74; Jason Frank, "Sympathy and Separation: Benjamin Rush and the Contagious Republic," *Modern Intellectual History* 6, no. 1 (2009): 27–57; Sari Altschuler, "From Blood Vessels to Global Networks of Exchange: The Physiology of Benjamin Rush's Early Republic," *Journal of the Early Republic* 32 (Summer 2012): 207–31. See also Eric Herschthal, "Antislavery Science in the Early Republic: The Case of Dr. Benjamin Rush," *Early American Studies* 15, no. 2 (Spring 2017): 274–307. For biographies of Rush, see Nathan Goodman, *Benjamin Rush: Physician and Citizen, 1746–1813* (Philadelphia: University of Pennsylvania Press, 1934); Donald D'Elia, *Benjamin Rush, Philosopher of the American Revolution* (Philadelphia: American Philosophical Society, 1974); David F. Hawke, *Benjamin Rush: Revolutionary Gadfly* (Indianapolis: Bobbs-Merrill, 1971); Carl Binger, *Revolutionary Doctor, Benjamin Rush, 1746–1813* (New York: W. W. Norton, 1966); Stephen Fried, *Rush: Revolution, Madness, and the Visionary Doctor Who Became a Founding Father* (New York: Crown, 2018). See also Lisbeth Haakonssen, *Medicine and Morals in the Enlightenment: John Gregory, Thomas Percival and Benjamin Rush* (Amsterdam: Rodopi, 1997).

35. Goodman, *Benjamin Rush*, chap. 2; Binger, *Revolutionary Doctor*, chaps. 2 and 3. Rush, "Travels through Life," in *Autobiography*, 54, 66–67.

36. Benjamin Rush, *An Address to the Inhabitants of the British Settlements, on the Slavery of the Negroes in America*, 2nd ed. (Philadelphia, 1773). This edition

includes the response he would write to a planter's critique, which Rush titled *A Vindication* and attached to the original *Address*. The *Vindication* will be discussed in greater detail shortly. For Rush's appointment to the chemistry professorship, see Goodman, *Benjamin Rush*, 128. For Benezet's role in his abolitionism, see Rush to Granville Sharp, May 1, 1773, in *Letters of Benjamin Rush*, ed. L. H. Butterfield, 2 vols. (Princeton: Princeton University Press, 1951), 1:80–81 (hereafter cited as *LBR*). See also Maurice Jackson, *Let This Voice Be Heard: Anthony Benezet, Father of Atlantic Abolitionism* (Philadelphia: University of Pennsylvania Press, 2009).

37. For environmentalist theories of race, see Bruce Dain, *A Hideous Monster of the Mind: American Race Theory in the Early Republic* (Cambridge: Harvard University Press, 2002), 1–39; Winthrop Jordan, *White over Black: American Attitudes toward the Negro, 1550–1812* (Chapel Hill: University of North Carolina Press, 1968), 287. See also Roxann Wheeler, *Complexion of Race: Categories of Difference in Eighteenth-Century British Culture* (Philadelphia: University of Pennsylvania Press, 2000).

38. Rush, *An Address to the Inhabitants of the British Settlements*, 2. He added: "We are to distinguish between an African in his own country, and an African in a state of slavery in America. Slavery is so foreign to the human mind, that the moral faculties, as well as those of understanding are debased, and rendered torpid by it." Rush to Granville Sharp, May 1, 1773, in "The Correspondence of Benjamin Rush and Granville Sharp, 1773–1809," ed. John A. Woods, *Journal of American Studies* 1, no. 1 (April 1967): 2. Rush to Franklin, May 1, 1773, in *PBF*, 20:192–93. Franklin to Rush, July 14, 1773, in *PBF*, 20:314–16. Richard Nisbet, *Slavery Not Forbidden by Scripture* (Philadelphia, 1773), 22–23, 21.

39. Rush, *An Address to the Inhabitants of the British Settlements*, 24–25, 20–21, 45. For the notion of "tropical exuberance," see Philip Curtin, *The Image of Africa: British Ideas and Action, 1780–1850* (Madison: University of Wisconsin Press, 1964), 62. For more on the lazy stereotype, see Klas Ronnback, " 'The Men Seldom Suffer a Woman to Sit Down': The Historical Development of the Stereotypes of the 'Lazy African,' " *African Studies* 73, no. 2 (2014): 211–27.

40. Nisbet, *Slavery Not Forbidden*, 27. Rush, *An Address to the Inhabitants of the British Settlements*, 40n–42n. Rush, "Diseases of Negroes," RFP, subseries VII, vol. 125. Rush to John Coakley Lettsom, April 21, 1788, in *Memoirs of John Coakley Lettsom*, ed. Thomas Pettigrew, 3 vols. (1817), 2:432. Lettsom to Rush, Aug. 10, 1788, RFP, subseries I, vol. 28. For "Dirteatis," which plantation physicians called "Cachexia Africana," see Rana Hogarth, *Medicalizing Blackness: Making Racial Difference in the Atlantic World, 1780-1840* (Chapel Hill: University of North Carolina Press, 2017), 81–103.

41. Rush, *An Address to the Inhabitants of the British Settlements*, 45, 19, 20, 19.

42. Thomas H. Broman, "The Medical Sciences," in *The Cambridge History of Science*, vol. 4, *Eighteenth-Century Science*, ed. Roy Porter (New York: Cambridge

University Press, 2003), 463–84; Binger, *Revolutionary Doctor*, 34–39, 88, 261–62; Goodman, *Benjamin Rush*, chap. 10.

43. Rush, "An Inquiry into the Natural History of Medicine among the Indians of North-America, and a Comparative View of Their Diseases and Remedies with Those of Civilized Nations," in Rush, *Medical Inquiries*, 2nd ed. (Philadelphia, 1789), 36, 19, 48–49, 20, 22.

44. Rush, "Commonplace Book," entry for June 17, 1799, in *Autobiography*, 246; Hawke, *Benjamin Rush*, 360–62.

45. Benjamin Rush, "On Account of the Influence of the Military and Political Events of the American Revolution upon the Human Body," in Rush, *Medical Inquiries*, 226–27. Benjamin Rush, "Diseases Caused by Government" (n.d.), RFP, subseries VII, vol. 158. There is unfortunately no date on many of the unpublished lectures. But given their close resemblance to many of the published medical lectures he began to print in the latter 1780s and 1790s, their close relation to political events at the time, and that he devoted most of his time to his professorial duties in the same period, they were probably conceived in the general time frame of the late 1780s and 1790s.

46. Rush, "Diseases Caused by Government."

47. Benjamin Rush, "On the Different Species of Mania" (n.d.), in *The Selected Writings of Benjamin Rush*, ed. Dagobert Runes (New York: Philosophical Library, 1947), 212–14.

48. David Brion Davis, *The Problem of Slavery in the Age of Emancipation* (New York: Alfred A. Knopf, 2014), 74–82, 83–104.

49. For political factions in the 1790s, see Stanley M. Elkins and Eric L. McKitrick, *The Age of Federalism* (New York: Oxford University Press, 1993), chaps. 7–9; Carol Berkin, *A Sovereign People: The Crises of the 1790s and the Birth of American Nationalism* (New York: Basic Books, 2017). For Jefferson, see Peter Onuf and Ari Helo, "Jefferson, Morality, and the Problem of Slavery," in Peter Onuf, *The Mind of Thomas Jefferson* (Charlottesville: University of Virginia Press, 2007).

50. Benjamin Rush, "An Account of the Progress of the Population, Agriculture, Manners, and Government in Pennsylvania, in a Letter to a Friend in England," in Rush, *Essays, Literary, Moral and Philosophical* (Philadelphia, 1798), 218–20.

51. Ibid. Benjamin Rush, "Diseases from the Different States of Society," RFP, subseries VII, vol. 158.

52. Rush, *An Address to the Inhabitants of the British Settlements*, 3–4.

53. Statistics from Nash and Soderlund, *Freedom by Degrees*, 18 (table 1–4); Richard Newman and James Mueller, introduction to *Antislavery and Abolition in Philadelphia: Emancipation and the Long Struggle for Racial Justice in the City of Brotherly Love*, ed. Richard Newman and James Mueller (Baton Rouge: Louisiana State University, 2011), 6; Julie Winch, "Self-Help and Self-Determination: Black Philadelphians and the Dimensions of Freedom," in *Antislavery and Abolition in*

Philadelphia, 75; Julie Winch, *Between Slavery and Freedom: Free People of Color in America from Settlement to the Civil War* (Lanham, Md.: Rowman & Littlefield, 2014), 39. The idea that the rise of a free Black population, not an enslaved one, provoked racial theorizing comes from Jordan, *White over Black*, 513–23.

54. Rael, *Eighty-Eight Years*, 126. One in nine free Blacks owned property in 1800; the remaining almost 90 percent were thus barred from voting. For statistics, see Nash and Soderlund, *Freedom by Degrees*, 175. See also Winch, *Between Slavery and Freedom*, 40–42; Richard Bell, *Stolen: Five Free Boys Kidnapped into Slavery and Their Astonishing Odyssey Home* (New York: Simon & Schuster, 2019).

55. Rush, "Of the Black Color of the Africans," RFP, subseries VII, vol. 161. In most other respects, this unpublished version is almost identical to the final lecture published in *Transactions*, which I will use for citation purposes from here on. For background on Samuel Stanhope Smith's "Essay on the Causes of Variety of Complexion and Figure in the Human Species" (Philadelphia, 1787), see Dain, *Hideous Monster of the Mind*, 40–45. Rush's essay on Blackness is often studied in relation to the "white negro" phenomenon. See Jordan, *White over Black*, 512–28; Melish, "Emancipation and the Embodiment of 'Race' "; Yokota, "Not Written in Black and White"; Yokota, *Unbecoming British*, 215–17; Martin, *The White African American Body*, 19–48.

56. On leprosy, see Warwick Anderson, "Leprosy and Citizenship," *Positions* 6, no. 3 (1998): 707–30. Jacques-Henri Bernardin de Saint-Pierre, *Studies of Nature*, trans. Henry Hunter (London, 1796), 254. Rush cites Bernardin by name, referring to this quote, though not quoting it directly; see Rush, "Observations Intended to Favour a Supposition," 291, as well as 292, 295. Rush to Thomas Jefferson, Feb. 4, 1797, in *The Papers of Thomas Jefferson*, ed. Barbara B. Oberg (Princeton: Princeton University Press, 2002), 29:284.

57. Rush, "Observations Intended to Favour a Supposition," 295.

58. Privately, Rush expressed disapproval of segregation. Seeing white and Black mourners together at a funeral gave him hope that "the partition wall which divide[s] the Blacks from the Whites will still be further broken down." Rush, "Commonplace Book," entry for June 18, 1792, in *Autobiography*, 221.

59. Rush, "Observations Intended to Favour a Supposition," 295–96. Moses Brown on Moss's blisters is quoted in Melish, "Emancipation and the Embodiment of 'Race,' " 227. For climate changing dark skin light, see Lee Alan Dugatkin, *Mr. Jefferson and the Giant Moose* (Chicago: University of Chicago Press, 2009), 27–28. For African albinism in the eighteenth century, see Andrew Curran, "Rethinking Race History: The Role of the Albino in the French Enlightenment Life Sciences," *History and Theory* 48, no. 3 (Oct. 2009): 151–79. For the reference Rush made to albinism among native Africans, see Rush, "Observations of the Black Color," 290–91. For Henry Moss, see Yokota, "Not Written in Black and White"; Harriet A. Washington, *Medical Apartheid: The Dark History of Medical Experimentation on*

Black Americans from Colonial Times to the Present (New York: Doubleday, 2006), 80–82, 88.

60. For recent studies of the 1793 yellow fever outbreak, see Thomas Apel, *Feverish Bodies, Enlightened Minds: Science and the Yellow Fever Controversy in the Early American Republic* (Stanford: Stanford University Press, 2016); Simon Finger, *The Contagious City: The Politics of Public Health in Early Philadelphia* (Ithaca: Cornell University Press, 2012), 120–62; Hogarth, *Medicalizing Blackness*, 25–40. For the politics beneath the contagionist-localist debate, see Katherine Arner, "Making Yellow Fever American: The Early American Republic, the British Empire and the Geopolitics of Disease in the Atlantic World," *Atlantic Studies* 7, no. 4 (Dec. 2010): 447–71; Jan Golinski, "Debating the Atmospheric Constitution: Yellow Fever and the American Climate," *Eighteenth-Century Studies* 49, no. 2 (Winter 2016): 149–65; Martin S. Pernick, "Politics, Parties, and Pestilence: Epidemic Yellow Fever in Philadelphia and the Rise of the First Party System," *William and Mary Quarterly* 29, no. 4 (Oct. 1972): 559–86. Only recently have scholars noted the influence of abolitionist politics on the yellow fever debates from the period. For examples, see Billy G. Smith, *Ship of Death: A Voyage That Changed the Atlantic World* (New Haven: Yale University Press, 2013), 169–74, 189–205, 247–48; Apel, *Feverish Bodies*, 127. For Rush's role in the epidemics, see Paul E. Kopperman, " 'Venerate the Lancet': Benjamin Rush's Yellow Fever Therapy in Context," *Bulletin of the History of Medicine* 78, no. 3 (Fall 2004): 539–74.

61. Colin Chisholm, *An Essay on the Malignant Pestilential Fever Introduced into the West Indian Islands . . .* (London, 1795), 82–84. For the British experience in Haiti and Toussaint-Louverture's exploitation of the disease environment, see John R. McNeill, *Mosquito Empires: Ecology and War in the Greater Caribbean, 1620–1914* (New York: Cambridge University Press, 2010), 244–48, 249–67.

62. E. H. Smith, "Medical Essays . . . On the Origin of the Pestilential Fever," *The Medical Repository* 1, no. 4 (May 1, 1798): 472, 492. Noah Webster, *A Brief History of Epidemic and Pestilential Diseases*, vol. 1 (Hartford, Conn., 1799), 294–95. Noah Webster to Benjamin Rush, Dec. 2, 1797, in *Letters of Noah Webster*, ed. Harry Warfel (New York: Library Publishers, 1953), 171–72.

63. Benjamin Rush, "Facts Intended to Prove the Yellow Fever Not Be Contagious, and Instances of Its Supposed Contagion Explained upon Other Principles," *Medical Repository* 6, no. 2 (July 1803): 167–68. Noah Webster wrote to the physician William Currie in December 1797 that the "philosopher and candid citizen will desert the indefensible ground of importation and acknowledge that *our climate* has obtained no decree of heaven exempting it from the operations of the general laws of nature"; Noah Webster to Dr. William Currie, Dec. 20, 1797, in *Letters of Noah Webster*, 164 (emphasis in original). For the influence of his students and Webster, see Arner, "Making Yellow Fever American," 457–58. For Rush's fear of implicating the nation as unhealthy, see Golinski, "Debating the Atmospheric Constitution," 150–51.

64. Benjamin Rush, *An Account of the Bilious Remitting Yellow Fever, as it Appeared in the City of Philadelphia, in the Year 1793* (Philadelphia: Tomas Dobson, 1794), 12–27. Richard Peters to Timothy Pickering, Oct. 22, 1793, *LBR*, 2:729–30n1. For Federalist antipathy toward Rush and the localist view, see Pernick, "Politics, Parties, and Pestilence," 569–70.

65. William Cobbett, *Rush-Light* (New York, 1800), 64, 77 (emphasis in original). For Cobbett's links to the Federalists, see Keith Arbour, "Benjamin Franklin as Weird Sister: William Cobbett and Federalist Philadelphia's Fears of Democracy" in *Federalists Reconsidered*, ed. Doron Ben-Atar and Barbara Oberg (Charlottesville: University Press of Virginia, 1998): 179–98. Cobbett's polemics against Rush led Rush to file and win a libel lawsuit in 1797. Though the case bankrupted Cobbett, he published one last lampoon of Rush in 1800 before leaving for England. Titled *Rush-Light*, it revealed how Rush's abolitionism and republicanism could be used to undermine the validity of his scientific oeuvre.

66. Benjamin Rush, "A Defense of Blood-Letting," in Rush, *Medical Inquiries and Observations*, 3rd ed., 4 vols. (Philadelphia, 1809), 4:379–80. This essay was first written in 1796.

67. Rush to Sharp, June 20, 1809, in "Correspondence of Benjamin Rush and Granville Sharp," 37. Rush to the Pennsylvania Abolition Society, n.d. (1794), in *LBR*, 2:754–56. For details of the eventual purchase of land, see Rush to PAS, n.d. (1794), in *LBR*, 2:755–56n. See also Goodman, *Benjamin Rush*, 303. For PAS freedom suits, see Newman, *Transformation of American Abolitionism*, 41–49.

68. Rush to John Adams, June 10, 1806, in *LBR*, 2:919. Rush to Adams, March 2, 1809, *LBR*, 2:997 (emphasis in original).

69. Benjamin Rush, "An Account of the Extraordinary Powers of Calculation, by Memory, Possessed by a Negro Slave, in Maryland, Communicated in a Letter from Dr. Rush, of Philadelphia, to a Gentleman in Manchester," *Universal Magazine of Knowledge and Pleasure* (Dec. 1788): 306. For another example, see Benjamin Rush, "Accounts of a Negro Practitioner of Physic, and a Self-taught Negro Calculator, by Benjamin Rush, M.D. Prof. of Chem . . . ." in William Dickson, *Letters on Slavery* (London, 1789), 184–87.

70. Ibram X. Kendi, *Stamped from the Beginning: The Definitive History of Racist Ideas in America* (New York: Nation Books, 2016), 94. For Banneker's wooden clock, see Anne Tyson Kirk, ed., *Banneker, the Afric-American Astronomer* (Philadelphia: Friends Book Association, 1884), 22. Banneker Astronomical Journal, Maryland Historical Society (hereafter cited as MDHS), item no. 2700. For the most authoritative account of Banneker's life, see Silvio Bedini, *The Life of Benjamin Banneker* (New York: Scribner, 1971). Bedini's biography includes transcriptions of many of the manuscript sources used in my account here.

71. Kirk, *Banneker, the Afric-American Astronomer*, 22 (emphasis in original). "A Sketch of Ellicott's Mills, and an Account of Benjamin Banneker Compiled from Remem-

brances of 1796," in *Selections from the Letters and Manuscripts of the Late Susanna Mason* (Philadelphia: Backliff & Jones, 1836), 243. Kirk, *Banneker, the Afric-American Astronomer*, 13, 53.

72. See, for instance, Richard Newman, " 'Good Communication Corrects Bad Manners': The Banneker-Jefferson Dialogue and the Project of White Uplift," in *Contesting Slavery: The Politics of Bondage and Freedom in the New American Nation*, ed. John Craig Hammond and Matthew Mason (Charlottesville: University of Virginia Press, 2011), 69–93; William Andrews, "Benjamin Banneker's Revision of Thomas Jefferson: Conscience versus Science in the Early American Antislavery Debate," in *Genius in Bondage: Literature of the Early Black Atlantic*, ed. Vincent Carretta and Philip Gould (Lexington: University Press of Kentucky, 2001), 218–41; Sinha, *Slave's Cause*, 144–46; Manisha Sinha, "To 'Cast Just Obliquy' on Oppressors: Black Radicalism in the Age of Revolution," *William and Mary Quarterly* 64, no. 1 (2007): 153.

73. Banneker, Astronomical Journal, page facing ephemeris for March 1799; page facing calculations for April 1799; page facing calculations for August 1802; Bedini, *Banneker*, 270.

74. Marion B. Stowell, *Early American Almanacs: The Colonial Weekday Bible* (New York: B. Franklin, 1976), x.

75. See William Pencak, "Poor Richard's Almanac," in Waldstreicher, *A Companion to Benjamin Franklin*, 275–89. See also Stowell, *Early American Almanacs*, 76–85; Patrick Spero, "The Revolution in Popular Publications: The Almanac and New England Primer, 1750–1800," *Early American Studies* 8, no. 1 (Winter 2010): 41–74.

76. Bedini, *Banneker*, 73, 77. Benjamin Banneker to George Ellicott, Oct. 13, 1789, VF, MDHS. To be sure, all the publishers Banneker approached had published antislavery works before this time. For the publishers Banneker initially approached, see Bedini, *Banneker*, 90–92.

77. Banneker to Andrew Ellicott, May 6, 1790, PAS Papers.

78. The only reason given for Hayes's decision comes from Joseph Townsend, the president of the Maryland Abolition Society, who mentioned in a letter to James Pemberton that Hayes rejected Banneker's letter because he already had a reliable ephemeris from Andrew Ellicott and thus "did not like to Change." Joseph Townsend to James Pemberton, Nov. 28, 1790, PAS Papers.

79. "George-Town, March 12," *General Advertiser* (Philadelphia), March, 19, 1791, 3. For evidence of his own astronomical observations during the survey, see Banneker, Astronomical Journal.

80. Kirk, *Banneker, the Afric-American Astronomer*, 37.

81. Joseph Townsend to James Pemberton, Nov. 14, 1790, PAS Papers. Pemberton's letter to Townsend has not been found, but this letter clearly implies that Pemberton asked him about Banneker, and that Townsend did not know about Banneker until that point. Elias Ellicott to James Pemberton, June 10, 1791, PAS Papers.

82. For authentication of Black authors, see Carretta and Gould, introduction to *Genius in Bondage*, 3. For the full document with verification signatures by eleven men, including George and Andrew Ellicott, see "Certificate in Regard to Genuineness of Banneker's Calculations," ca. Sept. 1791, PAS Papers. David Rittenhouse to James Pemberton, Aug. 6, 1791, PAS Papers. "Statement by William Waring, Aug. 16, 1791," PAS Papers. "Letter from Mr. James McHenry, Aug. 20, 1791," in *Benjamin Banneker's Pennsylvania, Delaware, Maryland and Virginia Almanack, and Ephemeris, for the Year of Our Lord, 1792* (Baltimore: Goddard and Angell, 1791), 3.

83. Banneker to Jefferson, Aug. 19, 1791, in *Papers of Thomas Jefferson* (hereafter cited as *PTJ*), ed. Charles Cullen (Princeton: Princeton University Press, 1986), 22:49–54. Thomas Jefferson, *Notes on the State of Virginia* (1787), ed. William Peden, 2nd ed. (Chapel Hill: University of North Carolina Press, 1982), 139. Banneker to Pemberton, Sept. 3, 1791, PAS Papers. Most recent work on Banneker focuses on this letter, which in my view overshadows Banneker's real contribution—his ephemeris. See Sinha, *A Slave's Cause*, 144–46; Newman, " 'Good Communication Corrects Bad Manners' "; Andrews, "Benjamin Banneker's revision of Thomas Jefferson."

84. Jefferson to Banneker, Aug. 30, 1791, in *PTJ*, 22:97–98. Jefferson to Condorcet, Aug. 30, 1791, in *PTJ*, 22:98–99. To be sure, Jefferson would, two decades later, again express doubts. See Jefferson to Barlow, Oct. 8, 1809, in *PTJ, Founders Online*, https://founders.archives.gov/documents/Jefferson/03-01-02-0461 (accessed June 11, 2017). Regarding Jefferson's being "rather old-fashioned," see Dain, *A Hideous Monster of the Mind*, 6. See also Robert P. Forbes, " 'The Cause of This Blackness': The Early American Republic and the Construction of Race," *American Nineteenth Century History* 13, no. 1 (March 2012): 65–94.

85. Banneker to Jefferson, Aug. 19, 1791, in *PTJ*, 22:97–98.

86. [Benjamin Rush], "A Plan for a Peace-Office, for the United States," in *Banneker's Almanack . . . for the Year 1793* (Philadelphia: Joseph Crukshank, 1792), 6–10. "Extract from Wilkinson's Appeal to England on Behalf of the Abused Africans," ibid., 26. The Baltimore edition of the 1793 almanac recycled famous epigrams from Franklin's *Poor Richard's Almanac*, such as "Time is money." See "Hints—by the Late Doctor Franklin," in *Banneker's Almanack . . . for the Year 1793*, 24.

87. *Banneker's Almanack . . . for the Year 1792*, 20. David Rittenhouse, "Dr. Rittenhouse's Oration on the Subject of Astronomy, Given before the American Philosophical Society, in 1775," reprinted in William Barton, ed., *Memoirs of the Life of David Rittenhouse, LLD. F.R.S.* (Philadelphia: Edward Parker, 1813), 566. *Banneker's Pennsylvania, Delaware, Maryland and Virginia Almanack, and Ephemeris, 1792*, 33 (emphasis in original).

88. *Banneker's Almanack . . . for the Year 1793*, 13. Another line noted that Egyptians invented "Letters" in 1822 BCE and centuries later "introduced [them] among the savages of Europe"; ibid., 7. For the possible African origins of Banneker's astro-

nomical interests, see Ron Eglash, "African Heritage of Benjamin Banneker," *Social Studies of Science* 27 (1997): 307–15; Charles A. Cerami, *Benjamin Banneker: Surveyor, Astronomer, Publisher, Patriot* (New York: Wiley & Sons, 2002), esp. 7, 15, 184–85, 217–19; Ellen Swartz, "Removing the Master Script: Benjamin Banneker 'Re-Membered,' " *Journal of Black Studies* 44, no. 1 (Jan. 2013): 31–49. For Molly Welsh, see Bedini, *Banneker*, 10–11, 17–18; Sandra W. Perot, "The Dairymaid and the Prince: Race, Memory and the Story of Banneker's Grandmother," *Slavery & Abolition* 38, no. 3 (2017): 445–58.

89. Kirk, *Banneker, the Afric-American Astronomer*, 9–11. For Bannaka's defiant character, see Bedini, *Banneker*, 17–18. For Dogon astronomy, see Laird Scranton, *The Science of the Dogon: Decoding the African Mystery Tradition* (Rochester, Vt.: Inner Traditions, 2006).

90. Banneker, Astronomical Journal.

91. Ibid., pages facing ephemeris for March 1799, April 1799, and August 1802. On the arson, see Bedini, *Banneker*, 270.

2 Full Steam Ahead

1. For British slave trade statistics, see Kenneth Morgan, *Slavery and the British Empire: From Africa to America* (New York: Oxford University Press, 2007), 12. For SEAST minutes, see entries for July 5, 1787, and Oct. 16, 1787, Fair Minute Books of the Committee for the Abolition of the Slave Trade, May 22, 1787–July 9, 1819, British Library, Add Mss. 21254–21256 (mf). On the medallion's creation, see J. R. Oldfield, *Popular Politics and British Anti-Slavery: The Mobilisation of Public Opinion against the Slave Trade, 1787–1807* (Manchester, U.K.: Manchester University Press, 1995), 155–84. See also Marcus Wood, *The Horrible Gift of Freedom: Atlantic Slavery and the Representation of Emancipation* (Athens: University of Georgia Press, 2010), 35–89; Martha Katz-Hyman, "Doing Good While Doing Well: The Decision to Manufacture Products That Supported the Abolition of the Slave Trade and Slavery in Great Britain," *Slavery & Abolition* 29, no. 2 (June 2008): 219–31.

2. Thomas Clarkson, *History of the Abolition of the Slave Trade*, 2 vols. (London, 1808), 2:191–92. For women's role in the early campaign, see Clare Midgley, *Women against Slavery: The British Campaigns, 1780–1870* (London: Routledge, 1992), 9–42; Charlotte Sussman, *Consuming Anxieties: Consumer Protest, Gender, and British Slavery, 1713–1833* (Stanford: Stanford University Press, 2000), chaps. 4–5. See also Oldfield, *Popular Politics and British Antislavery*, 155–84; J. R. Oldfield, "The London Committee and Mobilization of Public Opinion against the Slave Trade," *Historical Journal* 35, no. 2 (1992): 331–43.

3. For a brief discussion of Wedgewood's interest in chemistry, see Robin Reilly, *Josiah Wedgwood, 1730–1795* (London: Macmillan, 1992), 314–15. For the best study of

Wedgwood's skills in stoking consumer demand, see Neil McKendrick, John Brewer, and J. H. Plumb, *The Birth of a Consumer Society: The Commercialization of Eighteenth-Century England* (Bloomington: Indiana University Press, 1985), chaps. 1–4.

4. For the most useful studies of these men as a collective group, called the Lunar Society, see Robert Schofield, *The Lunar Society of Birmingham: A Social History of Provincial Science and Industry in Eighteenth-Century England* (Oxford: Clarendon Press, 1963). Schofield, however, pays little attention to abolitionism, except on pp. 59, 135, 213, and 356–57. See also Jennifer S. Uglow, *The Lunar Men: Five Friends Whose Curiosity Changed the World* (New York: Farrar, Straus and Giroux, 2002), which also has very little on abolitionism, except for pp. 55–56, 184, 258–62, and 410–14. For the most thorough study of Priestley, see Robert Schofield, *The Enlightenment of Joseph Priestley: A Study of His Life and Work from 1733 to 1773* (University Park: Pennsylvania State University Press, 1997), and Robert Schofield, *The Enlightened Joseph Priestley: A Study of His Life and Work from 1773 to 1804* (University Park: Pennsylvania State University Press, 2004). For Erasmus Darwin, see Maureen McNeil, *Under the Banner of Science: Erasmus Darwin and His Age* (Manchester, U.K.: Manchester University Press, 1987); Martin Priestman, *The Poetry of Erasmus Darwin: Enlightened Spaces, Romantic Times* (Burlington, Vt.: Ashgate, 2013); Patricia Fara, *Erasmus Darwin: Sex, Science, and Serendipity* (Oxford: Oxford University Press, 2012); Desmond King-Hele, *Doctor of Revolution: The Life and Genius of Erasmus Darwin* (London: Faber & Faber, 1977). For Wedgwood, see Reilly, *Josiah Wedgwood*, and McKendrick et al., *The Birth of a Consumer Society*, chaps. 1–4. There is a vast literature on Franklin, but particularly useful with regard to his scientific work is Joyce Chaplin, *The First Scientific American: Benjamin Franklin and the Pursuit of Genius* (New York: Basic Books, 2006).

5. Josiah Wedgwood to Matthew Boulton, Oct. 9, 1766, Letters of Josiah Wedgwood (hereafter cited as LJW), Wedgwood Museum Archives, Mss. 18130–25. Many of these letters are reprinted in *Correspondence of Josiah Wedgwood*, ed. Katherine Farrar, 3 vols. (Cambridge: Cambridge University Press, 2010) (hereafter cited as *CJW*); Joseph Priestley, *Experiments and Observations on Different Kinds of Air*, 3 vols., 2nd ed. (Birmingham: Thomas Pearson, 1790), 1:xxiii. For machines replacing slave labor, see Wedgwood to Anna Seward, Feb. 1788, *CJW*, 3:56.

6. Wedgwood to Thomas Bentley, March 15, 1768, LJW, Mss. 18193–25. Joseph Priestley to Wedgwood, Birmingham, May 26, 1781, Royal Society of London, Archives (hereafter cited as RS), Misc. Manuscripts, MM/5/2 (emphasis in original). For Wedgwood's youth and his experiments, see Reilly, *Josiah Wedgwood*, 2–3, 196–97, respectively. Excerpts from many of Priestley's letters to Wedgwood from the 1780s can be found in either *Scientific Correspondence of Joseph Priestley*, ed. Henry Carrington Bolton (New York, 1892), or *A Scientific Autobiography of Joseph Priestley: Selected Scientific Correspondence*, ed. Robert E. Schofield (Cambridge: MIT Press, 1966).

7. Joseph Priestley, *Memoirs of Dr. Joseph Priestley, to the Year 1795*, 2 vols. (London: J. Johnson, 1806–7), 1:97. There is a lively scholarly debate over the extent to which scientific knowledge and technological innovation propelled the Industrial Revolution. For scholars emphasizing the importance of advances in scientific and technological knowledge, see Joel Mokyr, *The Enlightened Economy: An Economic History of Britain, 1700–1850* (New Haven: Yale University Press, 2009), and Margaret C. Jacob, *The First Knowledge Economy: Human Capital and the European Economy, 1750–1850* (Cambridge: Cambridge University Press, 2013). For scholars who place greater emphasis on social factors, especially racial and gendered organization of labor, see Jan De Vries, *The Industrious Revolution: Consumer Behavior and the Household Economy, 1650 to the Present* (New York: Cambridge University Press, 2008), and Joseph E. Inikori, *Africans and the Industrial Revolution* (New York: Cambridge University Press, 2002).

8. "A Letter from a Gentleman at Paris to His Friend at Toulon, concerning a Very Extraordinary Experiment in Electricity," June 9, 1752, *Gentleman's Magazine*, June 1752, 264. "To Benjamin Franklin Esq; of Philadelphia, on his Experiments and Discoveries in Electricity," *Gentleman's Magazine*, Feb. 1754, 88. For Franklin's scientific career in this period, see Chaplin, *First Scientific American*, 73–74, 103–11, 116–17; James Delbourgo, *A Most Amazing Scene of Wonders: Electricity and Enlightenment in Early America* (Cambridge: Harvard University Press, 2006), 32–36. For his political career in England during this period, see George Goodwin, *Benjamin Franklin in London: The British Life of America's Founding Father* (New Haven: Yale University Press, 2016); Carla Mulford, *Benjamin Franklin and the Ends of Empire* (New York: Oxford University Press, 2015), chap. 6.

9. Joseph Priestley, *The History and Present State of Electricity, with Original Experiments, by Joseph Priestley*, 2nd ed. (London, 1769), xii. Benjamin Franklin, "Account of Dr. Priestley's New Experiments," *Papers of Benjamin Franklin*, 41 vols. (New Haven: Yale University Press, 1954–) (hereafter cited as *PBF*), 15:68. Priestley to Wedgwood, Birmingham, Nov. 30, 1780, RS, Mss. MM/5/1. Reilly, *Josiah Wedgwood*, 314–15.

10. Wedgwood to Boulton, Oct. 9, 1766. Steven Shapin, "The Image of the Man of Science," in *Cambridge History of Science*, vol. 4, *Eighteenth-Century Science*, ed. Roy Porter (New York: Cambridge University Press, 2003), 159–83.

11. Priestley, *History of Electricity*, xiv, xvii, iii–iv.

12. Barbara Solow, "Caribbean Slavery and British Growth: The Eric Williams Hypothesis," *Journal of Development Economics* 17 (1985): 113, 104–5. For science condoning slavery, see David Brion Davis, *Problem of Slavery in the Age of Emancipation* (New York: Alfred A. Knopf, 2014), 33–34; Robin Blackburn, *The American Crucible: Slavery, Emancipation and Human Rights* (London: Verso, 2011), 145–50.

13. Wedgwood to Thomas Bentley, July 18, 1766, *CJW*, 1:97. William Leybourne to Mr. Browhill, Dec. 5, 1776, LJW, Mss. 12731-L74. Darwin to Wedgwood, April 13, 1789, in *The Collected Letters of Erasmus Darwin*, ed. Desmond King-Hele, 2nd ed.

(New York: Cambridge University Press, 2007) (hereafter cited as *CLED*), 338. Company of Merchants Trading to Africa to Josiah Wedgwood II and Mr. Byerley, Oct. 1, 1802, LJW, Mss. 18160-L98; Longlands to Josiah Wedgwood II, 1802, LJW, Mss. 24973-L127.

14. Priestley, *Experiments and Observations on Different Kinds of Air*, 3 vols., 2nd ed. (Birmingham: Thomas Pearson, 1790), 1:xxii–xxiii. See also Priestley, *History of Electricity*, xviii, where he argued that advancements in science not only made mankind "more comfortable and happy," but also induced a benevolent cast of mind. For "contemplation of the works of God" see Priestley, *History of Electricity*, xviii. Priestley, *Experiments and Observations on Different Kinds of Air* (London: J. Johnson, 1775), 2:ix, xli. For how Enlightenment men of science found ways to accommodate religious belief and scientific ideas, see John Hedley Brooke, "Science and Religion," in *Cambridge History of Science*, vol. 4, *Eighteenth-Century Science*, 739–61. For Priestley's ability to mesh science with religion, see John G. McEvoy and J. E. McGuire, "God and Nature: Priestley's Way of Rational Dissent," *Historical Studies in Physical Sciences* 6 (1975): 325–404.

15. Here I build on the insights of Jan Golinski in *Science as Public Culture: Chemistry and Enlightenment in Britain, 1760–1820* (New York: Cambridge University Press, 1992), chaps. 3 and 4. Golinski uses Priestley as a case study to show how he developed a particular discursive style, one that emphasized "economy, modesty and liberality" (128), to gain credibility.

16. Priestley to Wedgwood, May 6, 1783, RS, Mss. MM/20/46. Priestley to Wedgwood, June 26, 1781, RS, Mss. MM/5/4. Priestley, *History of Electricity*, ix, iv. For the importance of active versus passive labor in the culture of scientific experimentation in the mid-eighteenth century, see Joyce Chaplin, "Benjamin Franklin's Discoveries: Science and Public Culture in the Eighteenth Century," *Agora* 46, no. 2 (2011): 14–22. For the divide between "hand" and "head" knowledge, see Chaplin, *The First Scientific American*, 9–38.

17. Joseph Priestley, *A Sermon on the Subject of the Slave Trade, Delivered to a Society of Protestant Dissenters at the New Meeting, in Birmingham* (Birmingham, U.K., 1788), 7. Priestley, *Experiments and Observations on Different Kinds of Air*, 3 vols., 2nd ed. (Birmingham: Thomas Pearson, 1790), 1:xxvi. Joseph Priestley, *Miscellaneous Observations Relating to Education. More Especially as It Respects the Conduct of the Mind*, 2nd ed. (Birmingham, U.K., 1788), 142.

18. Golinski, *Science as Public Culture*, 84–87.

19. Priestley, *Experiments and Observations on Different Kinds of Air* (London: J. Johnson, 1775), 2:xxviii, 3:xxx (emphasis in original). Joseph Priestley, *Experiments and Observations Relating to Various Branches of Natural Philosophy* (London, 1779), vi (emphasis in original). Lavoisier's creation of a new nomenclature for chemistry, and the new elemental basis that it represented, is traditionally called the Chemical Revolution, although some historians question whether any "revolution" really

occurred at all. For the traditional view of Lavoisier's importance to modern chemistry, see William H. Brock, *The Norton History of Chemistry* (New York: W. W. Norton, 1992), 99–121; for a more recent, skeptical account, see Hasok Chang, *Is Water H2O? Evidence, Realism and Pluralism* (New York: Springer, 2012). For Priestley's belief in gradual scientific progress, see John McEvoy, "Electricity, Knowledge, and the Nature of Progress in Priestley's Thought," *British Journal for the History of Science* 12, no. 1 (March 1979): 1–30.

20. Joseph Priestley, *An Essay on the First Principles of Government; and on the Nature of Political, Civil, and Religious Liberty* (London, 1768), 77–78 (emphasis in original), 124, 142, 153.

21. Priestley, *Sermon on the Subject of the Slave Trade*, xii, 28, 29.

22. Ibid., 3, 5, 8.

23. Ibid., 29–30 (emphasis in original).

24. Priestley, *History of Electricity*, x; Priestley, *Experiments and Observations on Different Kinds of Air*, 2nd ed. (Birmingham: Thomas Pearson, 1790), 1:xxiii.

25. Wedgwood to Boulton, Oct. 9, 1766.

26. Erasmus Darwin, *Botanic Garden* (London: J. Johnson, 1791), part 1 ("Economy of Vegetation"), canto I, ll. 289–92; part 2 ("Loves of the Plants"), canto II, ll. 95–98, 303–6, 85–94n; part 1, canto II, ll. 355–56, 367–70. For useful guides to Darwin's complex poem, see Fara, *Erasmus Darwin*; McNeil, *Under the Banner of Science*, 87–90; Priestman, *The Poetry of Erasmus Darwin*, chap. 4.

27. Lissa Roberts, "An Arcadian Apparatus: The Introduction of the Steam Engine into the Dutch Landscape," *Technology and Culture* 45, no. 2 (2004): 251–76.

28. Darwin, *Botanic Garden*, part 1, canto II, ll. 308, 315–16n.

29. Ibid., canto I, ll. 254–60n; part 2, canto II, ll. 85–104n.

30. Wedgwood to Seward, Feb. 1788, *CJW*, 3:56. Anna Seward to Wedgwood, Feb. 22, 1788, Wedgwood Correspondence (hereafter cited as WC), English MS 1110, ff. 39–43, John Rylands Library, University of Manchester. Wedgwood to James Watt, Etruria, Feb. 14, 1788, James Watt Papers, Birmingham Central Library, C 1/8 (mf) (emphasis in original).

31. Joseph Priestley, *Lectures on History and General Policy* (Dublin, 1788), 375, 380 (emphases in original).

32. See Sarah Knott, *Sensibility and the American Revolution* (Chapel Hill: University of North Carolina Press, 2009), 69–103.

33. Joseph Priestley, *Hartley's Theory of the Human Mind, on the Principle of the Association of Ideas* (London, 1775), xliii (emphasis in original). Erasmus Darwin, *Zoonomia: Part 1* (London, 1794), 480. Darwin, *Botanic Garden*, part 2, interlude between cantos II and III; part 1, canto II, ll. 315–16, 309.

34. Clarkson, *History of the Abolition of the Slave Trade*, 2:191–92.

35. Robert Whytt, *Observations on the Nature, Causes and Cure of Those Disorders Which Have Been Commonly Called Nervous, Hypochondriac or Hysteric*

(Edinburgh, 1765), 105. See also Londa Schiebinger, *The Mind Has No Sex? Women in the Origins of Modern Science* (Cambridge: Harvard University Press, 1989), 160–244.

36. Erasmus Darwin, *A Plan for the Conduct of Female Education* (1797; repr., Philadelphia, 1798), 50.

37. Darwin, *Botanic Garden*, part 2, canto III, ll. 431–32, 447–48.

38. Darwin, *A Plan for the Conduct of Female Education*, 9–10, 47. Priestley, *Miscellaneous Observations relating to Education*, 151–52. For excellent discussions of the way Darwin embedded patriarchal norms in his poetry, see Janet Browne, "Botany for Gentlemen: Erasmus Darwin and 'The Loves of the Plants,' " *Isis* 80, no. 4 (Dec. 1989): esp. 592–621. Fara, *Erasmus Darwin*, 95–98, 113.

39. Mary Wollstonecraft, *A Vindication of the Rights of Woman* (Philadelphia, 1792), xi. Darwin, *A Plan for the Conduct of Female Education*, 63.

40. For Rousseau and the "noble savage," see Mark Hulliung, *The Autocritique of Enlightenment: Rousseau and the Philosophes* (Cambridge: Harvard University Press, 1994), 156–200. For Wedgewood's busts of Linnaeus and Rousseau, see Wedgwood to Thomas Bentley, May 8, 1779, LJW, Mss. 18889-L26. Thomas Bentley, *Journal of a Visit to Paris, 1776*, ed. Peter France (Brighton, U.K.: University of Sussex Library, 1977), 59. For Bentley's influence on Wedgwood, see Reilly, *Josiah Wedgwood*, 31–32, 240–42.

41. Bentley, *Journal of a Visit to Paris*, 60, 63. Wedgwood to John Whitehurst, Oct. 19, 1775, LJW, Mss. 18193-25. Bentley, *Journal of a Visit to Paris*, 65. For geology and religious controversies, see Roy Porter, *The Making of Geology: Earth Science in Britain, 1660–1815* (New York: Cambridge University Press, 1977), 118–27.

42. Wedgwood to Bentley, Oct. 24, 1778, LJW, Mss. 18857-L26. Bentley, *Journal of a Visit to Paris*, 65. Darwin, *Botanic Garden*, part 1, canto II, ll. 65, 276–77, 272.

43. Darwin to Wedgwood, Feb. 22, 1789, WC, English MS 1110, ff. 47–48. Darwin to Wedgwood, June 22, 1789, WC, English MS 1110, ff. 49–50, 66–67. Darwin to Wedgwood, Derby, July 18, 1789, in *CLED*, 345–46. Wedgwood to Darwin, Oct. 1789, in *CJW*, 3:101–3; Wedgwood to Darwin, July 1789, *CJW*, 3:87–90. Darwin to Wedgwood, June 22, 1789 (all emphases in originals). The exact textual sources Wedgwood—or more likely, his chief artist, William Hackwood—drew on to design the slave are unknown. See Oldfield, *Popular Politics and British Anti-Slavery*, 156; Katz-Hyman, "Doing Good While Doing Well," 219–20; Wood, *Horrible Gift of Freedom*, 75–81.

44. Darwin to Wedgwood, June 21–27, 1789, in *CLED*, 342. Darwin to Wedgwood, June 10, 1790, in *CLED*, 366.

45. Robin Hallett, *The Penetration of Africa: European Enterprise and Exploration Principally in Northern and Western Africa up to 1815*, vol. 1 (London: Routledge, 1965), 273–75; Philip Curtin, *Image of Africa: British Ideas and Action, 1780–1850*, vol. 1 (Madison: University of Wisconsin Press, 1973), 109.

46. Clarkson to Wedgwood, Aug. 25, 1791, LJW, Mss. 24738-E32. Clarkson to Wedgwood, June 17, 1793, LJW, Mss. 24742-32. For scholarship on the founding of Sierra
Leone, see Seymour Drescher, *The Mighty Experiment: Free Labor versus Slavery
in British Emancipation* (New York: Oxford University Press, 2002), 88–100; Christopher Fyfe, *A History of Sierra Leone* (London: Oxford University Press, 1962),
1–65; Stephen J. Braidwood, *Black Poor and White Philanthropists: London's Blacks
and the Foundation of the Sierra Leone Settlement, 1786–1791* (Liverpool: Liverpool
University Press, 1994); Christopher L. Brown, "From Slaves to Subjects: Envisioning an Empire without Slavery, 1772–1834," in *The Black Experience and the Empire*, ed. Philip Morgan and Sean Hawkins (New York: Oxford University Press,
2004), 111–40; Isaac Land and Andrew M. Schocket, "New Approaches to the
Founding of the Sierra Leone Colony, 1786–1808," *Journal of Colonialism &
Colonial History* 9, no. 3 (Winter 2008).

47. Wedgwood to Bentley, May 20, 1767, LJW, Mss. 18146-25. Wedgwood to
Bentley, Dec. 15, 1777, in *CJW*, 2:283 (emphasis in original). Before 1783 Wedgwood had imported nearly five tons of Cherokee clay; see Reilly, *Josiah Wedgwood*,
160.

48. For a study of the twinned projects of Sierra Leone and New South Wales, see
Deirdre Coleman, *Romantic Colonization and British Anti-Slavery* (New York:
Cambridge University Press, 2004).

49. Josiah Wedgwood, "On the Analysis of a Mineral Substance from New South
Wales," *Philosophical Transactions* 80 (1790): 306. Priestley, *Sermon on the Slave
Trade*, 26. Darwin, *Botanic Garden*, part 1, canto II, ll. 310–16, n.

50. "The Loves of the Triangles, continued," *Anti-Jacobin, or Weekly Examiner* (April
23, 1798): canto I, ll. 133–34. Priestley, *Memoirs*, 2:118. For Priestley's life in America, see Jenny Graham, "Joseph Priestley in America," in *Joseph Priestley, Scientist,
Philosopher, and Theologian*, ed. Isabel Rivers and David L. Wykes (Oxford: Oxford
University Press, 2008), 203–30

51. Darwin to James Watt, Jan. 19, 1790, in *The Essential Writings of Erasmus Darwin*,
ed. Desmond King-Hele (London: MacGibbon & Kee, 1968), 37.

52. Erasmus Darwin, *Phytologia; or, The Philosophy of Agriculture and Gardening*
(London, 1800), viii, 526–27, 78. Erasmus Darwin, *The Temple of Nature* (London,
1803), canto IV, l. 66n.

53. Darwin, *The Temple of Nature*, canto IV, ll. 224–26. For a summary of Erasmus
Darwin's theory of evolution, see McNeil, *Under the Banner of Science*, 115–24. For
Erasmus Darwin's influence on Charles Darwin, see Adrian Desmond and James
Moore, *Darwin's Sacred Cause: How a Hatred of Slavery Shaped Darwin's Views on
Human Evolution* (Boston: Houghton Mifflin Harcourt, 2009).

54. Martin Priestman published the manuscript of Darwin's *Progress of Society* in
Priestman, *The Poetry of Erasmus Darwin*, "Appendix A: *The Progress of Society, or
the Temple of Nature*, by Erasmus Darwin," 259–82.

55. Darwin, *Temple of Nature*, canto IV, ll. 373–74, 399. He also mentioned slavery in this canto (ll. 73–74), suggesting it was on par with disease and war.

56. Thomas Malthus, *An Essay on the Principle of Population* (London, 1798), 44.

57. For Malthus's influence on Darwin, see Priestman, *Poetry of Erasmus Darwin*, 137; McNeil, *Under the Banner of Science*, 113–14. Seymour Drescher discusses Malthus's work in relation to the early abolitionist campaign in *Mighty Experiment*, 40–42.

58. For instance, in 1807 William Wilberforce wrote that to "emancipate [slaves] at once" would be the "grossest violation." See William Wilberforce, *Letter on the Abolition of the Slave Trade* (London, 1807), 258–59.

59. For the Adams administration's opening of diplomatic relations with Haiti, see Ronald Johnson, *Diplomacy in Black and White: John Adams, Toussaint Louverture, and Their Atlantic World Alliance* (Athens: University of Georgia Press, 2014). For the Haitian Revolution's leading a few white abolitionists to call for immediate emancipation within the United States, see James Alexander Dun, "American Reaction to the Haitian Revolution," in *Encyclopedia of African-American Culture and History*, ed. Colin Palmer, 6 vols., 2nd ed. (Detroit: Macmillan, 2006), 3:979. "Character of the Celebrated Black General Toussaint L'Ouverture," *Balance and Columbian Repository* (New York), July 16, 1801, 36. *Mercantile Advertiser* (New York), Oct. 5, 1802, 3 (emphasis in original).

60. For Priestley's letters responding to Rush's offers, see Priestley to Benjamin Rush, Sept. 14, 1794, in *Scientific Correspondence of Joseph Priestley*, 139–44. "A Letter from Republican Natives of Great Britain and Ireland," June 13, 1794, in Priestley, *Memoirs*, 2:253. William Cobbett, *Observations on the Emigration of Dr. Joseph Priestley, and on the Several Addresses Delivered to Him on His Arrival at New-York* (Philadelphia, 1794), 30–31.

61. Priestley, "Letter in Response to the Republican Natives of Great Britain and Ireland," June 13, 1794, in *Memoirs*, 2:254 (emphasis in original). Joseph Priestley, *Letters to the Inhabitants of Northumberland and Its Neighbourhood* (Northumberland, 1799), 6 (emphasis in original). For the attack Priestley responded to, see William Cobbett, "Priestley: Completely Defected," *Porcupine's Gazette*, Aug. 20, 1798, 2.

62. Priestley to John Wilkinson, Dec. 17, 1795, extract quoted in Priestley Papers, RS, MS/654 f. 33. For Priestley's "Servants alone" remark, see Henrietta Liston to her uncle, Germantown, Aug. 14, 1796, in "A Diplomat's Wife in Philadelphia: Letters of Henrietta Liston, 1796–1800," ed. Bradford Perkins, *William and Mary Quarterly* 11, no. 4 (Oct. 1954): 602. For Priestley's use of a slave, see Graham, "Joseph Priestley in America," 217.

63. Joseph Priestley, "Maxims of Political Arithmetic," first published in the *Aurora* (Pa.), Feb. 26, 27, 28, 1798; reprinted as an appendix to Priestley, *Letters to the Inhabitants of Northumberland*, 39.

64. Joseph Priestley, *A General History of the Christian Church: To the Fall of the Western Empire*, 2nd ed. (Northumberland, 1803–4), xiv. Thomas Jefferson to Joseph Priestley, Jan. 18, 1800, *Papers of Thomas Jefferson*, ed. Barbara B. Oberg (Princeton: Princeton University Press, 2004–), 31:319–23. For Jefferson's scientific ideas, see Timothy Sweet, "Jefferson, Science and Enlightenment," in *Cambridge Companion to Thomas Jefferson*, ed. Frank Shuffleton (New York: Cambridge University Press, 2009), chap. 6. For Jefferson and slavery, see Paul Finkelman, "Jefferson and Slavery: 'Treason against the Hopes of the World,' " in *Jeffersonian Legacies*, ed. Peter Onuf (Charlottesville: University Press of Virginia, 1993), 181–221.

65. Thomas Cooper to Benjamin Rush, Feb. 6, 1804, in *Scientific Correspondence of Joseph Priestley*, 162.

3 A Natural History of Sierra Leone

1. Henry Smeathman to George Cumberland, Paris, Aug. 31, 1783, Cumberland Papers, British Library, Mss. 36494, vol. IV, f. 131. For Lettsom's introduction, see John Coakley Lettsom to Benjamin Franklin, Aug. 2, 1783, in *Papers of Benjamin Franklin*, 41 vols. (New Haven: Yale University Press, 1954–) (hereafter cited as *PBF*), 40:426–27. For European knowledge of West Africa in this period, see Philip Curtin, *The Image of Africa: British Ideas and Action, 1780–1850* (Madison: University of Wisconsin Press, 1964), 3–27. See also Philip Curtin, " 'The White Man's Grave': Image and Reality, 1780–1850," *Journal of British Studies* 1, no. 1 (Nov. 1961): 94–110.

2. For prominent works on Sierra Leone's history and founding, some of which mention the scientific backgrounds of the colony's early advocates but do not explore them in any detail, see Christopher Fyfe, *A History of Sierra Leone* (London: Oxford University Press, 1962); Stephen J. Braidwood, *Black Poor and White Philanthropists: London's Blacks and the Foundation of the Sierra Leone Settlement, 1786–1791* (Liverpool: Liverpool University Press, 1994); Alexander X. Byrd, *Captives and Voyagers: Black Migrants across the Eighteenth-Century British Atlantic World* (Baton Rouge: Louisiana State University Press, 2008), 125–252; Cassandra Pybus, *Epic Journeys of Freedom: Runaway Slaves of the American Revolution and Their Global Quest for Liberty* (Boston: Beacon, 2006), 103–22, 139–56, 169–82; Padraic X. Scanlan, *Freedom's Debtors: British Antislavery in Sierra Leone in the Age of Revolution* (New Haven: Yale University Press, 2017); and Bronwen Everill, *Abolition and Empire in Sierra Leone and Liberia* (New York: Palgrave, 2013). To be sure, Starr Douglas has explored Smeathman's scientific work in Africa in relation to the rise of abolitionism, though she emphasizes the role that slavery, not antislavery, played in shaping his science. See Starr Douglas, "The Making of Scientific Knowledge in an Age of Slavery: Henry Smeathman, Sierra Leone and Natural History," *Journal of Colonialism and Colonial History* 9, no. 3 (2008), https://muse.

jhu.edu/article/255265. Deirdre Coleman has also explored Smeathman's African career in depth; see Deirdre Coleman, *Henry Smeathman, the Flycatcher: Natural History, Slavery, and Empire in the Late Eighteenth Century* (Liverpool: Liverpool University Press, 2018); Deirdre Coleman, *Romantic Colonization and British Anti-Slavery* (New York: Cambridge University Press, 2004), 28–62. See also Isaac Land and Andrew M. Schocket, "New Approaches to the Founding of the Sierra Leone Colony, 1786–1808," *Journal of Colonialism & Colonial History* 9, no. 3 (2008), https://muse.jhu.edu/article/255263. For antislavery support for colonization before the American Colonization Society's founding, see Matthew Spooner, " 'I Know This Scheme Is from God': Toward a Reconsideration of the Origins of the American Colonization Society," *Slavery & Abolition* 35, no. 3 (2014): 559–75; Beverly Tomek, " 'From Motives of Generosity, as Well as Self-preservation': Thomas Branagan, Colonization, and the Gradual Emancipation Movement," *American Nineteenth Century History* 6, no. 2 (2005): 121–47.

3. Samuel Hopkins to Granville Sharp, Jan. 15, 1789, in *Memoir of the Life and Character of Samuel Hopkins*, ed. Edwards A. Park, 2nd ed. (Boston, 1854), 140–41. Samuel Hopkins, "A Discourse upon the Slave Trade (1793)," ibid., 145.

4. Braidwood, *Black Poor and White Philanthropists*, 22–33. The Somerset case involved an enslaved Bostonian, James Somerset, who escaped from his owner while in England. Somerset's case was taken up by Granville Sharp, and the ruling, issued in 1772 by England's highest court, declared that slavery was contrary to English law. Though the ruling freed Somerset and set a precedent that slavery in England was not recognized by law, it was not an emancipation decree. The small number of slave owners in England could keep their slaves, and the ruling did not apply to the British Empire's colonies overseas. That said, the case was widely understood as a serious blow to slavery. See David Brion Davis, *The Problem of Slavery in the Age of Revolution, 1770–1823* (Ithaca: Cornell University Press, 1975), 480–82.

5. Katherine Paugh, *Politics of Reproduction: Race, Medicine, and Fertility in the Age of Abolition* (New York: Oxford University Press, 2017), 75. Granville Sharp, "Remarks on Case of John Hylas and His Wife Mary," in *Memoirs of Granville Sharp*, ed. Prince Hoare (London: H. Colburn, 1820), vi. For white attitudes toward Blacks in Britain during this period, see Anthony Barker, *The African Link: British Attitudes to the Negro in the Era of the Atlantic Slave Trade, 1550–1807* (London: Frank Cass, 1978), 157–93.

6. John Fothergill to Granville Sharp, 1769, in John Fothergill, *Chain of Friendship: Selected Letters with Introduction and Notes*, ed. Christopher C. Booth and Betsy C. Corner (Cambridge: Harvard University Press, 1971), 313. Fothergill to Sharp, Feb. 2, 1772, ibid., 375. Abolitionists who credited Fothergill with the original idea include Sharp, Carl Bernhard Wadström, and John Coakley Lettsom. See Sharp to Lettsom, Oct. 13, 1788, in *Memoirs of the Life and Writings of the Late John Coakley Lettsom*, ed. Thomas Pettigrew, 3 vols. (London, 1817), 2:236–37. Carl Bernhard

Wadström, *An Essay on Colonization, Particularly Applied to the Western Coast of Africa*, 2 vols. (London, 1794–95), 2:3. On King Agaja, see *Memoirs of John Fothergill*, ed. John Coakley Lettsom, 4th ed. (London: C. Dilly, 1786), 70. To be sure, other European thinkers had written about creating free Black colonies somewhere in West Africa before Fothergill. But Fothergill focused on the Sierra Leone region in particular, and his efforts to outfit a scientific expedition to the region would directly influence the actual colony's founding. For West African colonization plans before Sierra Leone, see Christopher L. Brown, "Empire without America: British Plans for Africa in the Era of the American Revolution," in *Abolitionism and Imperialism in Britain, Africa, and the Atlantic*, ed. Derek Peterson (Athens: Ohio University Press, 2010), 88–89. For the scholarly debate about Agaja's motives, see Robin Law, "Dahomey and the Slave Trade: Reflections on the Historiography of the Rise of Dahomey," *Journal of African History* 27, no. 2 (1986): 237–67.

7. John Fothergill to John Bartram, May 1, 1769, in Fothergill, *Chain of Friendship*, 302. Fothergill to Carl Linnaeus, April 4, 1774, ibid., 409. Much of Fothergill's correspondence with Americans in *Chain of Friendship* involves botanical exchanges. For more examples, see Fothergill to Humphry Marshall, March 2, 1767, ibid., 274–46; Fothergill to Bartram, Feb. 22, 1743/4, ibid., 83–86. For Fothergill's plant collection, see Corner and Booth, introduction to Fothergill, *Chain of Friendship*, 3–34. Fothergill and Collinson arranged for Franklin's electrical work to be published in London's *Gentleman's Magazine* in 1745, and Fothergill wrote the unsigned preface. See Margaret DeLacy, "John Fothergill (1712–1780)," *Oxford Dictionary of National Biography* (New York: Oxford University Press, 2004).

8. John Fothergill to Samuel Fothergill, June 9, 1772, in Fothergill, *Chain of Friendship*, 385. Thomas Clarkson, *The History of the Rise, Progress and Accomplishment of the Abolition of the African Slave-Trade*, 2 vols. (London: Longman, 1808), 1:169. For Benezet's solicitation of Fothergill, see *Memoirs of Fothergill*, 71n. For Samuel Fothergill's and Israel Pemberton's travels throughout the American colonies, see Davis, *Problem of Slavery in the Age of Revolution*, 230.

9. Anthony Benezet, *Some Historical Account of Guinea* (Philadelphia: Joseph Crukshank, 1771), 14 (emphases in original). See also: Maurice Jackson, *Let This Voice Be Heard: Anthony Benezet, Father of Atlantic Abolitionism* (Philadelphia: University of Pennsylvania Press, 2009), 78–79, 105–6. For Adanson's work and other key natural histories written before the 1780s, see Philip Curtin, *Image of Africa: British Ideas and Action, 1780–1850*, vol. 1 (Madison: University of Wisconsin Press, 1973), 12–16.

10. Joseph Banks to Thomas Coltman, March 1792, quoted in John Gascoigne, *Joseph Banks and the English Enlightenment: Useful Knowledge and Polite Culture* (New York: Cambridge University Press, 1994), 40. Gascoigne, the leading scholar on Banks, sums up Banks's attitude toward abolitionism as "rather ambivalent" (ibid., 41). For the long history of the Royal Society's debts to slaveholders, see James

Delbourgo, *Collecting the World: Hans Sloane and the Origins of the British Museum* (Cambridge: Harvard University Press, 2017), esp. chaps. 1–3; Kathleen S. Murphy, "Collecting Slave Traders: James Petiver, Natural History, and the British Slave Trade," *William and Mary Quarterly* 70, no. 4 (2013): 637–70; Douglas, "Making of Scientific Knowledge in an Age of Slavery." On the funders of Smeathman's 1771 voyage, see Smeathman's eulogy of John Fothergill, Oct. 19, 1782, in *The Works of John Fothergill, M.D.*, ed. John Coakley Lettsom, 3 vols. (London, 1784), 3:185–86.

11. Smeathman's eulogy of John Fothergill, 184–86. Fothergill to Carl Linnaeus, April 4, 1774, in Fothergill, *Chain of Friendship*, 409–10.

12. Joseph Banks described Fothergill's efforts to help him transplant the breadfruit in 1774. See *Memoirs of Fothergill*, xli–xlii. For the breadfruit initiative, see Richard Sheridan, "Captain Bligh, the Breadfruit and the Botanic Gardens of Jamaica," *Journal of Caribbean History* 23, no. 1 (1989): 28–50, and Anya Zilberstein, "Bastard Breadfruit and Other Cheap Provisions: Early Food Science for the Welfare of the Lower Orders," *Early Science and Medicine* 21, no. 5 (2016): 492–508. For amelioration campaigns, see J. W. Ward, *British West Indian Slavery, 1750–1834: The Process of Amelioration* (Oxford: Clarendon Press, 1988), and Christa Dierksheide, *Amelioration and Empire: Progress and Slavery in the Plantation Americas* (Charlottesville: University of Virginia Press, 2014).

13. Fothergill to John Ellis, Sept. 2, 1773, in *Works of John Fothergill*, 2:316, 325. For a general description of Fothergill's work on coffee cultivation, see *Memoirs of Fothergill*, 45. For Ellis's advocacy of transplanting breadfruit to the Caribbean, see John Ellis, *A Description of the Mangostan and the Bread-fruit* (London, 1775).

14. John Coakley Lettsom to Dr. Cuming, Oct. 20, 1787, *Memoirs of Lettsom*, 1:132, 2:36. *Memoirs of Fothergill*, 66–67.

15. Smeathman also spoke highly of the Liverpool slave trader Miles Barber in his private journal. See Henry Smeathman Journal, December 1771, Uppsala University Library, D.26. For details on marriages and the contract with William James, see Coleman, *Romantic Colonization*, 33. Smeathman to Lettsom, Oct. 15, 1785, *Memoirs of Lettsom*, 2:282, 284. For scholars who doubt his antislavery convictions, see Douglas, "The Making of Scientific Knowledge in an Age of Slavery"; Fyfe, *A History of Sierra Leone*, 22; Braidwood, *Black Poor and White Philanthropists*, 94.

16. Smeathman Journal, 1771. Smeathman to Marmaduke Tunstall, June 17, 1776, private collection of Michael Graves-Johnston, London. Smeathman to Tunstall, Tobago, May 28, 1777, private collection of Michael Graves-Johnston. Deirdre Coleman has recently published the manuscript in *Henry Smeathman, the Flycatcher*, Appendix, "Economy of a Slave Ship."

17. Smeathman Journal, 1771.

18. Ibid.

19. This description of African ethnography, and how it differed among pro- and antislavery men of science draws primarily from Curtin, *Image of Africa*, 28–57, and

Barker, *The African Link,* 179–93. For a fuller treatment of how understandings of race began to change in latter eighteenth-century Britain, see Roxann Wheeler, *Complexion of Race: Categories of Difference in Eighteenth-Century British Culture* (Philadelphia: University of Pennsylvania Press, 2000), 253–60.

20. Alexander Wilson, *Some Observations relative to the Influence of Climate on Vegetable and Animal Bodies* (London: T. Cadell, 1780), 279. For "tropical exuberance," see Klas Ronnback, "Enlightenment, Scientific Exploration and Abolitionism: Anders Sparrman's and Carl Bernhard Wadström's Colonial Encounters in Senegal, 1787–1788, and the British Abolitionist Movement," *Slavery & Abolition* 34, no. 3 (Sept. 2013): 425–45; Curtin, *Image of Africa,* 58–60.

21. Henry Smeathman to Dr. Thomas Knowles, ca. July 1783 (second letter), *New-Jerusalem Magazine* (1790): 286, 288.

22. Smeathman Journal, 1771.

23. Smeathman's eulogy of Fothergill, 194. Smeathman to Tunstall, June 17, 1776, private collection of Michael Graves-Johnston. For the introduction of African botanicals to the Americas, see Judith Carney and Richard Rosomoff, *In the Shadow of Slavery: Africa's Botanical Legacy in the Atlantic World* (Berkeley: University of California Press, 2009).

24. Smeathman to Tunstall, May 28, 1777, private collection of Michael Graves-Johnston. Henry Smeathman, "Some Account of the Termites, Which Are Found in Africa and Other Hot Climates, *Philosophical Transactions of the Royal Society of London* 71 (1781): 141.

25. Smeathman, "Some Account of the Termites," 145, 161 (emphasis in original). For Smeathman's parallel views on termites and Africans, see Coleman, *Romantic Colonization,* 28–62.

26. Smeathman's eulogy of Fothergill, 194. Fothergill to Carl Linnaeus, April 4, 1774, in Fothergill, *Chain of Friendship,* 409. Fothergill to John Ellis, July 14, 1774, in Fothergill, *Chain of Friendship,* 416–17.

27. Granville Sharp, "Diving Bell Drawing," ca. 1775, RS, MS L&P/8/122. For Sharp's attendance at Royal Society meetings, see Royal Society Minutes, Feb. 11, 1762, Nov. 11, 1762, RS, JBO/25/34–53. Braidwood notes that Smeathman hoped to use the profits from ballooning to fund the colony; see Braidwood, *Black Poor and White Philanthropists,* 7, 38. Henry Smeathman, "On a Flying Vessel," March 4, 1784, RS, AP/5/12. George Cumberland to Highman, Oct. 17, 1783, British Library, Additional MSS 36494, Cumberland Papers, f. 175v. Smeathman to Lettsom, July 16, 1784, in *Memoirs of Lettsom,* 2:271; Smeathman to Lettsom, Feb. 7, 1784, in *Memoirs of Lettsom,* 2:280.

28. Smeathman to Lettsom, Feb. 7, 1784, in *Memoirs of Lettsom,* 2:278. Here I draw on Deirdre Coleman's insights about Smeathman, ballooning, and slavery in *Romantic Colonization,* 49–56. For a recent general history of ballooning, science, and

Romantic poetry, see Richard Holmes, *Falling Upwards: How We Took to the Air* (New York: Pantheon, 2013). Shelley is quoted in Coleman, *Romantic Colonization*, 49. Smeathman Journal, 1771. Smeathman to Lettsom, Oct. 15, 1785, in *Memoirs of Lettsom*, 2:284.

29. Fyfe, *History of Sierra Leone*, 14–15; Braidwood, *Black Poor and White Philanthropists*, 93–102; Byrd, *Captives and Voyagers*, 126–38. Smeathman first began to advocate for an antislavery colony in 1783, as evidenced by his letters to Franklin and two letters to the Quaker abolitionist and physician Dr. Thomas Knowles. See Smeathman to Dr. Thomas Knowles, July 21, 1783, and ca. July 1783 (second letter), *New-Jerusalem Magazine* (1790): 279–93.

30. Granville Sharp to William Sharp, Jan. 1788, in *Memoirs of Granville Sharp*, 260–61. Braidwood, *Black Poor and White Philanthropists*, 101–2; Land and Schocket, "New Approaches to the Founding of the Sierra Leone Colony, 1786–1808."

31. Henry Smeathman, *Plan of a Settlement to Be Made Near Sierra Leone* (London, 1786), 18. For the Black settlers' role in shaping Smeathman's documents, see Byrd, *Captives and Voyagers*, 130–31.

32. Smeathman, *Plan of a Settlement*, title page. On Smeathman's ambiguity about whether slavery would be tolerated in the colony, see Braidwood, *Black Poor and White Philanthropists*, 93–102; Fyfe, *History of Sierra Leone*, 15; Byrd, *Captives and Voyagers*, 130–31.

33. Smeathman, *Plan of a Settlement*, 8–11.

34. A. Elliot to Granville Sharp, July 20, 1787, in *Memoirs of Granville Sharp*, 320–21. Elliot was a white teacher sent to instruct the settlers and indigenous children. For the initial group of settlers, see Braidwood, *Black Poor and White Philanthropists*, esp. 186–87, 194–95, 206; Fyfe, *History of Sierra Leone*, 16–28; Byrd, *Captives and Voyagers*, 200–243. Stephen Braidwood, like other scholars, emphasizes the death rate as a leading cause of the colony's early troubles. Though he also notes that the colony's projectors shared responsibility for setting high expectations, I extend his interpretation further and home in on the specific importance of the men of science who touted these views.

35. Mariola Espinosa, "The Question of Racial Immunity to Yellow Fever in History and Historiography," *Social Science History* 38, no. 3–4 (2014): 437–53; P. W. Hedrick, "Population Genetics of Malaria Resistance in Humans," *Heredity* 107, no. 4 (Oct. 2011): 283–304; Curtin, " 'The White Man's Grave': Image and Reality, 1780–1850."

36. Sharp to Franklin, Jan. 10, 1788 (unpublished), *Papers of Benjamin Franklin* online, franklinpapers.org (emphasis in original). Sharp to Lettsom, Oct. 13, 1788, in *Memoirs of Lettsom*, 2:239.

37. Wadström, *Essay on Colonization*, 2:221, 1:44. Lettsom to Dr. Cuming, Oct. 20, 1787, in *Memoirs of Lettsom*, 1:135.

38. Smeathman, *Plan of a Settlement*, 10. Lettsom to Dr. Cuming, Oct. 20, 1787, in *Memoirs of Lettsom*, 1:133. Wadström, *Essay on Colonization*, 2:227, 221. Braidwood, *Black Poor and White Philanthropists*, 183–85; Fyfe, *History of Sierra Leone*, 22–23.

39. Sharp to Lettsom, Oct. 13, 1787, in *Memoirs of Lettsom*, 2:246. Lettsom to Cuming, Oct. 20, 1787, in *Memoirs of Lettsom*, 1:136. Braidwood also cites Lettsom's quote, pointing out the irony; see Braidwood, *Black Poor and White Philanthropists*, 188. Wadström, *Essay on Colonization*, 1:35, 37

40. The Old Settlers at Sierra Leone to Granville Sharp, Sept. 3, 1788, in *Memoirs of Granville Sharp*, 331. Suzanne Schwarz, " 'A Just and Honourable Commerce': Abolitionist Experimentation in Sierra Leone in the Late Eighteenth and Early Nineteenth Centuries," *African Economic History* 45, no. 1 (2017): 6. Fyfe, *History of Sierra Leone*, 26–31. Byrd, *Captives and Voyagers*, 210–15.

41. Fyfe, *History of Sierra Leone*, 31–35; Pybus, *Epic Journeys of Freedom*, 103–22, 139–56, 169–82; Byrd, *Captives and Voyagers*, 177–99; Alan Gilbert, *Black Patriots and Loyalists: Fighting for Emancipation in the War of Independence* (Chicago: University of Chicago Press, 2013), 207–42.

42. "Minutes of the Evidence Taken before a Committee of the House of Commons, Being a Select Committee, Appointed on the 23rd Day of April 1790 . . . To Consider Further the Circumstances of the Slave Trade" (London, 1790), 19. Clarkson, *The History of the Rise, Progress and Accomplishment of the Abolition of the African Slave-Trade*, 1:488. Klas Ronnback is one of the few scholars to highlight the importance of men of science to the early abolitionist campaign, particularly between 1788 and 1792. He focuses on Wadström and his expedition partner Anders Sparrman; see Klas Ronnback, "Enlightenment, Scientific Exploration and Abolitionism: Anders Sparrman's and Carl Bernhard Wadström's Colonial Encounters in Senegal, 1787–1788, and the British Abolitionist Movement," *Slavery & Abolition* 34, no. 3 (Sept. 2013): 425–45.

43. Carl Wadström, *Observations on the Slave Trade and a Description of Some Part of the Coast of Guinea* (London: James Phillips, 1789), 53, iii, 6, vi.

44. Ibid., 31–33, 6 (emphasis in original), 38, 42.

45. "Minutes of the Evidence . . . Appointed on the 23rd Day of April 1790," 31–35.

46. Ibid., 33, 40, 35–36.

47. Benjamin Rush, "An Account of the Sugar Maple-Tree; Together with Observations upon the Advantages Both Public and Private of This Sugar, in a Letter to Thomas Jefferson," in *Transactions of the American Philosophical Society* 3 (1793): 79, 68. Rush cites James Bruce's *Travels to Discover the Source of the Nile, in the Years 1768, 1769, 1770, 1771, 1772, and 1773*, 5 vols. (1790). Wadström's *Observations on the Slave Trade* appeared in several editions in 1791 and 1792. For an example of one published in Philadelphia, see William Bell Crafton, *A Short Sketch of the Evidence for the Abolition of the Slave Trade, Delivered before a Committee of the House of Commons* (Philadelphia, 1792). For an example of an edition with

Wadström's evidence highlighted, see *An Abstract of the Evidence Delivered before a Select Committee of the House of Commons, in the Years 1790 and 1791*, 2nd ed. (Bury, U.K.: R. Hayworth, 1792), 25–26, 130–31, 220.

48. Wadström, *Observations on the Slave Trade*, 67. Klas Ronnback notes that he published a version in Swedish that included almost all the same depictions of West Africa's natural history as his abolitionist works. See Ronnback, "Enlightenment, Scientific Exploration and Abolitionism." For the revival of the abolitionist campaign by 1804, see Roger Anstey, *The Atlantic Slave Trade and British Abolition, 1760–1810* (Atlantic Highlands, N.J.: Humanities Press, 1975), 341–63.

49. Wadström, *Essay on Colonization*, 1:iii, 26, 43.

50. Ibid., 4 (emphasis in original), 9–17. John Matthews, *A Voyage to the River Sierra-Leone, on the Coast of Africa* (London: B. White, 1791), 86, 88–90.

51. Helen Maria Williams, "Memoirs of the Life of Charles Berns Wadström," *Monthly Magazine, or, British Register* (July 1799): 463.

52. On Afzelius, see Alexander P. Kup, ed. and trans., introduction to Afzelius, *Adam Afzelius: Sierra Leone Journal, 1795–1796* (Uppsala: Almqvist & Wiksell, 1967). Joseph Banks to Olof Swartz, Soho Square, Dec. 26, 1790, in *Scientific Correspondence of Sir Joseph Banks, 1765–1820*, ed. Neil Chambers, 6 vols. (London: Routledge, 2007), 4:25–26.

53. Adam Afzelius to "the Governor of Sierra Leone Concerning some Plants and Seeds sent home by the Ocean and the Army," Freetown, Nov. 19, 1794, Afzelius Correspondence, Uppsala University Library, D.26. On bioprospecting, see Londa Schiebinger, *Plants and Empire: Colonial Bioprospecting in the Atlantic World* (Cambridge: Harvard University Press, 2004), 73–104. For European reliance on African agricultural knowledge, see Judith Carney, *Black Rice: The African Origins of Rice Cultivation in the Americas* (Cambridge: Harvard University Press, 2001). On the reliance of European naturalists on enslaved and free people of African descent, see Londa Schiebinger, *Secret Cures of Slaves: People, Plants, and Medicine in the Eighteenth-Century Atlantic World* (Palo Alto: Stanford University Press, 2017); Susan Scott Parrish, *American Curiosity: Cultures of Natural History in the Colonial British Atlantic World* (Chapel Hill: University of North Carolina Press, 2006), 259–306; Delbourgo, *Collecting the World*, chaps. 1–3; Murphy, "Collecting Slave Traders."

54. Entries for May 22, 23, 1795, Jan. 4, 1796, and April 30, 1795, in Afzelius, *Sierra Leone Journal*, 16, 39, 12. On Tarleton Fleming, see entry for May 22–23, 1795, in Afzelius, *Sierra Leone Journal*, 17. According to Kup, Mrs. Logan was the wife of a Black settler, Mr. Logan, who was a Nova Scotian and formerly a slave in Virginia; see Afzelius, *Sierra Leone Journal*, 88–89n54r.

55. Entry for May 22–23, 1795, in Afzelius, *Sierra Leone Journal*, 17; for a brief description of Fleming and timber statistics, see ibid., 87n40r. Banks to Olof Swartz, Feb. 26, 1794, in *Scientific Correspondence of Sir Joseph Banks*, 4:273–74.

56. Entry for March 5, 1796, in Afzelius, *Sierra Leone Journal*, 103. Thomas Cooper to John Symmonds, April 12, 1796, in *Records of the African Association: 1788–1831*, ed. Robin Hallett (London: T. Nelson, 1964), 274–75. On go-betweens, see Simon Schaffer, Lissa Roberts, Kapil Raj, and James Delbourgo, eds., *The Brokered World: Go-betweens and Global Intelligence, 1770–1820* (Sagamore Beach, Mass.: Science History Publications, 2009).

57. Mungo Park, *Travels in the Interior Districts of Africa* (1799), ed. Kate Ferguson Marsters (Durham: Duke University Press, 2000), 263. For the African Association's views on slavery, see Hallett, *Records of the African Association*, 14–17. For a discussion of other expeditions during the 1790s and early 1800s under the Sierra Leone Company's sponsorship, see Schwarz, " 'A Just and Honourable Commerce,' " 1–45.

58. Adam Afzelius to "the Governor of Sierra Leone Concerning some Plants and Seeds."

59. Entries for Jan. 2, 9, and March 7, 1796, in Afzelius, *Sierra Leone Journal*, 38, 42, 105. "Afzelia africana," in *The Cabi Encyclopedia of Forest Trees* (Oxford: CABI, 2013); "Afzelia africana Sm.," *JStor Global Plants*, http://plants.jstor.org/stable /10.5555/al.ap.upwta.3_86 (accessed June 9, 2017).

60. Thomas Winterbottom, *An Account of the Native Africans in the Neighbourhood of Sierra Leone*, 2 vols. (London: C. Whittingham, 1803), 2:253. For Peruvian bark history, see Matthew James Crawford, *The Andean Wonder Drug: Cinchona Bark and Imperial Science in the Spanish Atlantic, 1630–1800* (Pittsburgh: University of Pittsburgh Press, 2016).

61. John Gray to Afzelius, July 19, 1798 (emphasis in original), and July 26, 1798, Afzelius Correspondence. James Edward Smith to Afzelius, Jan. 8, 1798, ibid. (emphasis in original). "Afzelia africana," *Transactions of the Linnaean Society in London* 4 (1798): 221.

62. John Clarkson to Afzelius, Sept. 9, 1793; John Prinsep to Afzelius, Feb. 17, April 25, and May 31, 1797, Afzelius Correspondence. For the African bark as a promising cure, see also Samuel Parker to Afzelius, 1796, ibid. Afzelius to Gov. of SL Company, Freetown, June 14, 1794, ibid. Afzelius to the Governor and Council of Sierra Leone, "An Account of the State of the Public Gardens before the Arrival of the French and After Their Departure to the Present-Time," Freetown, Nov. 27, 1794, ibid.

63. Afzelius to the Governor and Council of Sierra Leone, "An Account of the State of the Public Gardens," and John Sims to Afzelius, March 2, 1795, Afzelius Correspondence.

64. Cato Perkins and Isaac Anderson to John Clarkson, London, Oct. 30, 1793, in *"Our Children Free and Happy": Letters from Black Settlers in Africa in the 1790s*, ed. Christopher Fyfe (Edinburgh: University of Edinburgh Press, 1991), 40, 6. Luke Jordan and Isaac Anderson to directors of Sierra Leone Company, June 28, 1794, ibid., 42–43. Fyfe, *History of Sierra Leone*, 56.

65. Fyfe, *History of Sierra Leone*, 79–104.

66. "Substance of the Report Delivered by the Court of Directors of the Sierra Leone Company, to the General Court of Proprietors: on Thursday, March 27th, 1794" (London, 1794), 50, 166, 170. See also "Substance of the Report Delivered by the Court of Directors of the Sierra Leone Company, to the General Court of Proprietors: on Thursday, the 29th March, 1798" (London, 1798).

67. "Substance of the Report delivered by the Court of Directors, March 27th, 1794," 60. Curtin, *Image of Africa*, 137. "Art. 46. Substance of the Report Delivered by the Court of Directors of the Sierra-Leone Company to the General Court of Proprietors on Thursday, the 26th March, 1801," *Critical Review, or, Annals of Literature* 33 (1801): 118.

68. Stuart Menzies, "Thomas Masterman Winterbottom, MD, 1766–1859," in *Medicine in Northumbria: Essays on the History of Medicine in the North East of England*, ed. David Gardner-Medwin, Anne Hargreaves, and Elizabeth Lazenby (Newcastle-upon-Tyne: Pybus Society for the History and Bibliography of Medicine, 1993): 193–210. To be sure, Winterbottom's mother, Lydia Masterman, was the daughter of a local shipowner, who may well have outfitted ships for the slave trade.

69. Winterbottom, *Account of the Native Africans*, 1:28; 11, 55, 57, 93–97, 161, 220–21.

70. Winterbottom, *An Account of the Native Africans*, 1:ii, 7, 161, 166. Three prominent texts Winterbottom cited were Bruce's *Travels to Discover the Source of the Nile* (1790); Robert Norris's *Memoirs of the Reign of Bossa Ahadee, King of Dahomy* (1789); and William Snelgrave's *A New Account of Some Parts of Guinea, and the Slave-Trade* (1734).

71. For useful works on early abolitionism's influence on racial theories, see Barker, *The African Link*, 157–93; Wheeler, *Complexion of Race*, 253–60; Andrew S. Curran, *The Anatomy of Blackness: Science & Slavery in an Age of Enlightenment* (Baltimore: Johns Hopkins University Press, 2011), 204–15. For overviews of racial science at the turn of the nineteenth century, see Winthrop Jordan, *White over Black: American Attitudes toward the Negro, 1550–1812* (Chapel Hill: University of North Carolina Press, 1968), 491–509; David Bindman, *Ape to Apollo: Aesthetics and the Idea of Race in the 18th Century* (Ithaca: Cornell University Press, 2002), 201–21; Wheeler, *Complexion of Race*, 26–27, 250–51.

72. Petrus Camper, *The Works of the Late Professor Camper* (London, 1794), 32. Winterbottom, *An Account of the Native Africans*, 1:199. Blumenbach's quote was widely cited by abolitionists. See, for example, Ignatius Sancho, *Letters of the Late Ignatius Sancho*, ed. Joseph Jekyll (London, 1803), xi.

73. Winterbottom, *An Account of the Native Africans*, 1:197–98, 216–17.

74. Ibid., 2:1–2.

75. Ibid., 1:251–52, 257.

76. Entry for May 23, 1798, in Zachary Macaulay Journal (1793–1799), Huntington Library (mf). Fyfe, *History of Sierra Leone*, 99–103. Pybus, *Epic Journeys of Freedom*, 215.

77. Anstey, *The Atlantic Slave Trade and British Abolition*, 343–44.

78. Thomas Branagan, *Serious Remonstrances . . . with a Simplified Plan for Colonizing the Free Negroes* (Philadelphia, 1805), 64.

4 Trials in Freedom

1. Benjamin Silliman, "Origin and Progress of Chemistry," book 1, pp. 37–39. Silliman Family Papers (hereafter cited as SFP), Yale University, series III, reel 1 (mf). John C. Calhoun to Benjamin Silliman, March 26, 1818, in George Park Fisher, *Life of Benjamin Silliman*, 2 vols. (New York: Scribner, 1866), 1:288; Ebenezer Baldwin, *History of Yale College: From Its Foundation, A.D. 1700, to the Year 1838* (New Haven, 1841), 241. For Maclure's reliance on slaveholders, see Maclure, "Notes without Title relating to the West Indies," William Maclure Letters, 1796–1848, American Philosophical Society (mf), reel 513.4. For Silliman, see Chandos Michael Brown, *Benjamin Silliman: A Life in the Young Republic* (Princeton: Princeton University Press, 1989); John F. Fulton and Elizabeth H. Thomson, *Benjamin Silliman, 1779–1864: Pathfinder in American Science* (New York: Schuman, 1947). See also Craig Steven Wilder, *Ebony & Ivy: Race, Slavery, and the Troubled History of America's Universities* (New York: Bloomsbury, 2013), 1, 62, 133–34, 272, 285. For extensive excerpts from Silliman's journals and private correspondence, see Fisher, *Life of Silliman*. For the most recent biography of Maclure, see Leonard Warren, *Maclure of New Harmony: Scientist, Progressive Educator, Radical Philanthropist* (Bloomington: Indiana University Press, 2009). For Silliman's and Maclure's importance to promoting science in the early republic, see George H. Daniels, *American Science in the Age of Jackson* (New York: Columbia University Press, 1968), 10–13; Conevery Bolton Valencius, David I. Spanagel, Emily Pawley, Sara Sidstone Gronim, and Paul Lucier, "Science in Early America: Print Culture and the Sciences of Territoriality," *Journal of the Early Republic* 36, no. 1 (Spring 2016): 73–123.

2. For an excellent overview of the historiography on colonization, see Samantha Seeley, "Beyond the American Colonization Society," *History Compass* 14, no. 3 (March 2016): 93–104. For recent revisions of the ACS, see Eric Burin, *Slavery and the Peculiar Solution: A History of the American Colonization Society* (Gainesville: University Press of Florida, 2005); Matthew Spooner, " 'I Know This Scheme is from God': Toward a Reconsideration of the Origins of the American Colonization Society," *Slavery & Abolition* 35, no. 4 (2014): 559–75; Beverly Tomek, *Colonization and Its Discontents: Emancipation, Emigration, and Antislavery in Antebellum Pennsylvania* (New York: New York University Press, 2011); Beverly Tomek and

Matthew Hetrick, eds., *New Directions in the Study of African American Recoloni-zation* (Gainesville: University Press of Florida, 2017); Marie Tyler-McGraw, *An African Republic: Black & White Virginians in the Making of Liberia* (Chapel Hill: University of North Carolina Press, 2007); Nicholas Guyatt, " 'The Outskirts of Our Happiness': Race and the Lure of Colonization in the Early Republic," *Journal of American History* 95, no. 4 (March 2009): 986–1011. For Blacks' responses to coloni-zation, see Floyd J. Miller, *The Search for a Black Nationality: Black Emigration and Colonization, 1787–1863* (Urbana: University of Illinois Press, 1975); Ousmane K. Power-Greene, *Against Wind and Tide: The African American Struggle against the Colonization Movement* (New York: New York University Press, 2014), 17–45; Claude A. Clegg III, *The Price of Liberty: African Americans and the Making of Liberia* (Chapel Hill: University of North Carolina Press, 2004), 30–53; Sara Fan-ning, *Caribbean Crossing: African Americans and the Haitian Emigration Move-ment* (New York: New York University Press, 2015); James T. Campbell, *Middle Passages: African American Journeys to Africa, 1787–2005* (New York: Penguin, 2006). For Black Philadelphians rejecting the ACS, see Gary Nash, *Forging Free-dom: The Formation of Philadelphia's Black Community, 1720–1840* (Cambridge: Harvard University Press, 1988), 233–45; Julie Winch, *Philadelphia's Black Elite: Activism, Accommodation, and the Struggle for Autonomy, 1787–1848* (Philadelphia: Temple University Press, 1988), 27–47.

3. Patrick Rael, *Eighty-Eight Years: The Long Death of Slavery in the United States, 1777–1865* (Athens: University of Georgia Press, 2015), 136–37; Nash, *Forging Free-dom,* 157.

4. Manisha Sinha, *The Slave's Cause: A History of Abolition* (New Haven: Yale Uni-versity Press, 2016), 160–91; Rael, *Eighty-Eight Years,* 91–125. Beverly Tomek, "Seek-ing 'An Immutable Pledge from the Slave Holding States': The Pennsylvania Abolition Society and Black Resettlement," *Pennsylvania History* 75, no. 1 (Winter 2008): 46.

5. For the most critical study of Frances Wright and Nashoba, see Gail Bederman, "Revisiting Nashoba: Slavery, Utopia, and Frances Wright in America, 1818–1826," *American Literary History* 17, no. 3 (Autumn 2005): 438–59. See also Anne Taylor, *Visions of Harmony: A Study of Nineteenth Century Millenarianism* (New York: Oxford University Press, 1987), 164–75; Carol Morris Eckhardt, *Fanny Wright: Rebel in America* (Cambridge: Harvard University Press, 1984), 108–40; Warren, *Maclure of New Harmony,* 198–202.

6. Fisher, *Life of Silliman,* 1:57, 21–22. Entry for May 7, 1805, in Benjamin Silliman, "Travels in England, Holland, and Scotland, 1805–6," SFP, series III, reel 5 (emphasis in original). Brown, *Benjamin Silliman,* 33. John Noyes to Benjamin Silliman, March 16, 1802, SFP, series 1, box 11, folder 144.

7. Daniels, *American Science in the Age of Jackson,* 60–61. Silliman, "Birth Days—1829, 1830, 1831," SFP, series III, reel 2. Silliman, "Personal Record No.

XVII. Beginning April 8, 1860—Oct. 8?," SFP, series III, reel 5. Silliman, "Origin and Progress of Chemistry," book 1, pp. 37–39.

8. Fisher, *Life of Silliman*, 1:169–70, 215, 219.

9. Col. George Gibbs account books, 1813–38, Connecticut Historical Museum. The manuscript of Col. Gibbs's account book from his Long Island estate indicates that he paid at least five household "servants" a small wage as late as 1815. Their names were Hannibal, Joseph, Betsey, Peter, and Mary. Though it is impossible to know with certainty whether these "servants" were slaves, it is very likely. Consider: in New York, slavery was not completely abolished until 1827; *servants* was often used as a euphemism for slaves; these five individuals were paid much lower wages than the white laborers occasionally mentioned in Gibbs's account book; last, the absence of surnames, as well as the common slave name Hannibal, reflects common naming practices for enslaved people. For Newport demographics, see *1800 Census: Return of the Whole Number of Persons within the Several Districts of the United States* (1801), 2, 26. For "coloured servant," see Fisher, *Life of Silliman*, 1:219. For Gibbs's offer to Harvard, see Baldwin, *History of Yale College*, 241. For Rhode Island's role in the slave trade, see Jay Coughtry, *The Notorious Triangle: Rhode Island and the African Slave Trade, 1700–1807* (Philadelphia: Temple University Press, 1981); Rachel Chernos Lin, "The Rhode Island Slave-Traders: Butchers, Bakers and Candlestick-Makers," *Slavery & Abolition* 23, no. 3 (2002): 21–38.

10. Fisher, *Life of Silliman*, 1:257.

11. Barbara L. Narendra, "Benjamin Silliman and the Peabody Museum," *Discovery* 14, no. 2 (1979): 24. Baldwin, *History of Yale College*, 243. Ebenezer Baldwin, *Annals of Yale College*, 2nd ed. (New Haven: Noyes, 1838), 251. Brown, *Benjamin Silliman*, 263, 252. Silliman, "Origin and Progress of Chemistry," book 4, pp. 161–68 (emphasis in original); book 6, p. 199.

12. Narendra, "Benjamin Silliman and the Peabody Museum," 23–24. Baldwin, *Annals of Yale College*, 242. For the deep ties between Yale and slaveholder alumni in the antebellum period, see Garry Lacy Reeder, "Elms and Magnolias: Yale and the American South," exhibition at Sterling Library (1996), http://www.library.yale.edu/mssa/exhibits/elms/ (accessed Aug. 1, 2018). Fisher, *Life of Silliman*, 1:278, 259.

13. Benjamin Silliman, "Reminiscences of the Late Mr. Whitney, Inventor of the Cotton Gin," *American Journal of Science* 21, no. 2 (Jan. 2, 1832): 255–60. Daniels, *American Science in the Age of Jackson*, 20, 26, 44–56. Silliman, "Personal Record No. XVII. Beginning April 8, 1860—Oct. 8?"

14. For cotton statistics, see Edward Baptist, *The Half Has Never Been Told: Slavery and the Making of American Capitalism* (New York: Basic Books, 2014), 114 (table 4.1). Sven Beckert, in *Empire of Cotton: A Global History* (New York: Alfred A. Knopf, 2014), chap. 5, gives slightly smaller numbers: 1.5 million pounds in 1790, 36.5 million pounds in 1800. See also Angela Lakwete, *Inventing the Cotton Gin:*

Machine and Myth in Antebellum America (Baltimore: Johns Hopkins University Press, 2005).

15. Fisher, *Life of Silliman*, 1:255. Silliman, "Reminiscences of the Late Mr. Whitney," 259n, 265n.

16. Irving H. Bartlett, *John C. Calhoun: A Biography* (New York: W. W. Norton, 1994), 86–189.

17. John C. Calhoun to Silliman, May 13, 1822, SFP, General Correspondence, series II, reel 10. Calhoun to Silliman, March 26, 1818, in Fisher, *Life of Silliman*, 1:288. Fisher, *Life of Silliman*, 1:305. Daniels, *American Science in the Age of Jackson*, 18; Narendra, "Benjamin Silliman and the Peabody Museum," 24.

18. Andrew Jackson to Benjamin Silliman, July 14, 1821, SFP, General Correspondence, series II, reel 11. Robert Finley, an antislavery Northerner and future University of Georgia president, established the ACS in December 1816. To gain support for the idea, he held several meetings in Washington, D.C., with some of nation's leading slaveholders, including Andrew Jackson. Jackson's name was included on the original list of charter members, though he later claimed it was done without his consent. In 1830, as United States president, Jackson ended federal funding for the ACS, caving in to his proslavery base. See Burin, *Slavery and the Peculiar Institution*, 18, 22, 24; Nicholas Guyatt, *Bind Us Apart: How Enlightened Americans Invented Racial Segregation* (New York: Basic Books, 2016), 271.

19. Benjamin Silliman to Louis McLane, Washington D.C., May 28, 1833, in Benjamin Silliman, *Manual on the Cultivation of the Sugar Cane* (Washington, D.C., 1833), 5, 107. Silliman, "Origin and Progress of Chemistry," book 6, pp. 162–63. Fisher, *Life of Silliman*, 1:376. Jeremiah Day to Charles Shepard, Nov. 16, 1832, Shepard Papers (hereafter cited as SP), Amherst College Archives, series 2, box 3, folder 5. For a background on sugar, see Richard Follett, *The Sugar Masters: Planters and Slaves in Louisiana's Cane World, 1820–1860* (Baton Rouge: Louisiana State University Press, 2005), 17–22. See also James Walvin, *Sugar: The World Corrupted: From Slavery to Obesity* (New York: Pegasus, 2018), 29–50; Sidney Mintz, *Sweetness and Power: The Place of Sugar in Modern History* (New York: Viking, 1985).

20. Silliman, "Origin and Progress of Chemistry," book 6, pp. 162–68. Silliman to Charles Upham Shepard, "Memoranda," Nov. 7, 1832; Silliman to Shepard, Dec. 25, 1832; see also Silliman to Shepard, Jan. 9, 1833; Silliman to Shepard, Jan. 11, 1833; Silliman to Shepard, Jan. 16, 1833; Silliman to Shepard, Jan. 19, 1833, all in SP, series 2, box 3, folder 5.

21. Unidentified planter to Shepard, March 3, 1833; John Tenny to Shepard, Jan. 4, 1833; Stephen Henderson to Shepard, n.d., all in SP, series 2, box 3, folder 5.

22. Silliman, *Manual on the Cultivation of the Sugar Cane*, 71, 115–16. For Howard's vacuum pan, see Frederick Kurzer, "The Life and Work of Edward Charles Howard, FRS," *Annals of Science* 56 (1999): 113–41. For a discussion of planter re-

sistance to Howard's vacuum-pan technology, see R. Keith Aufhauser, "Slavery and Technological Change," *Journal of Economic History* 34, no. 1 (1974): 41–44. See also Maria Portuondo, "Plantation Factories: Science and Technology in Late-Eighteenth-Century Cuba," *Technology and Culture* 44, no. 2 (April 2003): 231–57.

23. Silliman, *Manual on the Cultivation of the Sugar Cane*, 120, 12–14, 96. Silliman, "Origin and Progress of Chemistry," book 6, p. 168. John Locke, "Art. X. On the Manufacture of Sugar from the River Maple," *American Journal of Science* 2, no. 2 (Nov. 1820): 263.

24. Benjamin Silliman, *Some of the Causes of National Anxiety* (New Haven, 1832), 8, 11–16. For Garrison's challenge to the earlier, first-wave abolitionist societies, see Richard Newman, *The Transformation of American Abolitionism: Fighting Slavery in the Early Republic* (Chapel Hill: University of North Carolina Press, 2002), 107–30; Sinha, *Slave's Cause*, 195–227.

25. *An Address to the Public by the Managers of the Colonization Society of Connecticut* (1828), 6–9 (emphases in original).

26. James Forten et al., "Education—An Appeal to the Benevolent," in *College for Coloured Youth: An Account of the New-Haven City Meeting and Resolutions* (New York: 1831), 3–5. For the Black college campaign, see Julie Winch, *Gentleman of Color: The Life of James Forten* (New York: Oxford University Press, 2002), 246–52.

27. William Maclure to Benjamin Silliman, Aug. 19, 1820, "William Maclure and Benjamin Silliman Correspondence, 1817–1827," American Philosophical Society, reel 283 (mf). Warren, *Maclure of New Harmony*, 7, 18–23, 51–71. Warren makes no mention of slave-grown tobacco, but evidence from a lawsuit filed against Maclure and his partners in 1796 indicates that one of the company's chief exports was Virginia tobacco. The lawsuit, Brydie's Executor v. Miller, Hart & Co., states that in the 1790s the firm made "frequent and heavy shipments of tobacco" from Virginia to the firm's base in London. The case made it to the Supreme Court in 1809. See *Reports of Cases Decided by the Honourable John Marshall*, ed. John W. Brockenbrough, 2 vols. (Philadelphia, 1837), 1:147–57. For Maclure's importance to developing the early American republic's scientific institutions, see Daniels, *American Science in the Age of Jackson*, 10–12; Valencius et al., "Science in Early America."

28. Warren, *Maclure of New Harmony*, 153–236; Donald E. Pitzer, "William Maclure's Boatload of Knowledge: Science and Education into the Midwest," *Indiana Magazine of History* 94, no. 2 (1998): 110–37.

29. Julian U. Niemcewicz, *Under Their Vine and Fig Tree: Travels through America in 1797–1799, 1805, with Some Further Account of Life in New Jersey*, ed. and trans. Metchie J. E. Budka (Elizabeth, N.J.: Grassmann, 1965), 46–47. Warren, *Maclure of New Harmony*, 20–21.

30. Maclure to Thomas Jefferson, Nov. 20, 1801, in *Papers of Thomas Jefferson* (hereafter cited as *PTJ*), ed. Barbara B. Oberg (Princeton: Princeton University Press, 2004–), 35:706–8. For the American approach to science and fact over theory, see

Daniels, *American Science in the Age of Jackson*, 63–84. See also Andrew J. Lewis, *A Democracy of Facts: Natural History in the Early Republic* (Philadelphia: University of Pennsylvania Press, 2011).

31. Maclure to Thomas Jefferson, Nov. 20, 1801. William Maclure, "Observations on the Geology of the United States, Explanatory of a Geological Map," *Transactions of the American Philosophical Society* 6 (1809): 411–28. William Maclure, *Observations on the Geology of the United States of America* (Philadelphia, 1817). Warren, *Maclure of New Harmony*, 60.

32. Maclure, "Notes without Title relating to the West Indies." For the Silliman edition, see William Maclure, "ART. XV. Observations on the Geology of the West India Islands, from Barbadoes to Santa Cruz, Inclusive," *American Journal of Science* (Jan. 6, 1818): 311.

33. Maclure, *Observations on the Geology of the United States*, 20, 106.

34. Ibid., 189.

35. Ibid., 98. For manipulation of antislavery laws, see John C. Hammond, *Slavery, Freedom, and Expansion in the Early American West* (Charlottesville: University of Virginia Press, 2007), 121–23.

36. Taylor, *Visions of Harmony*, 71–73, 94–97. See also Paul R. Bernard, "Irreconcilable Opinions: The Social and Educational Theories of Robert Owen and William Maclure," *Journal of the Early Republic* 8, no. 1 (1988): 21–44.

37. Frances Wright to Harriet and Julia Garnett, ca. Oct. 1820, in "The Nashoba Plan for Removing the Evil of Slavery: Letters of Frances and Camilla Wright, 1820–1829," ed. Cecilia Helena Payne-Gaposchkin, *Harvard Library Bulletin* 23, no. 3 (July 1975): 225. Frances Wright to Julia Garnett, Jan./Feb. 1825, ibid., 232–33. For Wright's views on America, see Eckhardt, *Fanny Wright*, 1–24, 108–40; Taylor, *Visions of Harmony*, 163–75.

38. Marquis de Lafayette to George Washington, Feb. 5, 1783, *Founders Online*, National Archives, https://founders.archives.gov/documents/Washington /99-01-02-10575 (accessed May 21, 2020). "Eighth Annual Report of the American Society for Colonizing the Free People of Colour of the United States" (Washington, D.C., 1825), 3. For Lafayette's colonization schemes, see Guyatt, *Bind Us Apart*, 205–10, 276–80. For Wright's relationship with Lafayette, see Eckhardt, *Fanny Wright*, 78–107.

39. Frances Wright to Julia Garnett, Nov. 14, 1824, in Payne-Gaposchkin, "The Nashoba Plan," 230. Frances Wright to Thomas Jefferson, July 26, 1825, *Founders Online*, https://founders.archives.gov/?q=%20Author%3A%22Wright%2C%20 Frances%22&s=1111311111&r=4&sr= (accessed May 21, 2020). See also Marquis de Lafayette to Thomas Jefferson, Oct. 1, 1824, *Founders Online*, https://founders. archives.gov/documents/Jefferson/98-01-02-4587 (accessed May 21, 2020).

40. Wright to Garnett, Nov. 14, 1824, in Payne-Gaposchkin, "The Nashoba Plan," 230. For immigration statistics to Haiti, see Ada Ferrer, "Haiti, Free Soil, and Antislavery

in the Revolutionary Atlantic," *American Historical Review* 117, no. 1 (2012): 58. See also Fanning, *Caribbean Crossing*, 41–77. For immigration to Liberia, 1820–33, see Burin, *Slavery and the Peculiar Solution*, 171 (table 2).

41. Robert Owen, "Mr. Owen's Second Discourse on a New System of Society," in *Owen's American Discourses* (London, 1825), 17. See also George Lockwood, *The New Harmony Communities* (Marion, Ind.: Chronicle, 1902), 87–89. Robert Owen, *Robert Owen's Opening Speech and His Reply to the Rev. Alex. Campbell* (Cincinnati, 1829), 189. Frances Wright to Julia Garnett, June 8, 1825, quoted in Taylor, *Visions of Harmony*, 166. For the science behind Owen's educational theories, see Cornelia Lambert, " 'Living Machines': Performance and Pedagogy at Robert Owen's Institute for the Formation of Character, New Lanark, 1816–1828," *Journal of the History of Childhood and Youth* 4, no. 4 (2011): 420–33. For the enslaved people's names, see Eckhardt, *Fanny Wright*, 119.

42. [Frances Wright], "A Plan: For the Gradual Abolition of Slavery in the United States," *New Harmony Gazette*, Oct. 1, 1825.

43. Ibid. George Lockwood, *The New Harmony Movement* (New York: D. Appleton, 1905), 83–90. Though plantation managers are often neglected in the history of scientific management, Caitlin Rosenthal argues that planters were in fact leading innovators. See Caitlin Rosenthal, *Accounting for Slavery: Masters and Management* (Cambridge: Harvard University Press, 2018).

44. Owen, "Mr. Owen's Second Discourse on a New System of Society," 22. Bederman, "Revisiting Nashoba," 450–51. Taylor, *Visions of Harmony*, 172–73. Pitzer, "William Maclure's Boatload of Knowledge."

45. William Maclure, "Fear, a Concomitant of Slavery," in William Maclure, *Opinions on Various Subjects*, 3 vols. (New Harmony, Ind., 1831–38), 2:84. William Maclure, "On the Effects of Representative Governments," Feb. 22, 1826, ibid., 1:4. For Pestalozzi's influence on Maclure, see Warren, *Maclure of New Harmony*, 72–98, 187–203; Bernard, "Irreconcilable Opinions."

46. William Maclure, "Education," Feb. 13, 1828, in Maclure, *Opinions on Various Subjects*, 1:44–49. William Maclure, "Maclure's Outline, or Course of Study, for the New Harmony Schools," in Lockwood, *The New Harmony Movement*, 236–37. Maclure also urged Silliman to send his own son to the New Harmony school, but Silliman "politely declined," he wrote in his journal. See Benjamin Silliman, "Origin and Progress of Chemistry," book 6, p. 162.

47. Frances Wright, "Frances Wright's Establishment for the Abolition of Slavery," *Genius of Universal Emancipation*, Feb. 24, 1827, 2. Maclure to Fretageot, Dec. 13, 1826, in *Partnership for Posterity: the Correspondence of William Maclure and Marie Duclos Fretageot, 1820–1833*, ed. Josephine Mirabella Elliott (Indianapolis: Indiana Historical Society, 1994), 438–39. Camilla Wright (dictated by Frances Wright) to Julia Garnett, Dec. 8, 1826, quoted in Taylor, *Visions of Harmony*, 173. "Communication from the Trustees of Nashoba," *Genius of Universal Emancipation*, April 14, 1827, 23.

48. Frances Wright, *Fanny Wright Unmasked by Her Own Pen* . . . (New York, 1830), 3–5 (originally written Dec. 4, 1827). Wright, "Wright's Establishment for the Abolition of Slavery," 2.

49. Wright, "Wright's Establishment for the Abolition of Slavery," 18, 2.

50. Maclure's 1827 will, in *Partnership for Posterity*, Appendix H, 1086.

51. Maclure, "On the Effects of Representative Governments," 1. Maclure, "Fear, a Concomitant of Slavery," 84. Maclure, "Effects of Slavery on the Education of Free Children," Jan. 16, 1828, in *Opinions on Various Subjects*, 1:43.

52. William Maclure, "The Effects Produced by Climate, on Different Forms of Government," March 1, 1826, in *Opinions on Various Subjects*, 1:7, 6. Wright, "A Plan."

53. James Richardson, "Nashoba Journal," June 1, 1827, quoted in Eckhardt, *Fanny Wright*, 143. Taylor, *Visions of Harmony*, 177–92; Bederman, "Revisiting Nashoba," 453.

54. Frances Wright, "Nashoba: Explanatory Notes Respecting the Nature and Objects of the Institution of Nashoba," *Genius of Universal Emancipation*, Feb. 23, 1828, 45–46; March 1, 1828, 52–53; March 8, 1828, 61–62. Other copies were printed simultaneously in the *New Harmony Gazette*, the *Correspondent*, and other periodicals. A version was published in 1830 as a single, separate pamphlet: *Fanny Wright Unmasked by Her Own Pen*; quotes on 6, 9. Taylor, *Visions of Harmony*, 177–80.

55. Wright to Maclure, Jan. 13, 1829, William Maclure Letters, 1796–1848, APS. For the Hall of Science see Eckhardt, *Fanny Wright*, 193–94.

56. Maclure to Fretageot, May 15, 1830, in *Partnership for Posterity*, 729. Thomas Say, *American Conchology; or, Descriptions of the Shells of North America* (New Harmony, Ind., 1830). Thomas Say to Charles Wilkes Short, New Harmony, March 1, 1831, Thomas Say Papers, APS, series 1. Say's wife, Lucy Sistaire, did much of the work alongside Say; see Say to Short, July 19, 1831; Say to Short, Jan. 14, 1832; Say to Short, March 5, 1833, all in Thomas Say Papers, APS, series 1.

57. For revivals of New Harmony's reputation, see Pitzer, "William Maclure's Boatload of Knowledge"; Warren, *Maclure of New Harmony*. For the two hundredth anniversary of New Harmony, in 2014, the Indiana Historical Society devoted an entire issue of one of its publications, *Traces of Indiana & Midwestern History*, to New Harmony. Its contributions to science featured prominently. See especially Ryan Rokicki, "Science in Utopia," and Donald Pitzer, "Why New Harmony Is World Famous," *Traces of Indiana & Midwestern History* 26, no. 2 (Spring 2014): 5–15, 51–55.

58. Maclure, "Slavery. Difference between the Slave-Holding and the Free States. Necessity of Emancipation," Sept. 28, 1835, in *Opinions on Various Subjects*, 3:111–12. Maclure, "Sugar Tax," ibid., 2:373. Maclure, "Slavery," ibid., 3:109–11.

5 The Technological Fix

1. For British antislavery efforts between 1807 and the 1830s, see Robin Blackburn, *The Overthrow of Colonial Slavery, 1776–1848* (New York: Verso, 1988), chaps. 8, 11; Seymour Drescher, *Abolition: A History of Slavery and Antislavery* (Cambridge: Cambridge University Press, 2009), chaps. 8–10; Howard Temperley, *British Antislavery, 1833–1870* (London: Longman, 1972), chap. 1; Kenneth Morgan, *Slavery and the British Empire* (New York: Oxford University Press, 2007), 172–98. For Caribbean amelioration see Christa Dierksheide, *Amelioration and Empire: Progress and Slavery in the Plantation Americas* (Charlottesville: University of Virginia Press, 2014), esp. chaps. 5–6; J. R. Ward, *British West Indian Slavery, 1750–1834: The Process of Amelioration* (Oxford: Oxford University Press, 1988).

2. Mr. Whitmore, "Proceedings of Second General Meeting of the Society," *Anti-Slavery Monthly Reporter* 1, no. 1 (June 1825): 6.

3. William Allen, *Life of William Allen*, 3 vols. (London: Charles Gilpin, 1846), 1:2. Entries for May 12, 1788, and Feb. 10, 1807, ibid., 1:3, 83–84. No modern biography of Allen exists. A few studies of Allen's pharmaceutical work provide the details discussed here: Geoffrey Tweedale, *At the Sign of the Plough: 275 Years of Allen & Hanburys and the British Pharmaceutical Industry, 1715–1990* (London: J. Murray, 1990), esp. chaps. 1–2; Briony Hudson, "William Allen: Anti-Slavery Campaigner," *Pharmaceutical Journal* 278 (March 24, 2007): 344–45. See also Wayne Ackerson, *The A.I. (1807–1827) and the Antislavery Movement in Great Britain* (Lewiston, N.Y.: E. Mellen, 2005). Allen's pharmaceutical company was purchased in 1958 by the drug company that became GlaxoSmithKline, which holds some of his private papers.

4. John Kizell to William Allen, July 30, 1812; William Allen to John Kizell, Oct. 30, 1812, African Correspondence, GlaxoSmithKline Archives (hereafter cited as GSK). For abolitionism reflecting well on business, see Hudson, "William Allen," 344–45. For background of the Friendly Society, see Christopher Fyfe, *A History of Sierra Leone* (London: Oxford University Press, 1962), 113.

5. Kevin Lowther, *The African American Odyssey of John Kizell: A South Carolina Slave Returns to Fight the Slave Trade in His African Homeland* (Columbia: University of South Carolina Press, 2011). Kizell's description of his capture is quoted ibid., 46–47. Lowther argues that Kizell's description of his capture for witchcraft is probably accurate, but he advises caution, given that more than a decade later Kizell offered a somewhat different account of his enslavement; ibid., 245–46n77. For African healing traditions in the Anglo-American Atlantic world, see Kenneth M. Bilby and Jerome S. Handler, "Obeah: Healing and Protection in West Indian Life," *Journal of Caribbean History* 38, no. 2 (2004): 153–83; Sharla M. Fett, *Working Cures: Healing, Health, and Power on Southern Slave Plantations* (Chapel Hill: University of North Carolina Press, 2002); Diana Paton and Maarit

Forde, eds., *Obeah and Other Powers: The Politics of Caribbean Religion and Healing* (Durham: Duke University Press, 2012); Randy M. Browne, "The 'Bad Business' of Obeah: Power, Authority, and the Politics of Slave Culture in the British Caribbean," *William and Mary Quarterly* 68, no. 3 (2011): 451–80.

6. William Allen to James Wise, March 28, 1814; James Wise to William Allen, July 3, 1814, African Correspondence, GSK.

7. Allen to Friendly Society, April 30, 1812; Friendly Society to Allen, June 20, 1813, African Correspondence, GSK. For enslaved people and natural curiosity collecting, see James Delbourgo, *Collecting the World: Hans Sloane and the Origins of the British Museum* (Cambridge: Harvard University Press, 2017), esp. chaps. 1–3; Kathleen Murphy, "Translating the Vernacular: Indigenous and African Knowledge in the Eighteenth-Century British Atlantic," *Atlantic Studies* 8, no. 1 (2011): 29–48; Londa Schiebinger, *Plants and Empire: Colonial Bioprospecting in the Atlantic World* (Cambridge: Harvard University Press, 2004); Molly Warsh, "Enslaved Pearl Divers in the Sixteenth Century Caribbean," *Slavery & Abolition* 31, no. 3 (Sept. 2010): 345–62.

8. Allen to Kizell, Oct. 30, 1812; Henry Savage to William Allen, May 29, 1815, African Correspondence, GSK. For examples of natural history–related items published in the AI reports, including descriptions of geography, climate, soil, and ethnography, see *Fourth Report of the Directors of the African Institution, Read at the Annual General Meeting of March 28, 1810*, 2nd ed. (London, 1814), Appendix R, 75–82; *Tenth Report of the Directors of the African Institution* (London, 1816), 34; *Eighteenth Report . . . of the African Institution* (London, 1824), Appendix O, 209–25.

9. Allen to Kizell, Oct. 30, 1812; Kizell to Allen, March 15, 1813, African Correspondence, GSK. On Kizell's lost journal, see Lowther, *The African American Odyssey of John Kizell*, 192–200.

10. Paul Cuffe to William Allen, Sierra Leone, April, 22, 1811, in *Captain Paul Cuffe's Logs and Letters, 1807–1817: A Black Quaker's Voice from within the Veil*, ed. Rosalind Cobb Wiggins (Washington, D.C.: Howard University Press, 1996), 119. For scholarship on Paul Cuffe, see Lamont Thomas, *Rise to Be a People: A Biography of Paul Cuffe* (Urbana: University of Illinois Press, 1986); James Sidbury, *Becoming African in America: Race and Nation in Early Black Atlantic* (New York: Oxford University Press, 2007), 145–79; James T. Campbell, *Middle Passages: African American Journeys to Africa, 1787–2005* (New York: Penguin Press, 2006), 16, 30–39, 60, 103, 110, 219, 319, 430–31; Jeffrey A. Fortin, "Cuffe's Black Atlantic World, 1808–1817," *Atlantic Studies* 4, no. 2 (Oct. 2007): 245–66.

11. Paul Cuffe, *A Brief Account of the Settlement and Present Situation of the Colony of Sierra Leone in Africa* (New York, 1812), 8–13. Fortin, "Cuffe's Black Atlantic World," 248.

12. Paul Cuffe to William Allen, June 12, 1812, African Correspondence, GSK. For Pemberton's solicitation of Cuffe and Cuffe's initial rejection of emigration

schemes in 1793, see Thomas, *Rise to Be a People*, 35, 40, 19–22. Though Cuffe began a school for Black and white children in the 1790s near his home in Westport, Mass., this was mainly for his nephews and the few black children who lived in the area. See Thomas, *Rise to Be a People*, 20.

13. Sarah Howard to Paul Cuffe, July 3, 1816; Cuffe to Sarah Howard, Oct. 3, 1816, in *Cuffe's Logs and Letters*, 420, 403. For Smeathman's *Plan of a Settlement* being among the books in the African Union Society's library, see Fortin, "Cuffe's Black Atlantic World," 249. For Cuffe's links to Newport and the African Union Society, see Thomas, *Rise to Be a People*, 20. For AI's editing Park's journal, discussed later in this chapter, see Kate Ferguson Marsters, introduction to Mungo Park, *Travels in the Interior Districts of Africa*, ed. Marsters (Durham: Duke University Press, 2000). For Black identity formation, see Sidbury, *Becoming African in America*; Patrick Rael, *Black Identity and Black Protest in the Antebellum North* (Chapel Hill: University of North Carolina Press, 2002).

14. Entry for July 1811, in *Life of William Allen*, 1:133. Cuffe, *A Brief Account*, 4. For performance and speech as a means of scientific legitimation, see Geoffrey Cantor, "The Rhetoric of Experiment," in *The Uses of Experiment: Studies in the Natural Sciences*, ed. David Gooding, Trevor Pinch, and Simon Schaffer (New York: Cambridge University Press, 1989), 159–79. See also James Delbourgo, "Introduction: The Far Side of the Ocean," in *Science and Empire in the Atlantic World*, ed. James Delbourgo and Nicholas Dew (London: Routledge, 2008), 6.

15. Entry for Aug. 3 (?), 1811, in *Cuffe's Logs and Letters*, 148. For a map of the abolition movement, see Thomas Clarkson, *The History of the Rise, Progress, & Accomplishment of the Abolition of the African Slave-Trade by the British Parliament*, 2 vols. (Philadelphia, 1808), 1:259. William Allen commissioned Clarkson to write anonymous reviews for the philanthropic journal Allen edited, the *Philanthropist*: "Setting aside its geological contents," Clarkson wrote privately to Allen, in reference to a natural history of Sweden that Allen gave him to review, "the book . . . is one of the dullest I have ever read." Thomas Clarkson to William Allen, July 4, 1813, Thomas Clarkson Letters, Library of the Society of Friends, London, Temp Mss 4/6/7. For background on Clarkson's "African box," see Jane Webster, "Collecting for the Cabinet of Freedom: The Parliamentary History of Thomas Clarkson's Chest," *Slavery & Abolition* 38, no. 1 (Dec. 14, 2016): 1–20; Marcus Wood, "Packaging Liberty and Marketing the Gift of Freedom: 1807 and the Legacy of Clarkson's Chest," in *The British Slave Trade: Abolition, Parliament and People*, ed. Stephen Farrell, Melanie Unwin, and James Walvin (Edinburgh: Edinburgh University Press, 2007), 209–23.

16. Cuffe to Allen, Sept. 23, 1813, and April 1, 1816, in *Cuffe's Logs and Letters*, 255–56, 409. Entry for July 1811, in *Life of Allen*, 1:139. Paul Cuffe, *Memoir of Captain Paul Cuffee* (York, 1812), 23. For the African roots of African American nautical navigation, see W. Jeffrey Bolster, *Black Jacks: African American Seamen in the Age of Sail*

(Cambridge: Harvard University Press, 1997), 45–51. For Native American nautical navigation, see Andrew Lipman, *Saltwater Frontier: Indians and the Contest for the American Coast* (New Haven: Yale University Press, 2015), 66–73.

17. Entries for May 13, 1811, May 15, 1811, and June 1811, in *Cuffe's Logs and Letters*, 122–23. Thomas, *Rise to Be a People*, 18–19.

18. Allen to Kizell, Aug. 29, 1811, African Correspondence, GSK. Petition by Cuffe to Congress, June 16, 1813, in *Cuffe's Logs and Letters*, 252–53. Allen to James Wise, March 28, 1814, African Correspondence, GSK. For William Allen's interest in agricultural chemistry, see entries for March 3 and March 14, 1809, in *Life of William Allen*, 1:111.

19. Allen to Friendly Society, March 25, 1814; Capt. T. P. Thompson to Allen, Fort St. George, India, July 18, 1815, African Correspondence, GSK. Entry for [fall] 1811, in *Life of William Allen*, 1:142. "Extract of a Letter from Dr. Roxburgh, of Calcutta, with a List of Seeds, and Directions for Cultivating the Sunn and Paat Plants," Aug. 23, 1809, in *Fourth Report of the Directors of the AI*, 70–71.

20. Entry for Aug. 28, 1811, in *Life of William Allen*, 1:141.

21. Cuffe to Allen, June 1812, African Correspondence, GSK. In August 1814 Cuffe wrote to Allen telling him that the House of Representatives rejected another petition he had sent asking for a special trading license; Cuffe to Allen, Aug. 10, 1814, African Correspondence, GSK. Friendly Society to Allen, June 20, 1813; Kizell to Allen, Feb. 14, 1814; Allen to Friendly Society, Aug. 11, 1815; Thomas Clarkson to James Wise, Feb. 25, 1819 (emphasis in original), all in African Correspondence, GSK.

22. Fyfe, *History of Sierra Leone*, 114. Allen to James Wise, Nov. 23, 1814, African Correspondence, GSK.

23. *Third Census of the United States, 1810*, vol. 3 (Washington, D.C., 1990), 1. Cuffe to Allen, April 22, 1811, in *Cuffe's Logs and Letters*, 119. For Cuffe's recruitment effort, see Thomas, *Rise to Be a People*, 101.

24. Fyfe, *History of Sierra Leone*, 114–15, 136–37. Philip Curtin, *The Image of Africa: British Ideas and Action, 1780–1850* (Madison: University of Wisconsin Press, 1964), 1:157–76.

25. *Seventh Report of the Directors of the AI, Read on 24th of March, 1813* (London, 1813), 29–30. Mungo Park, *Journal of a Mission to the Interior of Africa in the Year 1805* (N.p.: 1815), 45–46 (emphases in original).

26. Park, *Travels in the Interior Districts*, 263. For background on the 1815 AI edition and the Edwards version, see Marsters, introduction to Park, *Travels in the Interior Districts*, 1–28.

27. Park, *Journal of a Mission to the Interior of Africa*, 22–26, 30. *Ninth Report of the Directors of the AI* (London, 1815), "Appendix N. African Mode of Dying [*sic*] Cloth, Taken from Mr. Park's Journal," 135–36, 65–66.

28. *Tenth Report of the Directors of the AI*, 36. Curtin, *Image of Africa*, 1:164–69. Christopher L. Brown, "From Slaves to Subjects: Envisioning an Empire without

Slavery, 1772–1834," in *Black Experience and the Empire*, ed. Philip D. Morgan and Sean Hawkins (Oxford: Oxford University Press, 2004), 111–40; Christopher L. Brown, *Moral Capital: Foundations of British Abolitionism* (Chapel Hill: University of North Carolina Press, 2006); Padraic X. Scanlan, *Freedom's Debtors: British Antislavery in Sierra Leone in the Age of Revolution* (New Haven: Yale University Press, 2017); Bronwen Everill, *Abolition and Empire in Sierra Leone and Liberia* (New York: Palgrave, 2013); Richard Huzzey, *Freedom Burning: Anti-slavery and Empire in Victorian Britain* (Ithaca: Cornell University Press, 2012); Derek R. Peterson, ed., *Abolitionism and Imperialism in Britain, Africa, and the Atlantic* (Athens: Ohio University Press, 2010); Andrew Porter, "Trusteeship, Anti-slavery, and Humanitarianism," in *The Oxford History of the British Empire*, vol. 3, *The Nineteenth Century*, ed. Andrew Porter (New York: Oxford University Press, 1999), 198–221.

29. Thomas Edward Bowdich, *Mission from Cape Coast Castle to Ashantee with a Statistical Account of That Kingdom, and Geographical Notices of Other Parts of the Interior of Africa* (N.p., 1819), 338–39. For early failures and the implicit purpose of Bowdich's narrative, see Curtin, *Image of Africa*, 1:169; 151, 158–60, 168–70.

30. Alexander Gordon Laing, *Travels in the Timannee, Kooranko, and Soolima Countries, in Western Africa* (London: J. Murray, 1825), 379, 281–82, 223.

31. Thomas Edward Bowdich, *Essay on the Superstitions, Customs, and Arts Common to the Ancient Egyptians, Abyssinians, and Ashantees* (Paris, 1821), preface. *Royal Gazette; and Sierra Leone Advertiser* (Freetown), July 14, 1821, and June 8, 1822. Gregory A. Good, "Sabine, Sir Edward (1788–1883)," in *Oxford Dictionary of National Biography* (Oxford: Oxford University Press, 2004). Christopher Fyfe, "Denham, Dixon (1786–1828)," in *Oxford Dictionary of National Biography*.

32. Review of Bowdich's *Mission from Cape Coast Castle to Ashantee*, *Quarterly Review* (Jan. 1820): 301. Kenneth Macaulay, *The Colony of Sierra Leone Vindicated from the Misrepresentations of Mr. Macaulay of Glasgow* (London, 1827), 18. For travel narratives of the 1820s as sites in the struggle of representation over slavery, including a discussion of Kenneth Macaulay's role, see David Lambert, "Sierra Leone and Other Sites in the War of Representation over Slavery," *History Workshop Journal* 64 (Autumn 2007): 103–32. For suppressing slave trade as a bargaining chip, see Fyfe, *History of Sierra Leone*, 156–67. For African nations stalling colonization, see Tom C. McCaskie, "Cultural Encounters: Britain and Africa in the Nineteenth Century," in Morgan and Hawkins, *Black Experience and the Empire*, 166–93; Rebecca Shumway, "From Atlantic Creoles to African Nationalists: Reflections on the Historiography of Nineteenth-Century Fanteland," *History in Africa* 42 (June 2015): 139–64.

33. For the place of the East Indies in slavery debates in the 1820s, see Seymour Drescher, *The Mighty Experiment: Free Labor versus Slavery in British Emancipation* (New York: Oxford University Press, 2002), 114–17. Drescher downplays the influence of the East Indies in abolitionist debates in the 1820s. This results, I

believe, from an unwarranted focus on parliamentary debates, rather than on the literature produced by abolitionists themselves. For the most comprehensive scholarly works on Raffles and his time in Southeast Asia, see C. E. Wurtzburg, *Raffles of the Eastern Isles* (London: Hodder and Stoughton, 1954), and John S. Bastin, *The Native Policies of Sir Stamford Raffles in Java and Sumatra: An Economic Interpretation* (Oxford: Clarendon Press, 1957).

34. Thomas Stamford Raffles, *A History of Java* (London, 1817), dedication page, 79n. Raffles to William Wilberforce, Sept. 1819, in *Memoir of the Life and Public Services of Sir Thomas Stamford Raffles*, ed. Sophia Raffles, 2 vols. (London, 1835), 2:44. Raffles to Wilberforce, Oct. 23, 1817, in *Correspondence of William Wilberforce*, ed. Robert Isaac Wilberforce and Samuel Wilberforce, 2 vols. (London: J. Murray, 1840), 2:386. Wilberforce to Raffles, July 30, 1825, and Wilberforce to Raffles, Dec. 23, 1825, Raffles Family Collection, British Library, Mss. Eur D742/3 ff. 88–90, 95–96. For Raffles in Java, see Wurtzburg, *Raffles of the Eastern Isles*, 157–201. Raffles was actually elected to the Royal Society on March 20, 1817, and *A History of Java* came out one month later. But Raffles had already been in contact with Joseph Banks, who undoubtedly knew the book was forthcoming. See Wurtzburg, *Raffles of the Eastern Isles*, 417.

35. Raffles to East India Company in Bengal, Dec. 5, 1812, quoted in H. R. C. Wright, "Raffles and the Slave Trade at Batavia in 1812," *Historical Journal* 3, no. 2 (1960): 189. For weak enforcement of anti–slave trade policies, see ibid., 184–91. For slavery in Southeast Asia, see Anthony Reid and Jennifer Brewster, eds., *Slavery, Bondage, and Dependency in Southeast Asia* (New York: St. Martin's, 1983), esp. 1–47, 156–81; and S. Abeyasekere, "Slaves in Batavia," in Reid and Brewster, *Slavery, Bondage, and Dependency*, 286–314. See also Kerry Ward, "Slavery in Southeast Asia, 1420–1804," in *The Cambridge World History of Slavery*, vol. 3, AD 1420—AD 1804, ed. David Eltis and Stanley Engerman (New York: Cambridge University Press, 2011). For statistics see Abeyasekere, "Slaves in Batavia," 288–89, 301.

36. Raffles, *History of Java*, 1:76. I am paraphrasing the argument made in Gillen D'Arcy Wood, "The Volcano Lover: Climate, Colonialism, and the Slave Trade in Raffles's 'History of Java' (1817)," *Journal for Early Modern Cultural Studies* 8, no. 2 (2008): 33–55. See also Gillen D'Arcy Wood, *Tambora: The Eruption That Changed the World* (Princeton: Princeton University Press, 2014), 23–32.

37. For the irony of slavery deriving from the natural environment, I am paraphrasing Wood, "The Volcano Lover." For background on slavery in Southeast Asia, see Anthony Reid, introduction to Reid and Brewster, *Slavery, Bondage, and Dependency*, 1–43; Ward, "Slavery in Southeast Asia, 1420–1804."

38. Abeyasekere, "Slaves in Batavia," 296; Wood, "The Volcano Lover," 38–39.

39. Raffles, *History of Java*, 1:106–7, 29–31. See also Curtin, " 'The White Man's Grave': Image and Reality, 1780–1850," *Journal of British Studies* 1, no. 1 (Nov. 1961): 94–110.

40. Thomas Horsfield, *An Experimental Dissertation on the Rhus Vernix . . . Commonly Known in Pennsylvania by the Name of Poison-Ash* (Philadelphia, 1798) (emphasis in original). For a brief history of the Batavian Society, see "Article XII. Transactions of the Batavian Society of Art and Sciences," *Quarterly Journal of Science and the Arts* 7 (Jan. 4, 1817): 326. For Horsfield's background, see D. T. Moore, "Horsfield, Thomas (1773–1859)," in *Oxford Dictionary of National Biography*. See also Wurtzburg, *Raffles of the Eastern Isles*, 196, 198.

41. Raffles, *History of Java*, 1:44–49n; 1:36. Erasmus Darwin, *Botanic Garden* (London: J. Johnson, 1791), 2:26. See also John Bastin, "New Light on J. N. Foersch and the Celebrated Poison Tree of Java," *Journal of the Malaysian Branch of the Royal Asiatic Society* 58, no. 2 (1985): 25–44.

42. Raffles, *History of Java*, 1:131. Raffles to Somerset, Aug. 20, 1820, in *Memoir of the Life of Raffles*, 2:148–52. For naturalist backgrounds of abolitionists like the Duke of Somerset and Raffles's move to Bencoolen, see Wurtzburg, *Raffles of the Eastern Isles*, 417, 420, 422; 557–92.

43. *Eleventh Report of the Directors of the African Institution . . . 26th Day of March, 1817* (London, 1817), 78. Thomas Clarkson, *Thoughts on the Necessity of Improving the Condition of the Slaves in the British Colonies: With a View to Their Ultimate Emancipation* (London, 1823), 60–61. "New Work," *Anti-Slavery Monthly Reporter*, May 1828, 236.

44. For the best studies of amelioration, see Ward, *British West Indian Slavery*; Dierksheide, *Amelioration and Empire*. Though excellent, their studies focus on slaveholders, not abolitionists. For one study that does focus on planters' adoption of the plow, see J. W. Ward, "The Amelioration of British West Indian Slavery, 1750–1834: Technical Change and the Plough," *New West Indian Guide* 63, no. 1–2 (1989): 41–58. For the origins of agricultural science in Britain, see Richard Drayton, *Nature's Government: Science, Imperial Britain, and the "Improvement" of the World* (New Haven: Yale University Press, 2000), 50–81.

45. William Dickson, *Mitigation of Slavery: In Two Parts*, 2 vols. (London: R. and A. Taylor, 1814), 1:xi, 2:199. I arrive at the number twenty by combining the seventeen plow testimonials in *Mitigation*'s appendix (2:459–96) with the three references about plows included in the main text. The classic study that explains the social and economic divergence between European and non-European societies, and that focuses largely on European adoption of the plow, is Jack Goody, *Production and Reproduction: A Comparative Study of the Domestic Domain* (New York: Cambridge University Press, 1976). For a recent study that challenges the idea that hoe design had not improved on Anglo-American plantations, see Chris Evans, "The Plantation Hoe: The Rise and Fall of an Atlantic Commodity, 1650–1850," *William and Mary Quarterly* 69, no. 1 (Jan. 2012): 71–100. For Dickson's background, see: H. T. Dickinson, "Dickson, William," in *Oxford Dictionary of National Biography*.

46. Edward Long, *History of Jamaica*, 3 vols. (London, 1774), 2:436, 448. Dickson, *Mitigation of Slavery*, 2:464–67, 293, 289 (emphasis in original). For the original testimony, see "Mr. Ashley, concerning the Use of the Plough on His Estate in Jamaica," in *Reports of the Lords of the Committee Council Appointed for the Consideration of All Matters Relating to Trade and Foreign Plantations . . . Dated 11th of February, 1788* (London, 1789), part III, no. 9, pp. 279–80. Dickson, *Mitigation of Slavery*, 2:467–84.

47. Dickson, *Mitigation of Slavery*, 2:464–67; 1:xii; 2:471–74 (emphasis in original). For details on Steele, see Drescher, *Mighty Experiment*, 110–13.

48. Dickson, *Mitigation of Slavery*, 2:291–94.

49. Ibid., 293–301.

50. For attempts to use the plow in Jamaica and Antigua and why they failed, see Ward, *British West Indian Slavery*, 72, 80–95. For Jamaican planters hiring chemists, see ibid., 102. For breadfruit, see Richard Sheridan, "Captain Bligh, the Breadfruit and the Botanic Gardens of Jamaica," *Journal of Caribbean History* 23, no. 1 (Jan. 1989): 28–50. For Caribbean slaveholders' own use of the amelioration narrative, see Dierksheide, *Amelioration and Empire*, esp. 155–209; Justin Roberts, *Slavery and the Enlightenment in the British Atlantic, 1750–1807* (New York: Cambridge University Press, 2013).

51. I am indebted here to Chris Evan's essay on plantation hoe development, "The Plantation Hoe." For African-derived foods transported to the Caribbean, see Judith Carney and Richard Nicholas Rosomoff, *In the Shadow of Slavery: Africa's Botanical Legacy in the Atlantic World* (Berkeley: University of California Press, 2011). Ward, *British West Indian Slavery*, chap. 4, emphasizes that, of all the amelioration reforms planters attempted, the most effective was giving enslaved people more land to grow food for themselves. This would have meant that enslaved Africans' knowledge of how to grow indigenous African crops was even more essential to increases in productivity. The scholarly debate over African origins is in regard to rice; see Judith Carney, *Black Rice: The African Origins of Rice Cultivation in the Americas* (Cambridge: Harvard University Press, 2001), and a critique: David Eltis, Philip Morgan, and David Richardson, "Black, Brown, or White? Color-Coding American Commercial Rice Cultivation with Slave Labor," *American Historical Review* 115, no. 1 (Feb. 2010): 164–71. Evans, "The Plantation Hoe," applies this debate to the plantation hoe.

52. Blackburn, *Overthrow of Colonial Slavery*, 430, chap. 11, 429–32. For Heyrich and slave revolts, see Sinha, *The Slave's Cause*, 179–80, 196–97. For the Demerara revolt, see Michael Craton, *Testing the Chains: Resistance to Slavery in the British West Indies* (Ithaca: Cornell University Press, 1982), 267–70. For the broader effects of slave revolts on emancipation, see Vincent Brown, *Tacky's Revolt: The Story of an Atlantic Slavery War* (Cambridge: Harvard University Press, 2020).

53. Clarkson, *Thoughts on Improving the Condition of the Slaves*, 56 (emphasis in original). Whitmore, "Proceedings of Second General Meeting of the Society," 6. For the changes that occurred in the British antislavery movement between 1823 and 1838, see Blackburn, *Overthrow of Colonial Slavery*, chap. 11; Drescher, *Abolition*, chap. 9; and Temperley, *British Antislavery*, chap. 1.

54. "West Indian Statistics," *Anti-Slavery Monthly Reporter* 2, no. 2 (July 1827): 15 (emphasis in original). *Anti-Slavery Monthly Reporter* 2, no. 3 (Aug. 1827): 39–40. *Anti-Slavery Monthly Reporter* 1, no. 24 (May 1827): 385. For demographic decline, see Blackburn, *Overthrow of Colonial Slavery*, 424. For the increasing use of statistical data and mortality rates in the early nineteenth century, see Drescher, *Mighty Experiment*, 46–47.

55. Major Thomas Moody, "Separate Report of Major Thomas Moody, Royal Engineers," House of Commons, March 16, 1825, in *Parliamentary Papers: Related to Captured Negroes* (1825): 58, 49 (emphasis in original). Major Thomas Moody, "Second Part of Major Moody's Report Related to Captured Negroes," *Parliamentary Papers: Related to Captured Negroes* (1826): 38–40.

56. "East India Trade," *Anti-Slavery Monthly Reporter* 1, no. 22 (March 1827): 321n. For technology's fueling slavery's expansion, see Daniel Rood, *The Reinvention of Atlantic Slavery: Technology, Labor, Race, and Capitalism in the Greater Caribbean* (New York: Oxford University Press, 2017); Walter Johnson, *River of Dark Dreams: Slavery and Empire in the Cotton Kingdom* (Cambridge: Harvard University Press, 2013); R. Keith Aufhauser, "Slavery and Technological Change," *Journal of Economic History* 34, no. 1 (1974): 39–50.

57. *The Injurious Effects of Slave Labour: An Appeal to the Reason, Justice, and Patriotism of the People of Illinois*, ed. James Cropper (London, 1824), 6. "East India Trade," *Anti-Slavery Monthly Reporter* 1, no. 22 (March 1827): 322. *Anti-Slavery Monthly Reporter* 1, no. 24 (May 1827): 386.

58. Clarkson, *Thoughts on Improving the Condition of the Slaves* (1823), 56 (emphasis in original).

59. For background on the emancipation bill, see Drescher, *Mighty Experiment*, 121–22; Blackburn, *Overthrow of Colonial Slavery*, 453. For the 1831 revolt, also known as the Baptist War, see Craton, *Testing the Chains*, 291–322; Blackburn, *Overthrow of Colonial Slavery*, 432–33.

60. *Report from Select Committee on the Extinction of Slavery throughout the British Dominions . . . Ordered by the House of Commons, to be Printed, 11 August 1832* (London, 1833), 152, 37, 164, 345, 168.

61. For the end of slavery in the British Caribbean, see Morgan, *Slavery and the British Empire*, 190–98. For post-emancipation challenges in the British Caribbean, see Frederick Cooper, Thomas C. Holt, and Rebecca J. Scott, *Beyond Slavery: Explorations of Race, Labor, and Citizenship in Post-emancipation Societies* (Chapel Hill: University of North Carolina Press, 2000); Jeffrey R. Kerr-Ritchie, *Rites of August*

First: Emancipation Day in the Black Atlantic World (Baton Rouge: Louisiana State University Press, 2007); Natasha Lightfoot, *Troubling Freedom: Antigua and the Aftermath of British Emancipation* (Durham: Duke University Press, 2015). Gibson is quoted in Kerr-Ritchie, *Rites of August First*, 33.

62. For the challenges British abolitionists faced when having to defend emancipation's success in economic terms, see Drescher, *Mighty Experiment*, 158–230.

6 Antislavery in an Age of Science

1. Jonathan Baldwin Turner, *The Three Great Races of Men* (N.p., 1861), 29, 6–8, 43, 7, 101.

2. Ibid., 78. For hiding of runaways, see "Friend," to Jonathan Turner, Louisville, Ky., Sept. 11, 1842, Jonathan Baldwin Turner Papers, University of Illinois at Urbana-Champaign (hereafter cited as JBT-UI), box 1, folder 2. For his firing from Illinois College, see Mary Turner Carriel, *The Life of Jonathan Baldwin Turner* (Jacksonville, Ill., 1911), 53–64. For the Lincoln meeting, see Jonathan Baldwin Turner to Rhodolphia Turner, Sept. 19, 1862, in Carriel, *Life of Turner*, 275. For the diversity of antislavery views in this period, see Eric Foner, *Free Soil, Free Labor, Free Men: The Ideology of the Republican Party before the Civil War* (New York: Oxford University Press, 1970); Manish Sinha, *The Slave's Cause: A History of Abolition* (New Haven: Yale University Press, 2016), 357–58, 497–98, 578–83.

3. For abolitionism in the antebellum era, see Sinha, *Slave's Cause*, 195–227, 266–98, 461–99; Patrick Rael, *Eighty-Eight Years: The Long Death of Slavery in the United States, 1777–1865* (Athens: University of Georgia Press, 2015), 163–238; Richard Newman, *The Transformation of American Abolitionism: Fighting Slavery in the Early Republic* (Chapel Hill: University of North Carolina Press, 2002), 107–30; John L. Brooke, *"There Is a North": Fugitive Slaves, Political Crisis, and Cultural Transformation in the Coming of the Civil War* (Amherst: University of Massachusetts Press, 2019); Benjamin Quarles, *Black Abolitionists* (New York: Oxford University Press, 1969); James Brewer Stewart, *Holy Warriors: The Abolitionists and American Slavery*, rev. ed. (New York: Hill and Wang, 1997), 35–205; John Stauffer, *The Black Hearts of Men: Radical Abolitionists and the Transformation of Race* (Cambridge: Harvard University Press, 2002); W. Caleb McDaniel, *The Problem of Democracy in the Age of Slavery: Garrisonian Abolitionists & Transatlantic Reform* (Baton Rouge: Louisiana State University Press, 2013); Julie Roy Jeffrey, *The Great Silent Army of Abolitionism: Ordinary Women in the Antislavery Movement* (Chapel Hill: University of North Carolina Press, 1998); Shirley Yee, *Black Women Abolitionists: A Study in Activism, 1828–1860* (Knoxville: University of Tennessee Press, 1992); Richard Blackett, *Building an Antislavery Wall: Black Americans in the Atlantic Abolitionist Movement, 1830–1860* (Baton Rouge: Louisiana State University Press, 1983).

4. Sinha, *Slave's Cause*, 499.

5. On objectivity, see Lorraine Daston and Peter Galison, *Objectivity* (New York: Zone, 2007), 27–35. For professionalization and changes in nineteenth=century American science, see Paul Lucier, "The Professional and the Scientist in Nineteenth-Century America," *Isis* 100, no. 4 (Dec. 2009): 699–732; Robert V. Bruce, *The Launching of Modern American Science, 1846–1876* (New York: Alfred A Knopf, 1987); George H. Daniels, "The Process of Professionalization in American Science: The Emergent Period, 1820–1860," *Isis* 58, no. 2 (1967): 150–66; Nathan Reingold, "Definitions and Speculations: The Professionalization of Science in America in the Nineteenth Century," in *Science, American Style*, ed. Nathan Reingold (New Brunswick: Rutgers University Press, 1991), 24–53.

6. Frederick Douglass, *The Claims of the Negro, Ethnologically Considered* (Rochester, N.Y., 1854), 9.

7. Ibid. The trade-off African Americans faced when attacking ethnology has been studied by several historians. Mia Bay and Bruce Dain argue that by fighting ethnology in its own terms, Black writers reproduced the biases inherent in racial science; by contrast, Britt Rusert and Patrick Rael argue that Black abolitionists successfully reappropriated ethnology for their own ends. See Mia Bay, *The White Image in the Black Mind: African-American Ideas About White People, 1830–1925* (New York: Oxford University Press, 1999), 38–74; Bruce Dain, *A Hideous Monster of the Mind: American Race Theory in the Early Republic* (Cambridge: Harvard University Press, 2002), 227–63; Britt Rusert, *Fugitive Science: Empiricism and Freedom in Early African American Culture* (New York: New York University Press, 2017); Patrick Rael, "A Common Nature, a United Destiny: African American Responses to Racial Science from the Revolution to the Civil War," in *Prophets of Protest: Reconsidering the History of American Abolitionism*, ed. Timothy P. McCarthy and John Stauffer (New York: New Press, 2006), 183–99. See also Stephen Howard Browne, "Counter-Science: African American Historians and the Critique of Ethnology in Nineteenth-Century America," *Western Journal of Communication* 64, no. 3 (Summer 2000): 268–84.

8. For changes in U.S. racial science in early nineteenth century, see Dain, *Hideous Monster*, 197–226; William Stanton, *The Leopard's Spots: Scientific Attitudes toward Race in America, 1815–59* (Chicago: University of Chicago Press, 1960); George M. Fredrickson, *The Black Image in the White Mind: The Debate on Afro-American Character and Destiny, 1817–1914* (1971; repr., Middletown, Conn.: Wesleyan University Press, 1987), 71–164; Ann Fabian, *The Skull Collectors: Race, Science, and America's Unburied Dead* (Chicago: University of Chicago Press, 2010). It is worth noting that these developments were not unique to the United States, though national politics certainly shaped the way each nation used racial science. See James Poskett, *Materials of the Mind: Phrenology, Race, and the Global History of Science, 1815–1920* (Chicago: University of Chicago Press, 2019), 83–110. For

population statistics, see United States Census Office, *Population of the United States in 1860* (Washington, D.C.: National Archives and Records Service, 1967), ix.

9. For common depictions of polygenism as a "proslavery science," see Frederickson, *Black Image in the White Mind*, 71–96; Bay, *White Image in the Black Mind*, 42–44; Sinha, *Slave's Cause*, 309. For challenges to polygenism as a proslavery science, see Christopher Willoughby, "Pedagogies of the Black Body: Race and Medical Education in the Antebellum United States" (PhD diss., Tulane University, 2016), 70–71; Rusert, *Fugitive Science*, 11–12; Stephen Jay Gould, *The Mismeasure of Man* (New York: W. W. Norton, 1980), 74; Stanton, *Leopard's Spots*, 193; Rael, *Eighty-Eight Years*, 131. Suffice it to say that polygenism's appeal in the antislavery North is not widely appreciated beyond these few specialists.

10. Charles D. Meigs, *A Memoir of Samuel George Morton* (Philadelphia, 1851), 12. Samuel George Morton, *Memoir of William Maclure* (Philadelphia, 1841), 31. Morton to Silliman, Feb. 6, 1832, Silliman Family Papers (hereafter cited as SFP), Yale University Library, reel 12. Samuel George Morton, *Crania Americana* (Philadelphia: J. Dobson, 1839), 7, 3. For background on Morton, see Dain, *Hideous Monster*, 198–220; Fabian, *Skull Collectors*, 79–118.

11. Morton, *Crania Americana*, i, 475–76; George Combe, *Notes on the United States of North America during a Phrenological Visit in 1838–39–40*, 2 vols. (Philadelphia: Carey & Hart, 1841), 1:113–14.

12. "Review of Morton's *Crania Americana*," *American Journal of Science* 38, no. 2 (1840): 341. Morton to Silliman, March 31, 1840, SFP, reel 12. Samuel George Morton, "Hybridity in Animals, Considered in Reference to the Question of the Unity of the Human Species: Part 1," *American Journal of Science* 3, no. 7 (January 1847): 39–50; Samuel George Morton, "Hybridity in Animals, Considered in Reference to the Question of the Unity of the Human Species: Part 2," *American Journal of Science* 3, no. 8 (March 1847): 203–212. See also Edward Lurie, "Louis Agassiz and the Races of Man," *Isis* 45, no. 3 (1954): 232. Samuel George Morton, "Observations on the Size of the Brain in Various Races and Families of Man," *American Journal of Science* 9, no. 26 (March 1850): 246–49. Morton described Egyptians as part of the Caucasian race in *Crania Aegyptiaca* (Philadelphia: J. Penington, 1844). For the Mexican-American War and subsequent slavery debates, see James McPherson, *Battle Cry of Freedom: The Civil War Era* (New York: Oxford University Press, 1988), 47–77; Sean Wilentz, *The Rise of American Democracy: Jefferson to Lincoln* (New York: W. W. Norton, 2005), 602–32.

13. Louis Agassiz to his mother, quoted in Louis Menand, "Morton, Agassiz, and the Origins of Scientific Racism in the United States," *Journal of Blacks in Higher Education* 34 (Jan. 31, 2002): 112; and Louis Menand, *The Metaphysical Club* (New York: Farrar, Straus & Giroux, 1999), 105. For Agassiz's background, see Fabian, *Skull Collectors*, 112–15; Frederickson, *Black Image in the White Mind*, 74–76.

14. For "zoological province," see Louis Agassiz, cited in Josiah Nott, "An Examination of the Physical History of the Jews, in Its Bearings on the Question of the Unity of the Races," *Proceedings of the American Association for the Advancement of Science. Third Meeting, March 1850* (Charleston, S.C.: Walker and James, 1850), 107. Louis Agassiz, "The Diversity of Origin of Human Races," *Christian Examiner* 49 (1850): 110, 142, 145.

15. For background on Josiah Nott and George Gliddon's *Types of Mankind* (Philadelphia: J. B. Lippincott, 1854), see Fabian, *Skull Collectors*, 111–15; Frederickson, *Black Image in the White Mind*, 74–76. Proslavery Southerners did not unanimously embrace *Types of Mankind* or polygenism. Many proslavery Southerners abhorred polygenism's biblical heresy and instead used the Bible's story of the curse of Ham to justify slavery on racial grounds. See Michael O'Brien, *Conjectures of Order: Intellectual Life and the American South, 1810–1860*, 2 vols. (Chapel Hill: University of North Carolina Press, 2004), 1:249–50.

16. "Types of Mankind," *Medical Examiner* (Philadelphia) 114 (June 1854): 343; "Types of Mankind," *Scientific American*, July 1, 1854, 333. "Indigenous Races of the Earth," *Buffalo Medical Journal* 13, no. 2 (1857): 95. For northern religious critiques, see Stanton, *Leopard's Spots*, 193. For science and religion, see Stephen P. Weldon, "Science and Religion," in *Science and Religion: A Historical Introduction*, ed. Gary B. Ferngren, 2nd ed. (Baltimore: Johns Hopkins University Press, 2017), 3–22.

17. Jonathan Baldwin Turner, "Remarks on Prof. Silliman's Lectures," JBT-UI, 2/3. Turner, "The Millennium of Labor," *Transactions of the Illinois State Agricultural Society* 1 (1853–54): 59. Turner has received limited scholarly attention. For useful biographical studies, see Carriel, *Life of Turner*; and Judith Ann Hancock, "Jonathan Baldwin Turner (1805–1899): A Study of an Educational Reformer" (PhD diss., University of Washington, 1971).

18. Jonathan Baldwin Turner, "Facsimile Written by a Moor Son of the King of Timbuctoo," JBT-UI, 2/3. "The African Prince," *Connecticut Journal* (New Haven), Oct. 14, 1828, 2. For background on Ibrahima, see Terry Alford, *Prince among Slaves* (New York: Harcourt Brace Jovanovich, 1977); Sylviane Diouf, *Servants of Allah: African Muslims Enslaved in the Americas* (New York: New York University Press, 1998), 25–26.

19. Jonathan Baldwin Turner, "Outline of Lecture on Geology and Scripture," JBT-UI, 2/4. Carriel writes that he taught "all branches of knowledge in the curriculum, except chemistry": Carriel, *Life of Turner*, 46. Jonathan Baldwin Turner, "Historic Morgan," in Carriel, *Life of Turner*, 53. "Friend" to Turner, Louisville, Ky., Sept. 11, 1842, JBT-UI, 1/2. J. N. Coleman to the editor of the *Illinois Statesman*, May 10, 1843, JBT-UI, 1/2. Another factor in his firing was his unorthodox evangelical sermons. See Carriel, *Life of Turner*, 60–64.

20. Unknown correspondent to Turner, from Yale College, Feb. 7, 1837, JBT-UI, 1/2. Jonathan Baldwin Turner, "Equality in Society," June 4, 1843, JBT-UI, 2/4.

21. "National Compensation Emancipation Society," 1857, JBT-UI, 2/6. Silliman declined, saying that he was too frail to travel to the first convention, in Cleveland.

22. "The Compensated Emancipation Society," *Frederick Douglass' Paper*, Feb. 11, 1859, 3. See also "Compensated Emancipation," *National Anti-Slavery Standard* (New York), Sept. 5, 1857; "Burritt's Compensated Emancipation," *Liberator*, Sept. 4, 1857, 142; "Compensation Convention," *Liberator*, Feb. 11, 1859, 23. "Emancipation in Delaware," *Frederick Douglass' Paper*, April 15, 1859, 4.

23. "National Compensation Emancipation Society."

24. Jonathan Baldwin Turner, *Industrial Universities for the People* (Jacksonville, Ill., 1853), 42. Jonathan Baldwin Turner, "The Reign of War, of Words, and of Works," *Transactions of the Wisconsin State Agricultural Society* 4 (1854–57), 387–88, 379; Turner, "Millennium of Labor," 58, 55. Turner, "The Great Crisis; or, The True Issue between Freemen and Tyrants," ca. 1861, JBT-UI, 2/4. For southern resistance to federally funded agricultural colleges, see Sarah T. Phillips, "Antebellum Agricultural Reform, Republican Ideology, and Sectional Tension," *Agricultural History* 74, no. 4 (Autumn 2000): 799–822. Roger L. Geiger, *The History of American Higher Education: Learning and Culture from the Founding to World War II* (Princeton: Princeton University Press, 2015), 229–42.

25. Turner, "Reign of War," 387. For amoral motivations of free white laborers, see Foner, *Free Soil*, 11–39, 262.

26. Owen Lovejoy to Turner, June 2, 1856, and Lyman Trumbull to Turner, Oct. 19, 1857, JBT-UI, 1/2. For other antislavery Republicans who solicited Turner's advice on the land-grant college idea, see, Richard Yates to Turner, May, 25, 1856, Justin Morrill to Turner, Dec. 30, 1861, and Charles H. Howland to Turner, Jan. 7, 1865, all JBT-UI, 1/2.

27. Turner, "Reign of War," 376. Turner, "Millennium of Labor," 55, 54. Turner, *Three Great Races*, 24, 57.

28. Turner, *Three Great Races*, 24, 12, 52, 29 (emphasis in original), 31, 47, 43–44, 8.

29. Ibid., 88, 20, 47, 43, 60, 95.

30. Ibid., iii. John Kennicott to Turner, Highland Park, Ill., Jan. 28, 1863, Jonathan Baldwin Turner Papers, Abraham Lincoln Presidential Library (hereafter cited as JBT-ALPL), 2/1 (emphases in original). The lecture was almost certainly written before the summer of 1857; on Turner's copy of the "National Compensation Emancipation Society" pamphlet, he wrote that the call for the society came just "after my lecture on the three races." See "National Compensation Emancipation Society."

31. Thomas Ewbank, *Inorganic Forces Ordained to Supersede Human Slavery* (New York, 1860), 4, 8, 30, 32 (emphasis in original). For Ewbank's helping Turner publish his "State Agricultural University" plan, see Richard Yates to Turner, Washington, D.C., July 10, 1852, JBT-UI, 1/2.

32. Edward Bissell Hunt, *Union Foundations: A Study of American Nationality as a Fact of Science* (New York: D. Van Nostrand, 1863), "Note," excerpted as "Editor's Table: Indivisibility of the Nation," *Harper's New Monthly Magazine*, Feb. 1863, 413–18.

33. Hunt, *Union Foundations*, 51, 1, 49–55.

34. Abraham Lincoln, "Address on Colonization to a Deputation of Negroes," in *The Collected Works of Abraham Lincoln*, ed. Roy P. Basler et al., 7 vols. (New Brunswick: Rutgers University Press, 1953–55), 5:371. For Lincoln and colonization, see Kate Masur, *An Example for All the Land: Emancipation and the Struggle over Equality in Washington, D.C.* (Chapel Hill: University of North Carolina Press, 2010), 13–15, 35–39; Kate Masur, "The African American Delegation to Abraham Lincoln: A Reappraisal," *Civil War History* 56, no. 2 (2010): 117–44; Eric Foner, "Lincoln and Colonization," in *Our Lincoln: New Perspectives on Lincoln and His World*, ed. Eric Foner (New York: W. W. Norton, 2008), 135–66. For coal, Chiriquí, and Lincoln's scientific background, see Peter Shulman, *Coal & Empire: The Birth of Energy Security in Industrial America* (Baltimore: Johns Hopkins University Press, 2015), 92–135; Jason Emerson, *Lincoln the Inventor* (Carbondale: Southern Illinois University Press, 2009).

35. Turner to Rhodolphia Turner, Stone Hospital, Washington, D.C., Sept. 12, 1862, JBT-ALPL, 2/1. See also Turner to Rhodolphia Turner, Stone Hospital, Washington D.C., Sept. 19, 1862, in Carriel, *Life of Turner*, 275–76.

36. Douglass, *Claims of the Negro*, 9–10. Frederick Douglass to Anne Tyson Kirk, Washington, D.C., March 29, 1879, Maryland Historical Society Archives (hereafter cited as MDHS). Douglass to Kirk, Washington, D.C., March 4, 1878 (MDHS). For background on the *Claims of the Negro* lecture, see *Frederick Douglass Papers*, ed. John Blassingame and John McKivigan, series 1 (New Haven: Yale University Press, 1982), 2:498–99; Frederick Douglass, *Life and Times of Frederick Douglass*, series 2, vol. 3 (New Haven: Yale University Press, 2011), 293–94; Waldo E. Martin Jr., *The Mind of Frederick Douglass* (Chapel Hill: University of North Carolina Press, 1984), 212–34. For other useful discussions of *Claims of the Negro*, see Rusert, *Fugitive Science*, 13–14, 18, 41, 126–28; Jared Gardner, *Master Plots: Race and the Founding of an American Literature, 1787–1845* (Baltimore: Johns Hopkins University Press, 1998), 178–85; Bay, *White Image in the Black Mind*, 68–71.

37. "Colored Orphan's Asylum: Physician's Report," *Colored American*, Jan. 26, 1839. James McCune Smith, "The Destiny of the People of Color," in *The Works of James McCune Smith: Black Intellectual and Abolitionist*, ed. John Stauffer (New York: Oxford University Press, 2006), 53. J. R. L., "Ethnology," *National Anti-Slavery Standard*, Feb. 1, 1849, 142.

38. "Prof. Agassiz on the Origin of the Human Race," *National Anti-Slavery Standard*, June 27, 1850, 18. "Types of Mankind," *Liberator*, Oct. 13, 1854, 164. "Anniversary of the New York City Anti-Slavery Society," *Liberator*, May 19, 1854, 79; "Mr. Parker and Prof. Agassiz," *Liberator*, May 26, 1854, 83.

39. James McCune Smith to Gerrit Smith, March 31, 1855, in *Works of James McCune Smith*, 317. For Black resistance to white paternalism, see Rael, *Eighty-Eight Years*, 198–206; Quarles, *Black Abolitionists*, 49–50, 56, 235.

40. Douglass, *Claims of the Negro*, 34.

41. Ibid., 6. Stauffer, introduction to *Works of James McCune Smith*, xiii–xlii.

42. I have found no evidence of Smith's ethnological work being cited in the period's most commonly mentioned ethnological essays.

43. Edward Jarvis, "Insanity among the Coloured Population of the Free States," *American Journal of the Medical Sciences* (Jan. 1844): 75, 83. James McCune Smith, "Freedom and Slavery for Afric-Americans," *Liberator*, Feb. 23, 1844, in *Works of James McCune Smith*, 62, 65. For the 1840 census and slavery, see Harriet A. Washington, *Medical Apartheid: The Dark History of Experimentation on Black Americans from Colonial Times to the Present* (New York: Doubleday, 2006), 145–51; Martin Summers, " 'Suitable Care of the African When Afflicted with Insanity': Race, Madness, and Social Order in Comparative Perspective," *Bulletin of the History of Medicine* 84, no. 1 (Spring 2010): 58–91; Dain, *Hideous Monster*, 247–48; Robert Whitaker, *Mad in America: Bad Science, Bad Medicine, and the Enduring Mistreatment of the Mentally Ill* (New York: Basic Books, 2002), 171. For Smith's scientific achievements, see Stauffer, introduction to *Works of James McCune Smith*, xiv–xv. It should be noted that neither of his 1840s articles attacked ethnological science directly. In regard to Black scientific practitioners in the antebellum period, many African Americans did in fact work in fields that today would be considered medical or health care related: Black women as nurses, midwives, and healers, and Black men as barbers, nurses, healers, dentists, and physician's assistants. Yet because African Americans were denied access to white-run educational institutions and were largely self-taught, their knowledge did not possess the same level of public authority in the eyes of white society. For Black medical practitioners in the period, see Leslie Falk, "Black Abolitionist Doctors and Healers, 1850–1885," *Bulletin of the History of Medicine* 42, no. 2 (Summer 1980): 258–72.

44. James McCune Smith, "Civilization: Its Dependence on Physical Circumstances" (1859), *Anglo-African*, reprinted in *Works of James McCune Smith*, 246–63; James McCune Smith, "On the Fourteenth Query of Thomas Jefferson's Notes on Virginia" (1859), *Anglo-African*, reprinted in *Works of James McCune Smith*, 264–81. Smith wrote about ethnology in his introduction to Frederick Douglass's second autobiography, *My Bondage and My Freedom* (New York: Miller, Orton & Mulligan, 1855), 18–19.

45. Smith, "Fourteenth Query," 267–70, 278. Smith, "Civilization," 260.

46. Smith, "Civilization," 247.

47. Ibid., 250, 262. Smith, "Fourteenth Query," 274.

48. Smith, "Civilization," 261–62. James McCune Smith, "Annual Meeting of the Coloured Orphan Asylum," *National Anti-Slavery Standard*, Feb. 22, 1849, 154. James McCune Smith, "Lecture on the Haytien Revolutions" (1841) in *Works of James*

McCune Smith, 39. James McCune Smith, "The Destiny of the People of Color" (1843), in *Works of James McCune Smith*, 56–58. Bay and Dain also argue that Smith sometimes perpetuated racial stereotypes; see Bay, *White Image in the Black Mind*, 59–63; Dain, *Hideous Monster*, 255–64. For Smith's problematic gender assumptions, see: Bay, *White Image in the Black Mind*, 41.

49. Smith, "Destiny of the People of Color," 48, 59–60.

50. "Speech of Dr. Rock," *Liberator*, March 12, 1858, 42. "Selections: Speech of Rev. Theodore Parker," *National Anti-Slavery Standard*, Feb. 27, 1858, 1. For background on Rock, see Bay, *White Image in the Black Mind*, 56–57.

51. "Speech of Dr. Rock," 42.

52. Ibid.

53. Douglass, *Claims of the Negro*, 5, 20.

54. To be clear, many scholars of nineteenth-century U.S. agricultural history have studied the ways *proslavery* agricultural reformers, most famously Edmund Ruffin, used agricultural science to bolster slavery. For excellent works in this regard, see Steven Stoll, *Larding the Lean Earth: Soil and Society in Nineteenth-Century America* (New York: Hill and Wang, 2002); Courtney Fullilove, *The Profit of the Earth: The Global Seeds of American Agriculture* (Chicago: University of Chicago Press, 2017), 68–89; Alan L. Olmstead and Paul W. Rhode, *Creating Abundance: Biological Innovation and American Agricultural Development* (New York: Cambridge University Press, 2008), 98–154; William Mathew, *Edmund Ruffin and the Crisis of Slavery in the Old South: The Failure of Agricultural Reform* (Athens: University of Georgia Press, 1988). There is also a related debate about whether slavery did in fact impede the implementation of scientific agricultural reforms. See Avery O. Craven, *Soil Exhaustion as a Factor in the Agricultural History of Virginia and Maryland, 1606–1860* (Urbana: University of Illinois Press, 1926), 108–21; Eugene Genovese, *The Political Economy of Slavery: Studies in the Economy & Society of the Slave South* (1965; repr., Middletown, Conn.: Wesleyan University Press, 1989), 124–53; Joyce Chaplin, *An Anxious Pursuit: Agricultural Innovation and Modernity in the Lower South, 1730–1815* (Chapel Hill: University of North Carolina Press, 1993); A. Glenn Crothers, "Agricultural Improvement and Technological Innovation in a Slave Society: The Case of Early National Northern Virginia," *Agricultural History* 75, no. 2 (Spring 2001): 135–67. But to my knowledge, few scholars have studied the ways *abolitionists* used agricultural science to attack slavery. For a rare exception, see Sarah Phillips, "Antebellum Agricultural Reform, Republican Ideology, and Sectional Tension," *Agricultural History* 74, no. 4 (Autumn 2000): 799–822. For background on scientific agriculture in nineteenth-century America, see Benjamin R. Cohen, *Notes from the Ground: Science, Soil, and Society in the American Countryside* (New Haven: Yale University Press, 2009); Margaret Rossiter, *The Emergence of Agricultural Science: Justus Liebig and the Americans, 1840–1880* (New Haven: Yale University Press, 1975).

55. William Kenrick, "Alleged Effect of Slavery on Agriculture of Virginia," *New England Farmer & Horticultural Register* 18, no. 2 (July 17, 1839): 13. Turner, "The Great Crisis," ca. 1861, JBT-UI, 2/4. "Who Has Not Heard of McCormick, the Mighty Reaper," *New York Daily Tribune*, Sept. 16, 1857, 4. For other scholars who discuss northern agricultural journals' critiques of slavery, see Ada W. Dean, *An Agrarian Republic: Farming, Antislavery Politics, and Nature Parks in the Civil War Era* (Chapel Hill: University of North Carolina Press, 2015), 34–36; Phillips, "Antebellum Agricultural Reform."

56. Edward Everett, "Agriculture, as Concerned with Mechanics and Engineering," *Transactions of the Illinois State Agricultural Society* 1 (1853–54): 439. Ewbank, *Inorganic Forces*, 21, 24–25. Everett's conservative politics was typical of many elite Northerners, who often opposed slavery but sacrificed any scruples about slavery to the cause of national union. See Matthew Mason, *Apostle of Union: A Political Biography of Edward Everett* (Chapel Hill: University of North Carolina Press, 2016).

57. "Letter from Germantown, Pa.," *Cultivator* (Albany), Aug. 1849, 238 (emphasis in original). "Flax v. Cotton," *New York Daily Tribune*, July 12, 1851, 4. Turner, *Three Great Races*, 51. Hunt, *Union Foundations*, 54–55.

58. Ewbank, *Inorganic Forces*, 26. *Congressional Globe*, 32nd Cong., 1st sess., May 17, 1852, 588. "Flax v. Cotton."

59. The idea that science promoted the social good, of course, dates to the early modern period and the Enlightenment ideal of science. See Peter Harrison, "Science, Religion, and Modernity: Early Modern Science and the Idea of Moral Progress," in *British Abolitionism and the Question of Moral Progress in History*, ed. Donald A. Yerxa (Columbia: University of South Carolina Press, 2012), 139–53.

60. Entries for March 20 and May 19, 1841, John Pitkin Norton Diary, John Pitkin Norton Papers, Manuscripts and Archives, Yale University Library (hereafter cited as JPN), box 3, folder 18. Norton's diaries are a common source for scholars studying the *Amistad* rebellion, but no historian has explored how his scientific career interacted with his career as an antislavery activist. For prominent works mentioning Norton's role in the *Amistad* case, see David Brion Davis, *Inhuman Bondage: The Rise and Fall of Slavery in the New World* (New York: Oxford University Press, 2006), 24; Marcus Rediker, *The Amistad Rebellion: An Atlantic Odyssey of Slavery and Freedom* (New York: Viking, 2012), 30, 192. For background on Norton and agricultural chemistry, see Rossiter, *Emergence of Agricultural Science*, 92–126; Louis I. Kuslan, "The Founding of the Yale School of Applied Chemistry," *Journal of the History of Medicine and Allied Sciences* 24, no. 4 (Oct. 1969): 430–51.

61. Entries for March 18, Aug. 25, and March 27, 1841, Norton diary, JPN, 3/18, and Sept. 17, 1841, JPN, 4/19 (emphasis in original).

62. Entries for Sept. 13, April 6, and May 5, 1841, Norton diary, JPN, 3/18, and Nov. 14, 1841, JPN, 4/19. Letter to the editor of the *Journal & Courier*, ca. fall 1851, JPN, 8/2.

63. Leonard Scott & Co. to Norton, Feb. 19, 1850, JPN, 1/10. Henry Stephens, *The Farmer's Guide to Scientific and Practical Agriculture*, ed. John Pitkin Norton, 2 vols. (New York: L. Scott, 1850), 1:12.

64. Stephens, *Farmer's Guide*, 1:1, 12, 39. For cotton exports in 1850, see Federal Reserve Board, *Federal Reserve Bulletin, May, 1923* (Washington, D.C.: Government Printing Office, 1923), 567.

65. Stephens, *Farmer's Guide*, 2:64. Norton was closely involved in the design of the McCormick reaper. See Charles H. R. Rockwell to Norton, March 21, 1849, and Rockwell to Norton, June 2, 1849, JPN, 1/9. For the McCormick reaper's development in Virginia, see Daniel Rood, "An International Harvest: The Second Slavery, the Virginia-Brazil Connection, and the Development of the McCormick Reaper," in *Slavery's Capitalism: A New History of American Economic Development*, ed. Sven Beckert and Seth Rockman (Philadelphia: University of Pennsylvania Press, 2016), 87–104; Daniel Rood, *The Reinvention of Atlantic Slavery: Technology, Labor, Race, and Capitalism in the Greater Caribbean* (New York: Oxford University Press, 2017), 174–96.

66. John Pitkin Norton, "On the Soil, Considered Scientifically as Well as Practically: Letter II," *Southern Cultivator* 9, no. 10 (Oct. 1851): 146 (reprinted from *Soil of the South*). On Norton's soil analyses, see John Pitkin Norton, "On the Value of Results Obtained by Comparative Analysis of Soils," *Cultivator* (Nov. 1851): 8, 11. For the soil analysis craze in these years and Norton's lack of salary, see Rossiter, *Emergence of Agricultural Science*, 109–26. For the average Yale faculty salary in 1850, see *University Record* (Chicago: University of Chicago Press, 1900), 4:81. On *Soil of the South* compensation, see William Chambers to Norton, May 21, 1851, JPN, 1/13, and Norton to Chambers, June 11, 1851, JPN, 8/2.

67. Chambers to Norton, April 18, 1851, JPN, 1/13.

68. Anthony Benezet Allen to Norton, Sept. 4, 1851, JPN, 1/13 (emphasis in original). Norton to I. Delafield, Nov. 18, 1851, JPN, 8/2. Norton to Charles S. Peirce, Nov. 19, 1851, JPN, 8/2. For verbal commitments, see Norton to I. Delafield, Nov. 18, 1851. For letters with Robinson, see Norton to Solon Robinson, Jan. 5, 1852; Norton to Robinson, Jan. 5, 1852, JPS 8/2. For the abolitionist attack on Robinson, see "A Convert to Slavery," *National Era* (Washington, D.C.), Aug. 23, 1849, 134. For Robinson's proslavery article, see Solon Robinson, "Negro Slavery at the South," *De Bow's Review*, Sept. and Nov. 1849, 206–25, 379–89.

69. John Pitkin Norton, "Professor Norton's Letter," *Plow* (New York), Feb. 1852, 42–43 (emphasis in original). Norton to the editors of the *Plow*, Oct. 8, 1851, JPN, 9/2.

70. "The Great Exhibition," *Morning Chronicle*, June 9, 1851, 2. Edward McDermott, "On Claussen's Flax-Cotton," *Journal of the Royal Agricultural Society of England* 12 (1851): 241. "Cultivation, Preparation, and Manufacture of Flax and Hemp," *Morning Chronicle*, Nov. 14, 1850, 5. For Black abolitionists at the Great Exhibition, see Lisa Merrill, "Exhibiting Race 'Under the World's Huge Glass Case':

William and Ellen Craft and William Wells Brown at the Great Exhibition in Crystal Palace, London, 1851," *Slavery & Abolition* 33, no. 2 (2012): 321–36. For background on the Great Exhibition, see Jeffrey Auerbach, *The Great Exhibition of 1851: A Nation on Display* (New Haven: Yale University Press, 1999).

71. "The Great Exhibition," 2. John Ryan, *The Preparation of Long-Line Flax-Cotton, and Flax-Wool, by the Claussen Processes* (London, 1852), 95–100 (appendix B). For how Claussen's machine worked, see McDermott, "On Claussen's Flax-Cotton," 243. For abolitionism details discussed here, see Seymour Drescher, *Abolition: A History of Slavery and Antislavery* (New York: Cambridge University Press, 2009), 274, 293, 316. For East Indian indentured servant statistics, see G. W. Roberts and J. Byrne, "Summary Statistics on Indenture and Associated Migration Affecting the West Indies, 1834–1918," *Population Studies* 20, no. 1 (July 1966): 125–34.

72. Peter Claussen, *The Flax Movement: Its National Importance and Advantages*, 3rd ed. (London, 1851), 43, preface. The publication was also printed in New York.

73. "Chevalier Claussen's Preparation of Flax," *Morning Chronicle*, Jan. 19, 1852, 7.

74. McDermott, "On Claussen's Flax-Cotton," 236. "Claussen's Improvements in the Preparation and Bleaching of Flax," *Scientific American*, March 29, 1851, 220.

75. "Flax Cotton," *North Star*, Jan. 16, 1851, 2. "The Flax Movement: Its National Importance and Advantages, by Chevalier Claussen," *Anti-Slavery Reporter*, Oct. 1851, 166. "Anti-Slavery Items," *Anti-Slavery Reporter*, Oct. 1855, 224. This last article was originally published in a New York newspaper and printed in antislavery periodicals on both sides of the Atlantic, including *Frederick Douglass' Paper*, July 6, 1855, 1.

76. Editorial correspondence, "Glances at Europe, No. XII: The Flax-Cotton Revolution," *New York Daily Tribune*, June 23, 1851, 4. "Flax-Cotton and Linen," *New York Daily Tribune*, Aug. 25, 1851, 4. "Flax v. Cotton," *New York Daily Tribune*, July 12, 1851, 4. For Greeley's background, see Adam Tuchinsky, *Horace Greeley's "New-York Tribune": Civil War–Era Socialism and the Crisis of Free Labor* (Ithaca: Cornell University Press, 2009), 126–64.

77. "Science, Industry, and Invention," *New York Daily Tribune*, March 31, 1860, 5. "Science Is King," *New England Farmer* 13, no. 1 (Jan. 1861): 24 (emphasis in original). "Flax Cotton," *New England Farmer* 14, no. 8 (Aug. 1862): 346. Turner, *Three Great Races*, 101. Edward Everett, *An Address Delivered before the Union Agricultural Society of Adams, Rodman, and Loraine, Jefferson County, New York, 12 September 1861* (Cambridge, Mass., 1861), 16, 23.

78. "Flax and Sugar Beet: Letter to the Editor," *New York Daily Tribune*, Jan. 3, 1853, 3. "Flax v. Cotton," *New York Daily Tribune*, July 12, 1851, 4. The *Boston Journal* quotation is quoted in Civis, "The Flax-Cotton Dodge," *Liberator*, Oct. 18, 1861, 167.

79. For background on the free-produce movement, see Carol Faulkner, "The Root of the Evil: Free Produce and Radical Antislavery, 1820–1860," *Journal of the Early Republic* 27, no. 3 (Fall 2007): 377–405; Sinha, *Slave's Cause*, 178–79.

80. Jacob C. White Jr., "The Inconsistency of Colored People Using Slave Produce," Dec. 30, 1852, Leon Gardiner Collection of American Negro Historical Society Records, Historical Society of Pennsylvania (hereafter cited as HSP), box 5G, folder 28. For Crummell's and Garnet's free-produce lectures in England, see Sinha, *Slave's Cause*, 372–73.

81. "Free and Slave Produce," *New York Daily Tribune*, June 20, 1853, 5. For reprints of *New York Tribune* articles on Claussen's machine, see "The Flax-Cotton Revolution," *National Anti-Slavery Standard*, July 10, 1851, 28; "Flax in England," *National Anti-Slavery Standard*, Sept. 2, 1852, 60; "A New Supply of Cotton," *National Anti-Slavery Standard*, Sept. 2, 1852, n.p. "Cotton and Slavery—Flax," *National Anti-Slavery Standard*, June 17, 1854, 14. "Flax Cotton," *National Era*, Aug. 7 1851, n.p. "From the Yates County Whig, Penn Yan, N.Y.: Flax Cotton," *National Era,* March 10, 1853, n.p. Greeley briefly drew closer to radical abolitionism during the war years, as did many Northerners. By March 22, 1862, he had explicitly dismissed flax solutions, particularly when writers used them as an excuse not to fight the war and not to end slavery. He wrote: "Have we not already tried it—this expedient smothering fire with flax—and where are we now?" "Plain Why and Because," *New York Daily Tribune*, March 22, 1862, 4. But by 1872, Greeley had gone back to his more conservative views, running for president on a platform that rejected full Black citizenship.

82. "Flax Cotton," *North Star*, Jan. 16, 1851, 2. "Anti-Slavery Items," *Anti-Slavery Reporter*, Oct. 1855, 223–27. "Coffee Is Dethroned," *Frederick Douglass' Paper*, Aug. 4, 1854, 2.

83. Henry Bibb, "Flax Cotton," *Voice of the Fugitive* (Sandwich, C.W. [Ontario]), July 30, 1851, n.p. *Official Proceedings of the Ohio State Convention of Colored Freemen, Held in Columbus, Jan. 19–21, 1853* (Cleveland, 1853), 5. "The Flax Crop," *Christian Recorder* (Philadelphia), Sept. 5, 1853, 1.

84. Maria W. Stewart, "An Address Delivered at the African Masonic Hall, Boston, Feb. 27, 1833," *Liberator*, March 2, 1833, in *Maria W. Stewart: America's First Black Political Writer: Essays and Speeches*, ed. Marilyn Richardson (Bloomington: Indiana University Press, 1987), 57. For Sarah Mapps Douglass, see Rusert, *Fugitive Science*, 191–203. For integrationism, see Patrick Rael, *Black Identity and Black Protest in the Antebellum North* (Chapel Hill: University of North Carolina Press, 2002), 118–208; Sinha, *Slave's Cause*, 299–330. For free Black activists' emphasis on citizenship, see Christopher James Bonner, *Remaking the Republic: Black Politics and the Creation of American Citizenship* (Philadelphia: University of Pennsylvania Press, 2020); Martha S. Jones, *Birthright Citizens: A History of Race and Rights in Antebellum America* (New York: Cambridge University Press, 2018); Stephen Kantrowitz, *More Than Freedom: Fighting for Black Citizenship in a White Republic, 1829–1889* (New York: Penguin, 2012).

85. Jacob C. White Jr., "Address Read on the Reception of Governor Pollock at the Institute for Coloured Youth," May 24, 1855, Leon Gardiner Collection of

American Negro Historical Society Records, HSP, box 5G, folder 28. For the Sarah Mapps Douglass and Henry Black lectures, see Emma Jones Lapsansky, " 'Discipline to the Mind': Philadelphia's Banneker Institute, 1854–1872," *Pennsylvania Magazine of History and Biography* 117, no. 1–2 (Jan.–April 1993): 96–97. Other free Black intellectual societies established in the antebellum North include the Philomathean Society in New York, Philadelphia's Gilbert Lyceum, and the Afric-American Female Intelligence Society in Boston, where Maria W. Stewart gave her early speeches, among many others. See Richardson, introduction to *Maria W. Stewart*, 26; Lapsansky, " 'Discipline to the Mind,' " 94.

86. Charles L. Reason, "The Colored People's Industrial College," in *Autographs of Freedom*, ed. Julia Griffiths (Auburn, N.Y., 1854), 11–15. For Reason's background, see Delano Greenidge-Copprue, "Charles L. Reason," in *Encyclopedia of African American History, 1619–1895: From the Colonial Period to the Age of Frederick Douglass*, ed. Paul Finkelman (New York: Oxford University Press, 2008).

87. Martin Delany, *Official Report of the Niger Valley Exploring Party* (New York: T. Hamilton, 1861), 14, 58–61. For the Black emigrationist movement, see Sinha, *Slave's Cause*, 330–38, 372–73, 574–76; Rael, *Black Identity and Black Protest*, 209–36; Floyd Miller, *The Search for a Black Nationality: Black Emigration and Colonization, 1787–1863* (Urbana: University of Illinois Press, 1975); Ousmane K. Power-Greene, *Against Wind and Tide: The African American Struggle against the Colonization Movement* (New York: New York University Press, 2014); James T. Campbell, *Middle Passage: African American Journeys to Africa, 1787–2005* (New York: Penguin, 2006), 57–135.

88. Delany, *Official Report*, 11–12, 58. Martin Delany, "Comets," *Anglo-African Magazine*, Feb. 1859, 59–60. See also Rusert, *Fugitive Science*, 149–80. For background on Delany's African expedition, see Robert S. Levine, ed., *Martin R. Delany: A Documentary Reader* (Chapel Hill: University of North Carolina Press, 2003), 315–19.

89. Benjamin Coates to Frederick Douglass, Philadelphia, Jan. 1, 1851, *North Star*, Jan. 16, 1851, 3. See also Power-Greene, *Against Wind and Tide*, 161–79.

90. "The Free Produce Question," *Liberator*, March 1, 1850, 34. Here Garrison was speaking against Black abolitionists' recent endorsement of the free-produce movement, but his principled opposition to flax technologies followed the same logic. Garrison may have also dismissed scientific solutions for personal reasons. In the early 1850s he had a bitter falling-out with several Black abolitionists, which stemmed in part from Black abolitionists' frustration with Garrison's paternalism. See Rael, *Eighty-Eight Years*, 195–96; Sinha, *Slave's Cause*, 492–94.

91. Civis, "The Flax-Cotton Dodge," 167. "What Are We Going to Make?" *Atlantic Monthly*, June 1858, 95. "A Southerner on Steam Engines and Niggers," *Liberator*, July 23, 1858, 117. "The Free Produce Question."

Conclusion

1. Abraham Lincoln to Horace Greeley, Aug. 22, 1862, in *Collected Works of Abraham Lincoln*, ed. Roy P. Basler et al., 7 vols. (New Brunswick: Rutgers University Press, 1953–55), 5:388. Frederick Douglass, "The Mission of the War: An Address Delivered in New York, New York, on 13 January 1864," in *Frederick Douglass Papers*, ed. John Blassingame and John McKivigan, series 1 (New Haven: Yale University Press, 1982), 4:24. "Speech of Dr. Rock," *Liberator*, March 12, 1858, 42. For Lincoln's border strategy, see Eric Foner, *The Fiery Trial: Abraham Lincoln and American Slavery* (New York: W. W. Norton, 2010), 166–205; James McPherson, *Battle Cry of Freedom: The Civil War Era* (New York: Oxford University Press, 1988), 489–509, 557–63. For Lincoln and colonization, see Eric Foner, "Lincoln and Colonization," in *Our Lincoln: New Perspectives on Lincoln and His World*, ed. Eric Foner (New York: W. W. Norton, 2008), 135–66; Kate Masur, "The African American Delegation to Abraham Lincoln: A Reappraisal," *Civil War History* 56, no. 2 (2010): 117–44. For Black abolitionists and violence, see Kellie Carter Jackson, *Force and Freedom: Black Abolitionists and the Politics of Violence* (Philadelphia: University of Pennsylvania Press, 2019).

2. For recent scholarship underscoring Black agency in bringing about emancipation, see Manisha Sinha, *The Slave's Cause: A History of Abolition* (New Haven: Yale University Press, 2016); Vincent Brown, *Tacky's Revolt: The Story of an Atlantic Slave War* (Cambridge: Harvard University Press, 2020); David Williams, *I Freed Myself: African American Self-Emancipation in the Civil War Era* (New York: Cambridge University Press, 2014); David Brion Davis, *The Problem of Slavery in the Age of Emancipation* (New York: Alfred A. Knopf, 2014). For fugitive enslaved people's role in emancipation, see Richard J. M. Blackett, *The Captive's Quest for Freedom: Fugitive Slaves, the 1850 Fugitive Slave Law, and the Politics of Slavery* (Cambridge: Cambridge University Press, 2018); Andrew Delbanco, *The War before the War: Fugitive Slaves and the Struggle for America's Soul from the Revolution to the Civil War* (New York: Penguin, 2018), 354–87; Sinha, *Slave's Cause*, 421–60. For Black military service, see Douglas Egerton, *Thunder at the Gates: The Black Civil War Regiments That Redeemed America* (New York: Basic Books, 2016); McPherson, *Battle Cry of Freedom*, 563–65. For the transformation of northern opinion, see Allen C. Guelzo, *Fateful Lightning: A New History of the Civil War and Reconstruction* (New York: Oxford University Press, 2012), 313–22.

3. William P. Atkinson, *Classical and Scientific Studies, and the Great Schools of England: A Lecture Read before the Society of Arts of the Massachusetts Institute of Technology, Delivered on April 6, 1865* (Cambridge, Mass.: Sever and Francis, 1865), 66.

4. Frederick Douglass, *Life and Times of Frederick Douglass* (1881), in *Frederick Douglass Papers*, ed. John McKivigan, series 2 (New Haven: Yale University Press, 2012), 3:162, 350.

5. William F. Ogburn, *Technological Trends and National Policy, including the Social Implications of New Inventions* (Washington, D.C.: U.S. Government Printing Office, 1937), 261. Silas Bent, *Slaves by the Billion: The Story of Mechanical Progress Home* (New York: Longmans, 1938), jacket flap. Gerald Piel, "Ideas of Technology: Commentary," *Technology & Culture* 3, no. 4 (Autumn 1962): 465. For slavery and technology in American print culture, see Tim Armstrong, *The Logic of Slavery: Debt, Technology, and Pain in American Literature* (New York: Cambridge University Press, 2012), 70–99.

6. Edgar Zilsel, "The Sociological Roots of Science," *American Journal of Sociology* 47, no. 4 (Jan. 1942): 559. For background on Zilsel, see Wolfgang Krohn and Diedrick Raven, "The 'Zilsel Thesis' in the Context of Edgar Zilsel's Research Programme," *Social Studies of Science* 30, no. 6 (Dec. 2000): 925–33.

7. Eugene Genovese, *The Political Economy of Slavery: Studies in the Economy & Society of the Slave South* (1965; repr., Middletown, Conn.: Wesleyan University Press, 1989), 50.

8. Frederick Douglass, *The Claims of the Negro, Ethnologically Considered* (Rochester, N.Y., 1854), 20.